Concepts and Strategies in Plant Sciences

Series Editor

Chittaranjan Kole, Raja Ramanna Fellow, Government of India, ICAR-National Research Center on Plant Biotechnology, Pusa, Delhi, India

This book series highlights the spectacular advances in the concepts, techniques and tools in various areas of plant science. Individual volumes may cover topics like genome editing, phenotyping, molecular pharming, bioremediation, miRNA, fast-track breeding, crop evolution, IPR and farmers' rights, to name just a few. The books will demonstrate how advanced strategies in plant science can be utilized to develop and improve agriculture, ecology and the environment. The series will be of interest to students, scientists and professionals working in the fields of plant genetics, genomics, breeding, biotechnology, and in the related disciplines of plant production, improvement and protection.

Interested in editing a volume? Please contact Prof. Chittaranjan Kole, Series Editor, at ckoleorg@gmail.com

More information about this series at http://www.springer.com/series/16076

Célia Miguel · Tamas Dalmay ·
Inês Chaves
Editors

Plant microRNAs

Shaping Development and Environmental Responses

Springer

Editors
Célia Miguel
Faculdade de Ciências, BioISI—Biosystems & Integrative Sciences Institute
Universidade de Lisboa
Lisbon, Portugal

Tamas Dalmay
School of Biological Sciences
University of East Anglia
Norwich, UK

Inês Chaves
iBET and ITQB NOVA
Oeiras, Portugal

ISSN 2662-3188 ISSN 2662-3196 (electronic)
Concepts and Strategies in Plant Sciences
ISBN 978-3-030-35774-0 ISBN 978-3-030-35772-6 (eBook)
https://doi.org/10.1007/978-3-030-35772-6

This Springer imprint is published by the registered company Springer Nature Switzerland AG
The registered company address is: Gewerbestrasse 11, 6330 Cham, Switzerland

Preface

Since they were first described in 1993, microRNAs have been recognized as central players in the regulation of gene expression. This important class of small non-coding RNA molecules is present in widely diverse groups of organisms, including plants and animals. The investigation of microRNA roles in cellular processes and their regulatory functions in response to the external environment and in disease is a highly active and prolific research field.

In plants, there are a number of relatively well-characterized microRNAs which are present across different taxonomic groups. The roles of some of these conserved microRNAs in diverse biological processes, either related to development or interaction with external factors, have been already characterized in some detail, especially in model plant species such as *Arabidopsis thaliana*. However, a wealth of new sequencing data is being produced at an increasing pace. The release of a growing number of plant genome sequences and the relatively low cost of coding and non-coding transcriptome sequencing represent both an opportunity and a challenge. On the one hand, the amount of available information allows the discovery of novel, non-conserved microRNAs, in a wide range of plant species, with as yet unknown but potentially relevant functions. On the other hand, the identification and unequivocal annotation of such sequences are still a major challenge, and the criteria used by different research groups are not homogeneous or consensual. Significant advances in this area are expected to occur in the near future taking advantage of more advanced technologies for investigating regulatory processes *in vivo* and with cellular resolution.

In this book, we provide a state-of-the-art overview of the functions of microRNAs in the regulation of plant development and their responses to the surrounding environment. This overview is presented in the form of review chapters which are organized around four main subjects, including microRNA investigation and annotation, their regulatory roles in diverse developmental processes, and in response to abiotic and biotic factors. Each of the 11 chapters, authored by experts in the field, details a specific aspect of microRNA investigation, including a systematized revision of the key findings in that area or topic. The book will be useful to those interested in up-to-date knowledge about microRNAs in plants, including undergraduate and graduate students, teachers and researchers. Due to the promising applications of microRNAs in crop breeding and protection, the book will be valuable for scientists in academia and in the private sector as well.

Lisbon, Portugal Célia Miguel
Norwich, UK Tamas Dalmay
Oeiras, Portugal Inês Chaves

Contents

Editors and Contributors

About the Editors

Prof. Célia Miguel is Assistant Professor at the Faculty of Sciences, University of Lisbon, Portugal, and Senior Researcher at the BioISI (Biosystems Integrative Sciences Institute). She has been research group leader for more than 10 years, working on the molecular mechanisms that control key developmental processes in trees, and responses to environmental challenges. She has contributed to generate genomic resources in diverse woody plant species, including their small non-coding RNA transcriptomes, and her research has been published in over 50 papers in internationally recognized journals.

Prof. Tamas Dalmay is Head of School of Biological Sciences at the University of East Anglia, Norwich. His research group has been working on small non-coding RNAs since 2002 and published about 120 papers on the topic. He worked on small RNAs of many different species and contributed to the UEA workbench, a freely available bioinformatics platform to analyse next-generation sequencing data of small RNAs.

Dr. Inês Chaves is Senior Researcher at iBET and ITQB NOVA, Oeiras, Portugal. Her work is dedicated to the study of development and stress response of woody plants. Her approach is mainly biochemical, but since 2006 she has focused her attention on the role of microRNAs as well as other non-coding small RNAs during these processes. She is also involved in the development of bioinformatic tools to be used in model and non-model species. She has over 20 international publications.

Contributors

Ana Alves Faculdade de Ciências, BioISI—Biosystems & Integrative Sciences Institute, Universidade de Lisboa, Lisbon, Portugal

Susheel Sagar Bhat Department of Gene Expression, Faculty of Biology, Institute of Molecular Biology and Biotechnology, Adam Mickiewicz University, Poznan, Poland

Dawid Bielewicz Department of Gene Expression, Faculty of Biology, Institute of Molecular Biology and Biotechnology, Adam Mickiewicz University, Poznan, Poland

Claudio Bonghi Department of Agronomy, Food, Natural Resources, Animals and Environment, University of Padova Agripolis, Legnaro, Italy

Mireia Bundó Centre for Research in Agricultural Genomics (CRAG), CSIC-IRTA-UAB-UB, Campus Universitat Autònoma de Barcelona (UAB), Bellaterra (Cerdanyola del Vallés), Barcelona, Spain

Sonia Campo Centre for Research in Agricultural Genomics (CRAG), CSIC-IRTA-UAB-UB, Campus Universitat Autònoma de Barcelona (UAB), Bellaterra (Cerdanyola del Vallés), Barcelona, Spain

Monica Canton Department of Agronomy, Food, Natural Resources, Animals and Environment, University of Padova Agripolis, Legnaro, Italy

Inês Chaves iBET, Instituto de Biologia Experimental e Tecnológica, Oeiras, Portugal;
Instituto de Tecnologia Química e Biológica, Universidade Nova de Lisboa (ITQB NOVA), Oeiras, Portugal

Bruno Vasques Costa iBET, Instituto de Biologia Experimental .e Tecnológica, Oeiras, Portugal;
Faculdade de Ciências, BioISI—Biosystems & Integrative Sciences Institute, Universidade de Lisboa, Lisbon, Portugal

Tibor Csorba National Agricultural Research and Innovation Center, Agricultural Biotechnology Institute (NARIC-ABC), Gödöllő, Hungary

Silvia Farinati Department of Agronomy, Food, Natural Resources, Animals and Environment, University of Padova Agripolis, Legnaro, Italy

Cristian Forestan Department of Agronomy, Food, Natural Resources, Animals and Environment, University of Padova Agripolis, Legnaro, Italy

Carl Gunnar Fossdal Division of Forestry and Forest Resources, Norwegian Institute for Bioeconomy Research, Ås, Norway

Aleksandra Grabowska Department of Gene Expression, Faculty of Biology, Institute of Molecular Biology and Biotechnology, Adam Mickiewicz University, Poznan, Poland

Péter Gyula National Agricultural Research and Innovation Center, Agricultural Biotechnology Institute (NARIC-ABC), Gödöllő, Hungary

Ismanizan Ismail Institute of Systems Biology, Universiti Kebangsaan Malaysia, Bangi, Selangor, Malaysia;
Department of Biological Sciences and Biotechnology, Faculty of Science and Technology, Universiti Kebangsaan Malaysia, Bangi, Selangor, Malaysia

Artur Jarmolowski Department of Gene Expression, Faculty of Biology, Institute of Molecular Biology and Biotechnology, Adam Mickiewicz University, Poznan, Poland

Zofia Szweykowska Kulinska Department of Gene Expression, Faculty of Biology, Institute of Molecular Biology and Biotechnology, Adam Mickiewicz University, Poznan, Poland

Feng Li Key Laboratory of Horticultural Plant Biology (MOE), College of Horticulture and Forestry Sciences, Huazhong Agricultural University, Wuhan, China

Susana Lopes Instituto de Tecnologia Química e Biológica, Universidade Nova de Lisboa, Oeiras, Portugal;
BioISI—Biosystems & Integrative Sciences Institute, Faculdade de Ciências, Universidade de Lisboa, Lisbon, Portugal;
iBET, Instituto de Biologia Experimental e Tecnológica, Oeiras, Portugal

Célia Miguel Faculdade de Ciências, BioISI—Biosystems & Integrative Sciences Institute, Universidade de Lisboa, Lisbon, Portugal;
Instituto de Biologia Experimental e Tecnológica (IBET), Oeiras, Portugal

Ana Milhinhos Instituto de Tecnologia Química e Biológica, Universidade Nova de Lisboa, Oeiras, Portugal;
BioISI—Biosystems & Integrative Sciences Institute, Faculdade de Ciências, Universidade de Lisboa, Lisbon, Portugal

Andreia S. Rodrigues Instituto de Biologia Experimental e Tecnológica (IBET), Oeiras, Portugal;
Instituto de Tecnologia Química e Biológica, Universidade Nova de Lisboa, Oeiras, Portugal

Muhammad Sajad Institute of Systems Biology, Universiti Kebangsaan Malaysia, Bangi, Selangor, Malaysia;
Department of Plant Breeding and Genetics, University College of Agriculture and Environmental Sciences, The Islamiah University of Bahawalpur, Bahawalpur, Punjab, Pakistan

Abdul Fatah A. Samad Institute of Systems Biology, Universiti Kebangsaan Malaysia, Bangi, Selangor, Malaysia;
Department of Biosciences, Faculty of Science, Universiti Teknologi Malaysia, Skudai, Johor, Malaysia

Blanca San Segundo Centre for Research in Agricultural Genomics (CRAG), CSIC-IRTA-UAB-UB, Campus Universitat Autònoma de Barcelona (UAB), Bellaterra (Cerdanyola del Vallés), Barcelona, Spain;
Consejo Superior de Investigaciones Científicas (CSIC), Barcelona, Spain

Deepali Singh School of Biotechnology, Gautam Buddha University, Greater Noida, India

Aleksandra Smoczynska Department of Gene Expression, Faculty of Biology, Institute of Molecular Biology and Biotechnology, Adam Mickiewicz University, Poznan, Poland

Henrik Mihály Szaker National Agricultural Research and Innovation Center, Agricultural Biotechnology Institute (NARIC-ABC), Gödöllő, Hungary

György Szittya National Agricultural Research and Innovation Center, Agricultural Biotechnology Institute (NARIC-ABC), Gödöllő, Hungary

Guiliang Tang Department of Biological Sciences, Michigan Technological University, Houghton, MI, USA

Sachin Teotia Department of Biotechnology, Sharda University, Greater Noida, India

Serena Varotto Department of Agronomy, Food, Natural Resources, Animals and Environment, University of Padova Agripolis, Legnaro, Italy

Marcos Viejo Division of Forestry and Forest Resources, Norwegian Institute for Bioeconomy Research, Ås, Norway

Yudan Wang Key Laboratory of Horticultural Plant Biology (MOE), College of Horticulture and Forestry Sciences, Huazhong Agricultural University, Wuhan, China

Igor Yakovlev Division of Forestry and Forest Resources, Norwegian Institute for Bioeconomy Research, Ås, Norway

List of Figures

List of Tables

Part I
Studying microRNAs in Plants

Chapter 1
Regulation of Plant microRNA Biogenesis

Aleksandra Grabowska, Susheel Sagar Bhat, Aleksandra Smoczynska, Dawid Bielewicz, Artur Jarmolowski and Zofia Szweykowska Kulinska

Abstract miRNAs play important regulatory roles in various plant metabolic pathways. Similar to any key regulator, miRNA genes are under a variety of transcriptional and post-transcriptional controls. Another layer of this regulation is provided by the regulation of Microprocessor protein components as well as their interactions with other regulatory proteins. When in cytoplasm, miRNA mode of action and turnover is also influenced by various protein partners that either bind directly to the miRNA or interact with them indirectly. We put together currently available data to provide a comprehensive overview regarding miRNA biogenesis, mode of action, and turnover in plants.

Keywords microRNA · Biogenesis regulation · Microprocessor · Posttranscriptional *MIR* gene regulation · microRNA action and turnover · Plants

1.1 Introduction

microRNAs (miRNAs) are small (21–24 nucleotides) regulatory RNAs that are products of RNA Polymerase II transcription. miRNAs are transcribed as long precursors called primary-miRNAs (pri-miRNAs). pri-miRNAs contain a double-stranded hairpin loop region and are processed by miRNA biogenesis machinery to release miRNA/miRNA* duplexes (Xie 2005; Kim et al. 2011). One strand of the miRNA/miRNA* duplex is then incorporated into Argonaute 1 (AGO1) protein which becomes a part of RNA-induced silencing complex (RISC) that executes the post-transcriptional regulation of messenger RNAs (mRNAs) (Vaucheret et al. 2004; Baumberger and Baulcombe 2005; review: Vaucheret 2008). The regulation can be achieved in two different ways: cleavage of mRNAs or translational inhibition (Llave et al. 2002; Palatnik et al. 2007; Brodersen et al. 2008; Eamens et al. 2012).

Aleksandra Grabowska and Susheel Sagar Bhat contributed equally to this work.

A. Grabowska · S. S. Bhat · A. Smoczynska · D. Bielewicz · A. Jarmolowski · Z. S. Kulinska (✉)
Department of Gene Expression, Faculty of Biology, Institute of Molecular Biology and Biotechnology, Adam Mickiewicz University, Poznan, Poland
e-mail: zofszwey@amu.edu.pl

C. Miguel et al. (eds.), *Plant microRNAs*, Concepts and Strategies in Plant Sciences, https://doi.org/10.1007/978-3-030-35772-6_1

miRNA biogenesis in plants starts from the transcription of *MIR* genes by RNA Pol II (Lee et al. 2004; Xie 2005). *MIR* genes in plants are mostly located in intergenic regions and are independent transcriptional units (Reinhart et al. 2002). Similar to other RNA Pol II transcripts pri-miRNAs also carry a 5′ 7 methylguanosine cap and 3′ polyadenylated tail (Jones-Rhoades and Bartel 2004; Xie 2005).

The long pri-miRNA transcripts are known to generally fold over to form dsRNA structures containing imperfect hairpin loops that house miRNA/miRNA* duplex. Generally, one pri-miRNA gives rise to one mature miRNA species but 'polycistronic' pri-miRNAs (one pri-miRNA gives rise to more than one miRNA) have also been reported (Talmor-Neiman et al. 2006; Merchan et al. 2009; Zhang et al. 2009; Baldrich et al. 2016). miRNA biogenesis is also affected by general transcriptional regulating factors like the Mediator complex (Kim et al. 2011). Absence of Mediator complex leads to lower levels of primary and mature miRNAs owing to lower RNA Pol II occupancy on *MIR* genes. In addition, *MIR* gene promoters have motifs that allow for binding of various transcription factors, which in turn allow for the regulation of miRNA biogenesis according to various developmental and/or environmental cues (Megraw et al. 2006; Yamasaki et al. 2009; Yant et al. 2010; Rogers and Chen 2013; Wang and Perry 2013; Barciszewska-Pacak et al. 2015a; Stepien et al. 2017). Certain transcriptional factors are noteworthy in this regulation and are discussed further.

Negative on TATA less 2 (NOT2a) is a core member of the Carbon Catabolite Repression 4 (CCR4)-NOT complex. NOT2a and NOT2b (previously known as Vire2 interacting protein2) are homologous proteins that act as general transcription factors and associate with RNA Pol II to regulate RNA Pol II based transcription. NOT2a and NOT2b can form homo or heterodimers and interact with Dicer Like 1 (DCL1, discussed later), RNA Pol II and other miRNA biogenesis factors like Serrate (SE) and Cap Binding protein 80/20. Absence or lower levels of NOT2 proteins leads to decreased accumulation of both primary and mature miRNAs (Wang et al. 2013). Cell Division Cycle 5 (CDC5) is another transcription factor that associates with *MIR* gene promoters and positively regulates *MIR* gene transcription (Zhang et al. 2013). Other transcription factors that provide conditional regulation to *MIR* gene expression include, but are not limited to, auxin response factors (ARFs; provide auxin sensitivity) (Megraw et al. 2006), Squamosa Promoter binding protein Like 7 (SPL7; copper sensitivity) (Yamasaki et al. 2009), MYB2 transcriptional factor (phosphate sensitivity) (Baek et al. 2013), Apetala 2 (AP2) (Yant et al. 2010) and Fusca 3 (Wang and Perry 2013) for organ-specific miRNA expression.

miRNA biogenesis is also affected by processes like splicing and alternative polyadenylation (Bielewicz et al. 2013; Schwab et al. 2013; Knop et al. 2017). It has been shown that presence of introns stimulates miRNA biogenesis when miRNA is present in the exonic region, an effect that is enhanced by the presence of active 5′ splice site (5′ ss) rather than splicing itself. In contrast, biogenesis of selected intron-derived miRNAs is enhanced when 5′ ss is inactive or splicing is inhibited. These effects have been annotated to the association between U1 small nuclear ribonucleoprotein (U1 snRNP, component of the spliceosome complex) and Microprocessor components (SE) and Cap Binding Complex (CBC).

pri-miRNAs are processed by the RNase III endonuclease type enzyme DCL1 (Reinhart et al. 2002; Park et al. 2005). DCL1 cleavage is dependent on structural features of pri-miRNAs, especially the stem loop (Song et al. 2010; Werner et al. 2010). The imperfect pairing of the stem loop below the miRNA/miRNA* duplex plays an important role in the DCL1 cleavage (Bologna et al. 2009). pri-miRNAs are processed by DCL1 in a two-step process. In the first step, DCL1 cleaves pri-miRNAs and a precursor-miRNA (pre-miRNA) containing the stem loop (carrying miRNA/miRNA*) with 2nt 3′ overhang and 5′ phosphate group is released (Kurihara and Watanabe 2004). The second cleavage step then releases the miRNA/miRNA* duplex. The precision of this cleavage is also dependent on whether the processing happened from base to loop or loop to base. This bi-directional activity is also attributed to the heterogeneity of pri-miRNA structures (Bologna et al. 2009, 2013; Song et al. 2010; Werner et al. 2010). DCL1 activity also requires several other proteins for proper functioning. Hyponastic Leaves 1/Double stranded RNA Binding protein 1 (HYL1/DRB1) is a double-stranded RNA binding protein that is thought to probably bind the miRNA/miRNA* double-stranded region and guide proper pri-miRNA cleavage (Kurihara, 2005). Similarly, zinc finger protein SE, binds single-stranded RNA regions of pri-miRNAs and helps in proper positioning of pri-miRNAs at the catalytic site of DCL1 (Lobbes et al. 2006; Laubinger et al. 2008). Another protein Tough (TGH) binds single-stranded RNA and is shown to promote DCL1 activity (Ren et al. 2012b). All these protcins, DCL1, HYL1, SE, and TGH are known to interact physically. The plant Microprocessor is largely considered to be formed by DCL1, HYL1, and SE; while the inclusion of TGH in the Microprocessor complex is not yet definitive.

Recently, Chromatin Remodeling Factor (CHR2) has also been shown to be involved in miRNA biogenesis (Wang et al. 2018b). CHR2 is a member of the SWI/SNF chromatin remodeling complex and has ATPase activity. Wang and colleagues show that CHR2 positively affects *MIR* gene transcription and hence leads to higher levels of pri-miRNAs, but when it associates with SE it remodels pri-miRNAs in a way that they are no more suitable substrates for DCL1 mediated cleavage. This remodeling of pri-miRNAs is a result of non-canonical RNA helicase activity of CHR2. The processing of pri-miRNAs thus can also be affected by any modifications that can result in altered structures of pri-miRNAs or can be identified by some specific proteins. Methylation of adenosine at N6 position (m^6A) is one such modification that has been shown to positively affect miRNA production in animals (Alarcón et al. 2015). A recent *Arabidopsis* based study showed that mRNA adenosine methylase (MTA), catalytic component of m^6A methyltransferase complex, binds to and methylates pri-miRNAs and affects their processing. MTA also interacts with RNA Pol II and TGH and thus the possibility that it may also affect *MIR* gene transcription cannot be ruled out (Bhat et al. 2019). Evidence of another regulatory step came in the form of retrograde signaling from chloroplast (Fang et al. 2019). Fang and colleagues showed that nuclear Exoribonuclease 2 (XRN2) degrades pri-miRNAs and it is inhibited by 3′-phosphoadenosine 5′-phosphate (PAP). The levels of PAP are further influenced by tocopherols produced in chloroplast. Another example of retrograde signaling that regulates miRNA biogenesis came from a study done on barley (Świda-Barteczka et al. 2018). Whirly 1, a DNA/RNA binding protein, that is localized in nucleus as well as chloroplast was shown to positively affect biogenesis of selected miRNAs under light stress.

After miRNA/miRNA* duplex is released following DCL1 cleavage, it undergoes 2′-*O*-methylation at the 3′ ends by HUA1 enhancer1 (HEN1) (Yang et al. 2006). This modification stabilizes the miRNA/miRNA* duplex and prevents it from degrading. In absence of HEN1 and hence the 3′ end methylation, Hen1 suppressor1 (HESO1) uridylates the 3′ ends of miRNA/miRNA* duplex leading to their degradation (Wang et al. 2018a).

In the final stage, the mature miRNA strand of miRNA/miRNA* is selectively loaded into the effector protein of the RISC: AGO1 (Vaucheret et al. 2004; Baumberger and Baulcombe 2005). The strand selectivity is facilitated by HYL1 (Eamens et al. 2009) and the miRNA* strand is degraded. Traditionally, the loading was thought to occur in the cytoplasm after the miRNA/miRNA* strand gets exported from the nucleus by an exportin protein Hasty (HST) (Park et al. 2005). Recently, there is increasing evidence of RISC assembly happening in the nucleus (Bologna et al. 2018). Once loaded, RISC then executes post-transcriptional gene regulation either via cleaving the target mRNA or by translational inhibition. Cleavage by AGO1 requires an almost perfect complementarity between the miRNA and target mRNA. Cleavage triggered by miRNAs produces fragments of target mRNAs with exposed hydroxyl and phosphate groups on the 5′ and 3′ ends leading to their degradation. This is the most prevalent pathway of miRNA directed gene regulation in plants. On the other hand, some miRNAs, and a fraction of AGO1 have been shown to be associated with polysomes. It has been shown that binding of RISC can sterically hinder ribosome binding to the mRNA, thus preventing translation (Brodersen et al. 2009). In an alternate pathway, homologues of HYL1, DRB2, 3 and 5 are involved in the miRNA mediated translational inhibition of target mRNAs (Eamens et al. 2012; Reis et al. 2015a). Notably, DRB2 was found to affect the biogenesis of a certain subset of miRNAs, based on the structural characteristics of pri-miRNAs. The miRNAs derived from such DRB2 associated pri-miRNAs have been suggested to associate with AGO family members (possibly other than AGO1) and initiate translational inhibition. Furthermore, some miRNAs may even work using both pathways. Studies have shown that several factors like the location of RISC binding on mRNA and the place where miRNA and its cognate mRNA are expressed might also affect this selection. Figure 1.1 summarizes major steps of miRNA biogenesis in plants.

A myriad of factors affects miRNA biogenesis at various different steps and provide tight regulation for the process. Considering that miRNAs play a crucial role in plant survival this regulation plays an equally important role. Thus, often the rate of *MIR* gene transcription may or may not be reflected in the final miRNA level. We throw some more light on these regulatory networks and describe them in detail below.

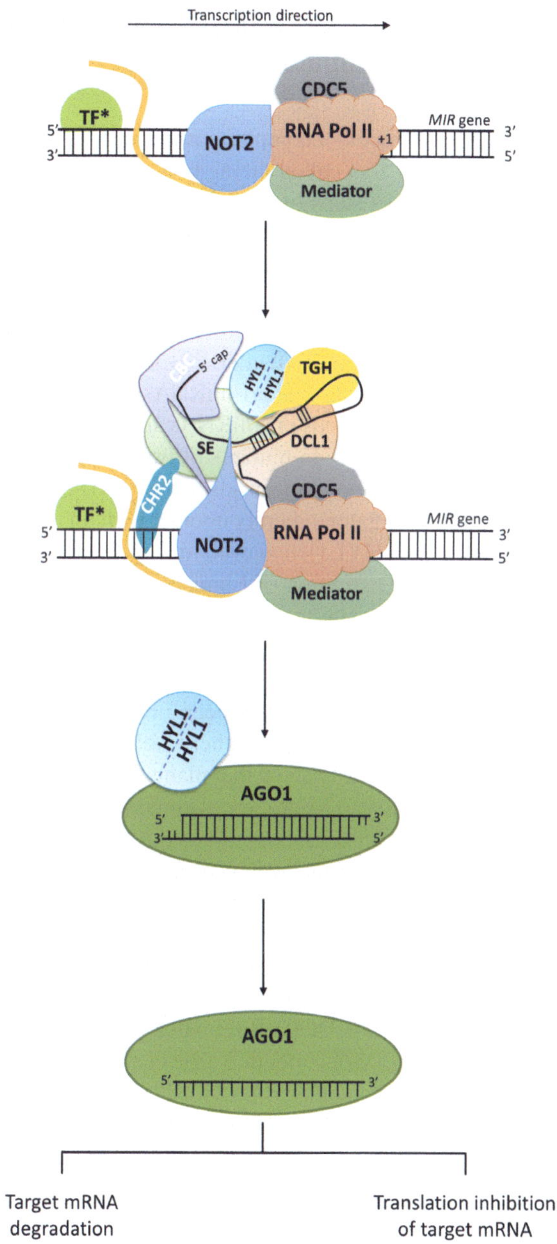

Fig. 1.1 Major proteins involved in miRNA biogenesis and action: from transcription to RISC formation. RNA Pol II transcribes *MIR* genes with factors like Mediator complex, Cell Division Cycle 5 (CDC5) and Negative on TATA less 2a (NOT2a)/2b positively affecting transcription. TF* = developmental or environmental stimuli dependent transcription factors. Other miRNA biogenesis related proteins like Serrate (SE), Dicer Like 1 (DCL1), Tough (TGH), Hyponastic Leaves 1 (HYL1), Cap Binding Complex (CBC) and Chromatin Remodeling Factor (CHR2) then facilitate proper processing of primary transcripts. HYL1 assisted strand selection into AGO1 results in the formation of active RISC

1.2 Regulation of microRNA Biogenesis Via Regulation of Microprocessor Complex

In wild type plants, two Microprocessor complex components (DCL1 and HYL1) are localized in specialized nuclear foci called dicing bodies (D-bodies) where according to current knowledge the first steps of pri-miRNA processing occur (Fang and Spector 2007). Cellular mislocalization of particular component of Microprocessor complex will affect miRNA biogenesis. Currently, several factors are considered to be important for D-body formation. For example, in a knock-out mutant of MOS2 (Modifier of Snc1), which is an RNA-binding protein, HYL1 localization was found relatively homogeneous in the nucleus (Wu et al. 2013b). Mislocalization of HYL1 in *mos2-2* mutant results in accumulation of pri-miRNAs and reduced level of mature miRNAs. It is proposed that MOS2 binds pri-miRNAs and promotes their processing via their recruitment to dicing bodies. In mutants of two elongator subunits (*elp2-2* and *elp5-1*) a reduced number of D-bodies were also observed (Fang et al. 2015). Fang and colleagues showed that Elongator subunits interact with all core components of Microprocessor (SE, DCL1, and HYL1), however, in *elp2-2* and *elp5-1* mutants only DCL1 localization is affected. Disruption of Elongator complex results in reduced level of mature miRNA and, opposite to *mos2-2,* reduced level of pri-miRNAs as well (Wu et al. 2013b; Fang et al. 2015). Interestingly, in plants lacking NOT2a and NOT2b proteins, components of transcriptional machinery of *MIR* genes, increased number of nuclear foci containing DCL1 has been observed. However, increased number of dicing bodies is not reflected in a higher level of mature miRNAs (Wang et al. 2013). These observations clearly show that plants regulate miRNA biogenesis through regulation of accessibility to Microprocessor components.

The core components of Microprocessor complex (DCL1 and SE) are tightly regulated in plants at the posttranscriptional level by negative feedback loops and also at the posttranslational level by phospho-regulation of HYL1. Posttranscriptional regulation of DCL1 transcript abundance was first described by Carrington and coworkers in 2003 (Xie et al. 2003). Xie and colleagues showed that DCL1 transcript, which is relatively low-abundant in wild type plants, is downregulated by miR162. This feedback loop allows to increase the DCL1 level when miRNA biogenesis pathway is suppressed for example during virus infection (Xie et al. 2003). Moreover, further fine-tuning of DCL1 transcript abundance is provided by the production of miR838, which is encoded in 14th intron of DCL1 and negatively regulates the transcript level of DCL1 (Rajagopalan et al. 2006).

SE transcript level is also regulated by a negative feedback loop. This loop depends on miR863 which decreases SE expression via translation inhibition (Niu et al. 2016). Interestingly, another main factor from miRNA biogenesis pathway (AGO1) is also regulated via posttranscriptional negative feedback loop. AGO1 transcript is a target of mRNA cleavage mediated by miR168 (Vaucheret et al. 2004). The third core component of Microprocessor HYL1 is regulated via posttranslational modification: phosphorylation. Phosphorylation status of HYL1 affects its cellular localization, stability and also activity (Manavella et al. 2012; Su et al. 2017; Achkar et al. 2018).

Currently, four proteins were found to be mainly involved in the phospho-regulation of HYL1: two kinases (Mitogen-Activated Protein Kinase 3 (MPK3) and SNF1-Related Protein Kinase Subfamily 2 (SnRK2)) and two phosphatases (C-Terminal Domain Phosphatase-Like 1 (CPL1) and Protein Phosphatase 4 (PP4)) (Raghuram et al. 2015; Yan et al. 2017) (Fig. 1.2). Different HYL1 phosphorylation patterns might have different consequences for miRNA expression. For example, in knock-out *MPK3* mutant, HYL1 protein accumulates, and consequently the expression of mature miRNAs is significantly higher than in wild-type plants. However, in a *snrk2.2/3/6* triple mutant, the levels of HYL1 and mature miRNAs were found to be decreased (Yan et al. 2017). MPK3 phosphorylation site of HYL1 resides in the first dsRBD (dsRNA Binding Domain) in the 'TP' motif (Bhagat et al. 2018) whereas SnRK2 phosphorylation site(s) is still not identified. The functional specificity of SnRK2 on regulation of HYL1 is difficult to reveal because SnRK2 subfamily consists of 10 redundant members (SnRK2.1-10) (Yan et al. 2017).

The detailed mechanism of how phosphatases regulate HYL1 is not entirely understood. It was demonstrated that CPL1 which dephosphorylates HYL1 protein also interacts with SE in vivo (Jeong et al. 2013). It was predicted that HYL1 contains seven serine residues which might be potentially phosphorylated (Manavella et al. 2012). Manavella and colleagues showed that two serine residues, which are

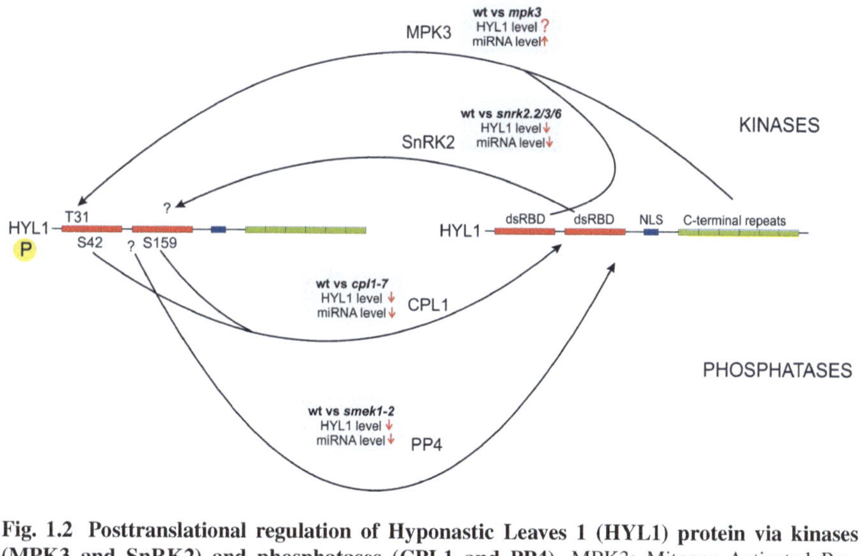

Fig. 1.2 Posttranslational regulation of Hyponastic Leaves 1 (HYL1) protein via kinases (MPK3 and SnRK2) and phosphatases (CPL1 and PP4). MPK3: Mitogen-Activated Protein Kinase 3, SnRK2: SNF1-Related Protein Kinase Subfamily 2, CPL1: C-Terminal Domain Phosphatase-Like1 and PP4: Protein Phosphatase4. HYL1 protein structure: dsRBD - double-stranded RNA binding domain, NLS—nuclear localization signal, T31—threonine; S42 and S159—serines. (?)—indicates that precise phosphorylation sites of HYL1 are unknown; wt—wild type plants. Arrows point to the particular or unknown serine, threonine or tyrosine residues that may be subjected to phosphorylation/dephosphorylation

in double-stranded RNA binding motifs, are especially important for HYL1 function. They demonstrated that expression of phosphomimicry allele of HYL1 (S42D or S159D) in *hyl1-2* mutant did not recover wild type plant phenotype. Authors demonstrated that hyperphosphorylated HYL1 is inactive in miRNA biogenesis. However, Achkar and colleagues showed that this hyperphosphorylated HYL1 can create a nuclear reserve pool of HYL1 (Manavella et al. 2012; Achkar et al. 2018). Quick dephosphorylation of HYL1, upon changes in environmental conditions, can rapidly restore HYL1 activity and miRNA production. PP4 is a highly conserved serine/threonine protein phosphatase among eukaryotes and the PP4 phosphatase complex consists of catalytic and regulatory subunits (Gingras et al. 2005; Kataya et al. 2017; Su et al. 2017). In *A. thaliana*, the catalytic subunits are encoded by two genes (*PPX1* and *PPX2*), and the proteins share 93% sequence identity and play redundant biological functions (Pujol et al. 2000; Wang et al. 2019). Regulatory subunit of PP4 phosphatase, in *A. thaliana*, is SMEK1/PSY2 which was recently also named PP4R3A (Wang et al. 2019). Currently three models of action of PP4 complex in miRNA biogenesis regulation were proposed. First model proposed by Su and colleagues assumes that reduced level of miRNA in a *smek1-2* mutant is a result of accelerated degradation of hyperphosphorylated HYL1 (Su et al. 2017). Second and third model was proposed recently by Wang et al. (2019). The second model assumes that SMEK1 recruits RNA Pol II to *MIR* genes to promote their transcription and at the same time dephosphorylates HYL1 to facilitate pri-miRNA processing (Wang et al. 2019). Wang and colleagues took also into consideration a third model, that PP4 might dephosphorylate chromatin-associated proteins which enables RNA Pol II and HYL1 to access the chromatin (Wang et al. 2019). Additionally, apart from phospho-regulation, HYL1 degradation is also regulated by yet unidentified cytoplasmic protease. Although protease is still not identified in *Arabidopsis*, it is known that RING-finger E3 ligase COP1 (Constitutive Photomorphogenic 1) indirectly protects HYL1 during daytime by inhibiting HYL1 protease. During the night, COP1 remobilizes to the nucleus, allowing the protease to degrade HYL1 which causes a reduction of correctly processed miRNAs (Cho et al. 2014).

1.3 Regulation of microRNA Biogenesis Via pri-miRNA Splicing, Alternative polyA Site Selection and miPEPs

The majority of *MIR* genes in *A. thaliana* represent independent transcription units (67%), while the remaining miRNAs are present within genes encoding either proteins or long non-coding RNAs (Brown et al. 2008; Shomron and Levy 2009). *MIR* genes vary in length (from 319 bp for *MIR165a* to 4975 bp for *MIR472*) and often contain introns. Thus, miRNAs can be derived either from exons (exonic miRNAs) or introns (intronic miRNAs). This complex structure of *MIR* genes allows the regulation of miRNA biogenesis via alternative splicing (AS), alternative polyA site selection and the presence of miRNA-encoded peptides (miPEPs). Below we discuss this regulatory network in detail.

Studies revealed that the presence of introns and active 5′ ss stimulates the production of miRNAs derived from exonic regions of intron-containing pri-miRNAs (Bielewicz et al. 2013; Schwab et al. 2013). In two independent studies, it was observed that the production of miR163 was positively affected by the presence of intron and active 5′ ss. As expected, removal of the intron or inactivation of 5′ ss leads to a decrease in miR163 production and similar results were obtained for miR161 and miR172a.

In the case of miRNAs derived from introns, splicing inhibition or inactivation of 5′ ss leads to an increase in miRNA production as was demonstrated by Knop and colleagues (Knop et al. 2017). Interestingly, when the stem-loop structure (which contains miRNA) was moved into the exon, mutation of the 5′ ss resulted in a decrease in the miRNA level, an effect that was previously observed for exonic miRNAs. The inactivation of 5′ ss was also accompanied by increased proximal polyA site selection that is located downstream of stem loop structure (Knop et al. 2017).

Knop and colleagues attributed the observations described above to the interactions of SE (Microprocessor component) with U1 snRNP (spliceosome component) and CBC (Knop et al. 2017). U1 snRNP binds to the 5′ ss of all U2 type introns and is necessary for the action of spliceosome (Mount et al. 1983; Rosbash and Seraphin 1991). In the case of intronic miRNAs it is proposed that in wild type plants, binding of CBC to the cap structure of primary transcripts, and binding of U1 snRNP to the 5′ ss leads to limited access of SE to Microprocessor complex that is located downstream of 5′ ss. On the other hand, when 5′ ss is inactivated and U1snRNP cannot bind the transcript, SE is freely available for Microprocessor, leading to upregulation of miRNA levels. In the case of exonic miRNAs, the components of Microprocessor are recruited and bind to the hairpin-loop structure immediately after CBC binding. U1 snRNP binds to the active 5′ ss and stabilizes SE and Microprocessor interactions leading to increased exonic miRNA production (Bielewicz et al. 2013; Knop et al. 2017) (Fig. 1.3).

miRNA response to environmental changes has been well documented (Allen et al. 2004; Kruszka et al. 2012; Barciszewska-Pacak et al. 2015b). Interestingly, Jia and Rock reported that abscisic acid (ABA) regulates the accumulation of selected miRNAs by affecting pri-miRNA alternative splicing (Jia and Rock 2013). MiR846 and miR842 are encoded within one transcription unit but are generated from alternatively spliced isoforms. In response to ABA treatment, the level of miR846 was decreased which was accompanied by the decreased level of alternatively spliced isoform from which miR846 is generated. Another example shows that heat stress affects the level of miRNA by alternative splicing inhibition. High temperature increased the accumulation of miR402 which correlates with the inhibition of miRNA-carrying intron splicing and higher selection of proximal polyadenylation site (located downstream of stem loop structure) (Knop et al. 2017).

These experiments showed that the presence of introns, their splicing and active 5′ ss's play a crucial role in the regulation of pri-miRNA processing pathway as well as in the accumulation of proper amount of miRNAs in the plant cell.

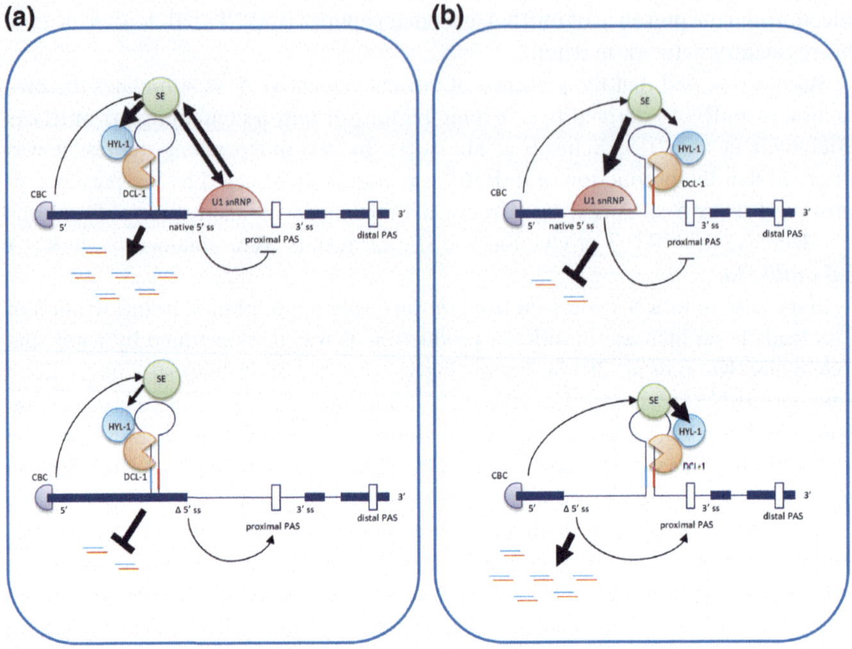

Fig. 1.3 A model of the interactions between the 5′ ss, Microprocessor and the U1 snRNP of selected exonic (A) and intronic (B) microRNA in plants. CBC: Cap Binding Complex, SE: Serrate; HYL1: Hyponastic Leaves 1; DCL1: Dicer-Like 1. Boxes represent exons and lines depict introns. Light blue and red parts of the stem-loop structure mark miRNA and miRNA*, respectively. Thick arrows show strong interactions, while thin arrows show weak interactions. The no-headed arrows point to the inhibitory effect. Open boxes mark proximal and distal polyA sites (PAS). Native and Δ5′ ss mean active and inactive 5′ ss, respectively

In the case of mRNA processing, it is known that alternative splicing and alternative polyadenylation site selection within the pre-mRNA can lead to production of many mRNA isoforms with different stability or protein-coding capacities (Shen et al. 2008; de Lorenzo et al. 2017). Recent studies revealed the presence of open reading frames (ORFs) in plant pri-miRNAs as well. These ORFs, located at the 5′ end of pri-miRNAs, encode miRNA-encoded peptides (miPEPs) (Lauressergues et al. 2015). It was shown that miPEP171b derived from pri-miRNA171b in *Medicago truncatula* and miPEP165a encoded by pri-miRNA165a in *A. thaliana* enhance the accumulation of their corresponding miRNAs. To investigate the function of miPEP171b, transformation of tobacco leaves to express native and mutant forms of pri-miRNA171b was carried out. The expression of the native pri-miR171b resulted in the large amount of mature miR171b, whereas the expression of a mutant, in which ORF was deleted, produced decreased amount of miR171b. What is more, overexpression of miPEP171b in *M. truncatula* roots led to increased accumulation of endogenous miR171 (Lauressergues et al. 2015).

Altogether, these data show that maturation of pri-miRNAs is a very dynamic process. Splicing, polyA site selection, and miPEPs presence strongly affect the level of mature miRNA.

1.4 microRNA Action and Turnover in the Cytoplasm

1.4.1 Export of miRNAs from Nucleus to the Cytoplasm

Cytoplasm is the main place of miRNA action (Lund et al. 2004; Park et al. 2005; Rogers and Chen 2013). As it was mentioned in the introduction, miRNA/miRNA* duplexes are cut out from pre-miRNA in the nucleus, protected from degradation by 2′-O-methylation and exported to the cytoplasm where they can be incorporated into multiprotein complex RISC to guide post-transcriptional regulation of their target genes (Park et al. 2005; Voinnet 2009). HST protein is responsible for export of miRNA/miRNA*duplexes from nucleus to the cytoplasm in Ran GTP-dependent manner. Park et al., in 2005 examined the levels of miRNAs in the three *HST* mutants (*hst-1, hst-6, hst-3). Hst-1* and *hst-6* are putative null alleles and *hst-3* bears 3-nt deletion in the first exon of *HST* gene which blocks the interaction between truncated HST and AtRAN1 (Ran-GTP) as it was shown in yeast two-hybrid system (Bollman et al. 2003). Accumulation of miRNAs in the *hasty* mutants, especially *hst-3* is reduced for several but not all miRNAs. For instance, miRNA172 that acts primarily in translation repression, is not affected, which suggests the existence of a pool of miRNAs that are exported in Hasty-independent manner (Park et al. 2005). Indeed, study by Bologna et al. show that AGO1 mediates export of miRNAs as well, as AGO1 contains functional nuclear localization signal (NLS) and nuclear export signal (NES) in N-terminus and shuttles between nucleus and cytoplasm (Bologna et al. 2018). Moreover, authors substituted four hydrophobic residues of NES signal for alanines creating GFP-AGO1^{mNES} mutant line displaying only nuclear localization of AGO1. Using these lines and plants containing cytosolic-dominant form of GFP-AGO1 RNA immunoprecipitation (RIP) was conducted and 18-35-nt-long sRNAs were subjected to Illumina-HiSeQ. More than 90% of cargo loaded by AGO1 in both mutants was 21-nt long miRNAs. Most importantly cytosolic-dominant AGO1 carried the same miRNA cohorts as its nuclear equivalent.

1.4.2 microRNA Strand Selection in RISC Complex

In the cytoplasm, one of the strands from miRNA/miRNA* duplex is selected to be incorporated into AGO protein to facilitate target gene regulation (Iki et al. 2010). Eamens and colleagues assessed the effects of 5′ terminus stability within miRNA/miRNA* duplex by introducing series of artificial miRNA/miRNA* duplexes with different 5′ terminus thermostability. Parallelly, they introduced sense or antisense version of the target transgene, β-glucuronidase (GUS). Authors observed preferential retention within AGO1 of a sRNA strand that displayed lower thermostability at the 5′ terminus coupled with down-regulation of corresponding target. Previous studies by Mi and colleagues; and Takeda and colleagues also show the preference of AGO proteins toward the strand with particular nucleotide at the 5′ end. AGO1 favors incorporation of miRNAs with uridine in the first position in 5′ terminus, while AGO2 and AGO4 recruit small RNAs with a 5′ terminal adenosine and AGO5 is associated with miRNA strands with cytosine in 5′ end (Mi et al. 2008; Takeda et al. 2008; Eamens et al. 2009). Altogether there are 10 AGO proteins in *Arabidopsis thaliana.* All recruit small RNAs and take part in regulation of expression of their target genes (Qu et al. 2008; Zhu et al. 2011; Jouannet et al. 2012; Hernandez-Lagana et al. 2016). AGO1 however, forms in the cytoplasm a complex with a dimer of HSP90 protein that binds ATP and causes conformational changes in AGO1 that allow for mature miRNA strand to be incorporated (Iki et al. 2010). Next, hydrolysis of ATP causes dissociation of AGO1 from HSP90 dimer and in consequence passenger strand is removed. Recent studies suggest the involvement of HYL1 and the CPL1 phosphatase in strand selection. Analyses of *hyl1* and *cpl1* mutants revealed accumulation of miRNA* strands in these plants (Eamens et al. 2009; Manavella et al. 2012). Generally the passenger strand (miRNA*) is subjected to degradation, but there are more and more reports concerning accumulation of some of miRNA* (Devers et al. 2011; Peng et al. 2011; Jia et al. 2014; Aceto et al. 2014; Yan et al. 2015; Zhang et al. 2016). Several studies have demonstrated that many miRNA*s have abundance similar to or even higher than their corresponding mature miRNAs in specific biological processes, in specific tissues and during development. For instance, miR156a*, miR164b*, and miR535* of *Betulalu minifera*; miR82* of *Morus notabilis* in leaf tissue; miR166*, miR159*, and miR171* of *Orchis italica*; miR171c*, miR369a*, and miR2612a*/b* of *Medicago truncatula*; and miR169 h*, miR408* and miR529a* of rice were all more abundant than their corresponding mature miRNAs (Devers et al. 2011; Peng et al. 2011; Jia et al. 2014; Aceto et al. 2014; Yan et al. 2015; Zhang et al. 2016).

1.4.3 Stabilization of microRNA in Cytoplasm

One of the processes that impacts stability of plant miRNAs is 3′truncation, uridylation, and adenylation. Generally, miRNA/miRNA* duplexes are protected from degradation by 2′-O-methylation mediated by HEN1 methyltransferase (Li et al. 2005; Yu et al. 2005). However, plant miRNAs also undergo common post-transcriptional modifications such as uridylation and adenylation (Lu et al. 2009; Wyman et al. 2011). The enzyme responsible for 3′ adenylation of miRNAs remains unidentified but it is known that in Arabidopsis both HESO1 and UTP:RNA Uridyltransferase1 (URT1) exhibit nucleotidyl transferase activities and uridylate unmethylated miRNAs leading them to degradation (Zhao et al. 2012). What is more, HESO1 prefers miRNAs with uracyl on 3′ terminus while URT1 favors A-ending miRNAs as a substrate (Ren et al. 2012a; Zhao et al. 2012; Tu et al. 2015; Wang et al. 2015). Methylation of miRNAs does not inhibit the function of Small RNA Degrading Nuclease (SDN), a 3′–5′ exonuclease that is responsible for the 3′ truncation of mature miRNAs. Following truncation further degradation by other, yet unidentified enzymes, of plant miRNAs takes place (Ramachandran and Chen 2008; Yu et al. 2017). Also, HESO1, URT1, and SDN1 can operate on AGO bound miRNAs (Zhao et al. 2012; Tu et al. 2015). Another layer of regulation of miRNA stability is the ability of naturally occurring molecular sponges to sequester miRNAs. In *Arabidopsis*, miR399 is regulated by a native transcript with a miR399 binding site from the *IPS1* (*Induced by Phosphate Starvation 1*) locus (Franco-Zorrilla et al. 2007). A 3-nt bulge at the cleavage site within the *IPS1* transcript abolishes miR399 mediated cleavage. *IPS1* transcript mimics target sequence and bounds miR399 reducing its activity by triggering miRNA degradation (Franco-Zorrilla et al. 2007). Genome-wide bioinformatics analyses indicate that many transcripts can serve as potential endogenous molecular sponges to regulate miRNA activity (Ivashuta et al. 2011; Wu et al. 2013a). Figure 1.4 depicts cytoplasmic miRNA degradation pathways.

1.4.4 microRNA Mode of Action: mRNA Cleavage or Translational Inhibition?

miRNA-mediated regulation of gene expression is achieved through cleavage of targeted mRNA or translational repression (Chen 2009; Voinnet 2009; Rogers and Chen 2013). miRNA-mediated slicing of target mRNA occurs in specific cleavage site that is located between residues paired to nucleotides 10 and 11 counting from 5′end of miRNA sequence (Llave et al. 2002). AGO protein contains PIWI domain that is responsible for the cleavage. PIWI domain forms an RNase H—like fold and exhibits endonuclease activity; this activity has been demonstrated for *Arabidopsis* AGO1, AGO2, AGO4, AGO7, and AGO10 (Mi et al. 2008; Montgomery et al. 2008; Takeda et al. 2008; Maunoury and Vaucheret 2011; Ji et al. 2011; Zhu et al. 2011). Eamens and colleagues suggested that DRB3 and DRB5 proteins are required for cleavage of miRNA target mRNAs and other DRB family members (DRB1 and DRB2) are involved in proper processing of miRNA precursors (Eamens et al. 2009, 2012). Introduction of construct with *PRI-MIR159B* delivering artificial miRNA that targets Phytoene Desaturase (PDS) gene into the wild type and *drb1*, *drb2*, *drb3*, *drb5*, and triple mutant *drb235* resulted in photo-bleached phenotype only in wild type plants. However, the processing of pri-miRNA was defective only in *drb1*, *drb2*, and *drb235* transgenic lines suggesting that DRB3 and DRB5 proteins play significant role in facilitating miRNA-mediated cleavage (Eamens et al. 2012). In *Arabidopsis*, Exoribonuclease 4 (XRN4), a 5′ to 3′ exonuclease, is responsible for degrading the 3′ fragments of cleaved mRNA (Souret et al. 2004). 5′ fragments of cleaved mRNA are

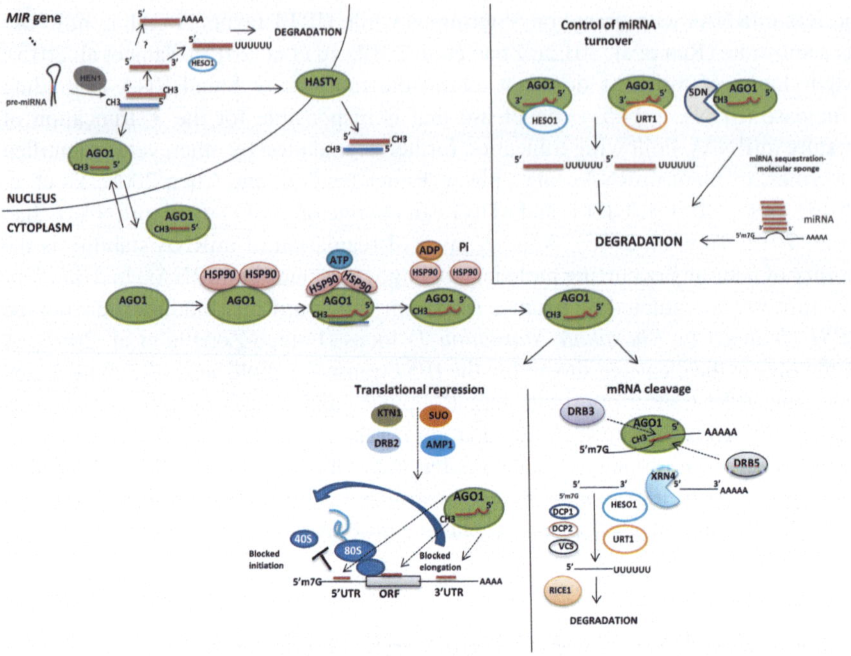

Fig. 1.4 miRNA action and turnover in the cytoplasm. miRNA is exported from the nucleus to the cytoplasm via Hasty (HST) and Argonaute 1 (AGO1) activities. In the cytoplasm, AGO1 forms a complex with a dimer of Heat Shock Protein 90 (HSP90) protein and miRNA strand is incorporated into AGO1 protein when ATP binds to a dimer of HSP90. Upon hydrolysis of ATP, passenger strand is released and degraded. miRNA loaded into AGO1 protein mediates target mRNA-cleavage or translational repression. Double-Stranded RNA Binding 3 (DRB3) and Double-Stranded RNA Binding 5 (DRB5) are involved in miRNA-mediated mRNA cleavage and Exoribonuclease 4 (XRN4) exonuclease is responsible for degrading 3′ fragments of cleaved mRNA. 5′ mRNA fragments are decapped by Decapping 1 (DCP1), Decapping 2 (DCP2) and Varicose (VCS) and 3′ ends are polyuridylated by HEN1 Suppressor 1 (HESO1) and UTP:RNA Uridylyltransferase 1 (URT1). RISC Interacting Clearing 3′-5′ Exoribonuclease 1 (RICE1) is responsible for the degradation of 5′ fragments of cleaved mRNA. miRNA can also regulate the expression of target on translational level. Double-Stranded RNA Binding 2 (DRB2), SUO, Katanin 1 (KTN1) and Altered Meristem Program 1 (AMP1) proteins are involved in miRNA-mediated inhibition of translation. Turnover of miRNAs is controlled by polyuridylation via HESO1 and URT1 activity as well as function of Small RNA Degrading Nuclease (SDN) exonucleases (see inset part of the figure). Sequestration of miRNAs to molecular sponges mimicking target sequence also triggers miRNA degradation

decapped by Decapping 1 (DCP1), Decapping 2 (DCP2), and Varicose (VCS) (Motomura et al. 2012; Maldonado-Bonilla 2014). HESO1 and URT1 polyuridylate the 3′ ends of 5′ fragments in vivo and in vitro (Ren et al. 2012a; Wang et al. 2015). RISC Interacting Clearing 3′–5′ Exoribonuclease 1 (RICE1) is responsible for the degradation of uridylated 5′ mRNA fragments in *Arabidopsis*, because these uridylated fragments are over accumulated in plants expressing a catalytically inactive RICE1 (Zhang et al. 2017). The cytoplasmic exosome may also play a role, as its cofactor's subunits, Superkiller 2 (SKI2), SKI3, and SKI8 are required for the degradation of RISC generated 5′ mRNA fragments (Branscheid et al. 2015). miRNA-mediated translational repression is less frequently observed. Well known examples of this process are AP2 and Squamosa Protein Binding Protein-Like 3 (SPL3) proteins regulated by miR172 and miR156/7, respectively. Overexpression of miR172 and miR156/7 did not cause changes in the level of AP2 and SPL3 transcripts but amounts of proteins were affected (Aukerman and Sakai 2003; Chen et al. 2004; Gandikota et al. 2007). DRB2 gene is strongly connected to miRNA-mediated translational regulation. Studies from 2015 by Reis and colleagues revealed strong accumulation of miRNA targets on protein level coupled with unchanged mRNA levels in *drb2* mutant. Similar correlation was observed in the case of triple mutant (*drb235*) but not in *drb3* and *drb5*. In *drb1* transgenic line both protein and mRNA levels of miRNA targets were elevated. These findings suggest that DRB1 and DRB2 play a role in determining miRNA mode of action but DRB2 is a sole DRB family member required for miRNA-guided translational inhibition (Reis et al. 2015b). Other proteins may be involved in regulation of miRNA-mediated translational repression. Mutations in Katanin 1(KTN1) connected to severing microtubules, GW-repeat protein SUO and ER membrane protein Altered Meristem Program 1 (AMP1) cause the alteration of miRNA-mediated translational repression and accumulation of target transcript in polysomal fractions (Brodersen et al. 2008; Yang et al. 2012; Li et al. 2013). In the mutant with knock-down of *AMP1* the expression of miRNAs involved in translational regulation was not changed, however accumulation of their corresponding proteins was higher with no effects on mRNA level. Using as an example miRNA398 and its target Cu/Zn Superoxide Dismutase (CDS2) authors assessed the levels of newly translated protein (labeling with 35S-Methionine) and compared them between wild type and mutant plants with knock-down of *AMP1*. They observed the accumulation of CDS2 even when the mutants were subjected to conditions causing up-regulation of miR398 (copper deprivation) proving the importance of AMP1 for miRNA-mediated translational repression (Li et al. 2013). Figure 1.4 presents the regulation pathways of miRNA-guided mRNA cleavage and/or translation inhibition.

1.5 Conclusions

miRNAs are small but incredibly important elements of plant metabolic pathways, thanks to their role in the posttranscriptional gene regulation. Accordingly, the production and action of miRNAs themselves need to be tightly controlled for proper functioning of plant cells. Both, over and underexpression of miRNAs are detrimental for plants. Overexpression of miR319 has been shown to cause excessive cell division while underexpression causes arrest of growth at seedling stage (Palatnik et al. 2003). Similarly, overexpression of miR156a leads to delayed flowering (Wu and Poethig 2006); miR396a/b overexpression leads to reduced cell number (due to over repression of Growth Response Factor (GRF)) while also providing drought tolerance (Liu et al. 2009) and overexpression of miR165 causes loss of shoot apical meristem (SAM) and defects in vascular growth. Owing to the importance of miRNAs in plant development and response to environmental cues, their production must be under tight regulatory control. Based on the knowledge so far, miRNA production is affected by many direct or indirect factors ranging from the regulation of Microprocessor proteins and proteins encoded by *MIR* genes themselves to pri-miRNA processing regulation. With more studies focusing on miRNA expression and regulation, much more information will come forth and very likely reveal even more complex network of pathways regarding their biogenesis.

Acknowledgements This work was supported by the NCN grants: 2017/, 27/N/NZ1/0020 to SSB 2016/23/D/NZ1/00152 to DB, 2016/23/N/NZ1/00005 to AS, 2013/10/A/NZ1/00557 to AJ and 2016/23/B/NZ9/00862 to Z S-K.

References

Aceto S, Sica M, De Paolo S et al (2014) The analysis of the inflorescence miRNome of the *Orchid Orchis italica* reveals a DEF-like MADS-box gene as a new miRNA target. PLoS ONE 9:e97839

Achkar NP, Cho SK, Poulsen C et al (2018) A quick HYL1-dependent reactivation of microRNA production is required for a proper developmental response after extended periods of light deprivation. Dev Cell 46:236–247.e6. https://doi.org/10.1016/j.devcel.2018.06.014

Alarcón CR, Lee H, Goodarzi H et al (2015) N6-methyladenosine marks primary microRNAs for processing. Nature 519:482–485. https://doi.org/10.1038/nature14281

Allen E, Xie Z, Gustafson AM et al (2004) Evolution of microRNA genes by inverted duplication of target gene sequences in *Arabidopsis thaliana*. Nat Genet 36:1282–1290. https://doi.org/10.1038/ng1478

Aukerman MJ, Sakai H (2003) Regulation of flowering time and floral organ identity by a microRNA and its APETALA2-like target genes. Plant Cell 15:2730 LP–2741. https://doi.org/10.1105/tpc.016238

Baek D, Park HC, Kim MC, Yun D-J (2013) The role of Arabidopsis MYB2 in miR399f-mediated phosphate-starvation response. Plant Signal Behav 8:e23488. https://doi.org/10.4161/psb.23488

Baldrich P, Hsing Y-IC, San Segundo B (2016) Genome-wide analysis of polycistronic microRNAs in cultivated and wild rice. Genome Biol Evol 8:1104–1114. https://doi.org/10.1093/gbe/evw062

Barciszewska-Pacak M, Milanowska K, Knop K et al (2015a) Arabidopsis microRNA expression regulation in a wide range of abiotic stress responses. Front Plant Sci 6:1–14. https://doi.org/10.3389/fpls.2015.00410

Barciszewska-Pacak M, Milanowska K, Knop K et al (2015b) Arabidopsis microRNA expression regulation in a wide range of abiotic stress responses. Front Plant Sci 6:410. https://doi.org/10.3389/fpls.2015.00410

Baumberger N, Baulcombe DC (2005) Arabidopsis ARGONAUTE1 is an RNA Slicer that selectively recruits microRNAs and short interfering RNAs. Proc Natl Acad Sci 102:11928–11933. https://doi.org/10.1073/pnas.0505461102

Bhagat PK, Verma D, Badmi R, Sinha AK (2018) Dynamic regulation of HYL1 provides new insights into its multifaceted role in Arabidopsis. bioRxiv 396861. https://doi.org/10.1101/396861

Bhat SS, Bielewicz D, Grzelak N et al (2019) mRNA adenosine methylase (MTA) deposits m6A on pri-miRNAs to modulate miRNA biogenesis in *Arabidopsis thaliana*. bioRxiv 557900. https://doi.org/10.1101/557900

Bielewicz D, Kalak M, Kalyna M et al (2013) Introns of plant pri-miRNAs enhance miRNA biogenesis. EMBO Rep 14:622–628. https://doi.org/10.1038/embor.2013.62

Bollman KM, Aukerman MJ, Park M-Y et al (2003) HASTY, the Arabidopsis ortholog of exportin 5/MSN5, regulates phase change and morphogenesis. Development 130:1493–1504

Bologna NG, Mateos JL, Bresso EG, Palatnik JF (2009) A loop-to-base processing mechanism underlies the biogenesis of plant microRNAs miR319 and miR159. EMBO J 28:3646–3656. https://doi.org/10.1038/emboj.2009.292

Bologna NG, Schapire AL, Zhai J et al (2013) Multiple RNA recognition patterns during microRNA biogenesis in plants. Genome Res 23:1675–1689. https://doi.org/10.1101/gr.153387.112

Bologna NG, Iselin R, Abriata LA et al (2018) Short article nucleo-cytosolic shuttling of ARGONAUTE1 prompts a revised model of the plant microRNA short article nucleo-cytosolic shuttling of ARGONAUTE1 prompts a revised model of the plant microRNA pathway. Mol Cell 69:1–11. https://doi.org/10.1016/j.molcel.2018.01.007

Branscheid A, Marchais A, Schott G et al (2015) SKI2 mediates degradation of RISC 5′-cleavage fragments and prevents secondary siRNA production from miRNA targets in Arabidopsis. Nucleic Acids Res 43:10975–10988. https://doi.org/10.1093/nar/gkv1014

Brodersen P, Crete P, Robaglia C et al (2009) Biochemical evidence for translational repression by arabidopsis microRNAs. Plant Cell Online 21:1762–1768. https://doi.org/10.1105/tpc.108.063412

Brodersen P, Sakvarelidze-Achard L, Bruun-Rasmussen M et al (2008) Widespread translational inhibition by plant miRNAs and siRNAs. Science 320:1185–1190. https://doi.org/10.1126/science.1159151

Chen X (2009) Small RNAs and their roles in plant development. Annu Rev Cell Dev Biol 25:21–44. https://doi.org/10.1146/annurev.cellbio.042308.113417

Brown JWS, Marshall DF, Echeverria M (2008) Intronic noncoding RNAs and splicing. Trends Plant Sci 13:335–342. https://doi.org/10.1016/j.tplants.2008.04.010

Chen J, Li WX, Xie D et al (2004) Viral virulence protein suppresses RNA silencing-mediated defense but upregulates the role of microRNA in host gene expression. Plant Cell 16:1302–1313. https://doi.org/10.1105/tpc.018986

Cho SK, Ben Chaabane S, Shah P et al (2014) COP1 E3 ligase protects HYL1 to retain microRNA biogenesis. Nat Commun 5:5867

de Lorenzo L, Sorenson R, Bailey-Serres J, Hunt AG (2017) Noncanonical alternative polyadenylation contributes to gene regulation in response to hypoxia. Plant Cell 29:1262–1277. https://doi.org/10.1105/tpc.16.00746

Devers EA, Branscheid A, May P, Krajinski F (2011) Stars and symbiosis: microRNA- and microRNA*-mediated transcript cleavage involved in arbuscular mycorrhizal symbiosis. Plant Physiol 156:1990 LP–2010. https://doi.org/10.1104/pp.111.172627

Eamens AL, Smith NA, Curtin SJ, et al (2009) The Arabidopsis thaliana double-stranded RNA binding protein DRB1 directs guide strand selection from microRNA duplexes The Arabidopsis thaliana double-stranded RNA binding protein DRB1 directs guide strand selection from microRNA duplexes. 2219–2235. https://doi.org/10.1261/rna.1646909

Eamens AL, Kim KW, Waterhouse PM (2012) DRB2, DRB3 and DRB5 function in a non-canonical microRNA pathway in Arabidopsis thaliana. Plant Signal Behav 7:1224–1229. https://doi.org/10.4161/psb.21518

Fang Y, Spector DL (2007) Identification of nuclear dicing bodies containing proteins for microRNA biogenesis in living Arabidopsis plants. Curr Biol 17:818–823. https://doi.org/10.1016/j.cub.2007.04.005

Fang X, Cui Y, Li Y, Qi Y (2015) Transcription and processing of primary microRNAs are coupled by Elongator complex in Arabidopsis. Nat Plants 1:15075. https://doi.org/10.1038/nplants.2015.75

Fang X, Zhao G, Zhang S et al (2019) Chloroplast-to-nucleus signaling regulates microRNA biogenesis in Arabidopsis. Dev Cell 48:371–382.e4. https://doi.org/10.1016/j.devcel.2018.11.046

Franco-Zorrilla JM, Valli A, Todesco M et al (2007) Target mimicry provides a new mechanism for regulation of microRNA activity. Nat Genet 39:1033–1037. https://doi.org/10.1038/ng2079

Gandikota M, Birkenbihl RP, Hohmann S et al (2007) The miRNA156/157 recognition element in the 3′ UTR of the Arabidopsis SBP box gene SPL3 prevents early flowering by translational inhibition in seedlings. Plant J 49:683–693. https://doi.org/10.1111/j.1365-313X.2006.02983.x

Gingras A-C, Caballero M, Zarske M et al (2005) A novel, evolutionarily conserved protein phosphatase complex involved in cisplatin sensitivity. Mol Cell Proteomics 4:1725–1740. https://doi.org/10.1074/mcp.M500231-MCP200

Hernandez-Lagana E, Rodriguez-Leal D, Lua J, Vielle-Calzada J-P (2016) A multigenic network of ARGONAUTE4 clade members controls early megaspore formation in Arabidopsis. Genetics 204:1045–1056. https://doi.org/10.1534/genetics.116.188151

Iki T, Yoshikawa M, Nishikiori M et al (2010) In vitro assembly of plant RNA-induced silencing complexes facilitated by molecular chaperone HSP90. Mol Cell 39:282–291. https://doi.org/10.1016/j.molcel.2010.05.014

Ivashuta S, Banks IR, Wiggins BE et al (2011) Regulation of gene expression in plants through miRNA inactivation. PLoS ONE 6:e21330. https://doi.org/10.1371/journal.pone.0021330

Jeong IS, Aksoy E, Fukudome A et al (2013) Arabidopsis C-terminal domain phosphatase-like 1 functions in miRNA accumulation and DNA methylation. PLoS ONE 8:e74739. https://doi.org/10.1371/journal.pone.0074739

Ji L, Liu X, Yan J et al (2011) ARGONAUTE10 and ARGONAUTE1 regulate the termination of floral stem cells through two microRNAs in Arabidopsis. PLoS Genet 7:e1001358. https://doi.org/10.1371/journal.pgen.1001358

Jia F, Rock CD (2013) Jacalin lectin At5g28520 is regulated by ABA and miR846. Plant Signal Behav 8:e24563. https://doi.org/10.4161/psb.24563

Jia L, Zhang D, Qi X et al (2014) Identification of the conserved and novel miRNAs in *Mulberry* by high-throughput sequencing. PLoS ONE 9:e104409. https://doi.org/10.1371/journal.pone.0104409

Jones-Rhoades MW, Bartel DP (2004) Computational identification of plant microRNAs and their targets, including a stress-induced miRNA. Mol Cell 14:787–799. https://doi.org/10.1016/j.molcel.2004.05.027

Jouannet V, Moreno AB, Elmayan T et al (2012) Cytoplasmic Arabidopsis AGO7 accumulates in membrane-associated siRNA bodies and is required for ta-siRNA biogenesis. EMBO J 31:1704–1713. https://doi.org/10.1038/emboj.2012.20

Kataya ARA, Creighton MT, Napitupulu TP et al (2017) PLATINUM SENSITIVE 2 LIKE impacts growth, root morphology, seed set, and stress responses. PLoS ONE 12:e0180478. https://doi.org/10.1371/journal.pone.0180478

Kim YJ, Zheng B, Yu Y et al (2011) The role of Mediator in small and long noncoding RNA production in *Arabidopsis thaliana*. EMBO J 30:814–822. https://doi.org/10.1038/emboj.2011.3

Knop K, Stepien A, Barciszewska-Pacak M et al (2017) Active 5′ splice sites regulate the biogenesis efficiency of Arabidopsis microRNAs derived from intron-containing genes. Nucleic Acids Res 45:2757–2775. https://doi.org/10.1093/nar/gkw895

Kruszka K, Pieczynski M, Windels D et al (2012) Role of microRNAs and other sRNAs of plants in their changing environments. J Plant Physiol 169:1664–1672. https://doi.org/10.1016/j.jplph.2012.03.009

Kurihara Y (2005) The interaction between DCL1 and HYL1 is important for efficient and precise processing of pri-miRNA in plant microRNA biogenesis. RNA 12:206–212. https://doi.org/10.1261/rna.2146906

Kurihara Y, Watanabe Y (2004) Arabidopsis micro-RNA biogenesis through Dicer-like 1 protein functions. Proc Natl Acad Sci U S A 101:12753–12758. https://doi.org/10.1073/pnas.0403115101

Laubinger S, Sachsenberg T, Zeller G et al (2008) Dual roles of the nuclear cap-binding complex and SERRATE in pre-mRNA splicing and microRNA processing in *Arabidopsis thaliana*. Proc Natl Acad Sci 105:8795–8800. https://doi.org/10.1073/pnas.0802493105

Lauressergues D, Couzigou J-M, Clemente HS et al (2015) Primary transcripts of microRNAs encode regulatory peptides. Nature 520:90–93. https://doi.org/10.1038/nature14346

Lee Y, Kim M, Han J et al (2004) microRNA genes are transcribed by RNA polymerase II. EMBO J 23:4051–4060. https://doi.org/10.1038/sj.emboj.7600385

Li J, Yang Z, Yu B, et al (2005) Methylation Protects miRNAs and siRNAs from a 3′- end uridylation activity in Arabidopsis. 15:1501–1507. https://doi.org/10.1016/j.cub.2005.07.029

Li S, Liu L, Zhuang X et al (2013) microRNAs inhibit the translation of target mRNAs on the endoplasmic reticulum in Arabidopsis. Cell 153:562–574. https://doi.org/10.1016/j.cell.2013.04.005

Liu D, Song Y, Chen Z, Yu D (2009) Ectopic expression of miR396 suppresses GRF target gene expression and alters leaf growth in Arabidopsis. Physiol Plant 136:223–236. https://doi.org/10.1111/j.1399-3054.2009.01229.x

Llave C, Xie Z, Kasschau KD, Carrington JC (2002) Cleavage of Scarecrow-like mRNA targets directed by a class of Arabidopsis miRNA. Science 297:2053–2056. https://doi.org/10.1126/science.1076311

Lobbes D, Rallapalli G, Schmidt DD et al (2006) SERRATE: a new player on the plant microRNA scene. EMBO Rep 7:1052–1058. https://doi.org/10.1038/sj.embor.7400806

Lu S, Sun Y-H, Chiang VL (2009) Adenylation of plant miRNAs. Nucleic Acids Res 37:1878–1885. https://doi.org/10.1093/nar/gkp031

Lund E, Guttinger S, Calado A et al (2004) Nuclear export of microRNA precursors. Science 303:95–98. https://doi.org/10.1126/science.1090599

Maldonado-Bonilla LD (2014) Composition and function of P bodies in *Arabidopsis thaliana*. Front Plant Sci 5:201. https://doi.org/10.3389/fpls.2014.00201

Manavella PA, Hagmann J, Ott F et al (2012) Fast-forward genetics identifies plant CPL phosphatases as regulators of miRNA processing factor HYL1. Cell 151:859–870. https://doi.org/10.1016/j.cell.2012.09.039

Maunoury N, Vaucheret H (2011) AGO1 and AGO2 act redundantly in miR408-mediated Plantacyanin regulation. PLoS ONE 6:e28729. https://doi.org/10.1371/journal.pone.0028729

Megraw M, Baev V, Rusinov V et al (2006) microRNA promoter element discovery in Arabidopsis. RNA 12:1612–1619. https://doi.org/10.1261/rna.130506

Merchan F, Boualem A, Crespi M, Frugier F (2009) Plant polycistronic precursors containing non-homologous microRNAs target transcripts encoding functionally related proteins. Genome Biol 10:R136. https://doi.org/10.1186/gb-2009-10-12-r136

Mi S, Cai T, Hu Y et al (2008) Sorting of small RNAs into Arabidopsis argonaute complexes is directed by the 5′ terminal nucleotide. Cell 133:116–127. https://doi.org/10.1016/j.cell.2008.02.034

Montgomery TA, Howell MD, Cuperus JT et al (2008) Specificity of ARGONAUTE7-miR390 interaction and dual functionality in TAS3 trans-acting siRNA formation. Cell 133:128–141. https://doi.org/10.1016/j.cell.2008.02.033

Motomura K, Le QT, Kumakura N et al (2012) The role of decapping proteins in the miRNA accumulation in *Arabidopsis thaliana*. RNA Biol 9:644–652. https://doi.org/10.4161/rna.19877

Mount SM, Pettersson I, Hinterberger M et al (1983) The U1 small nuclear RNA-protein complex selectively binds a 5′ splice site in vitro. Cell 33:509–518. https://doi.org/10.1016/0092-8674(83)90432-4

Niu D, Lii YE, Chellappan P et al (2016) miRNA863-3p sequentially targets negative immune regulator ARLPKs and positive regulator SERRATE upon bacterial infection. Nat Commun 7:11324. https://doi.org/10.1038/ncomms11324

Palatnik JF, Allen E, Wu X et al (2003) Control of leaf morphogenesis by microRNAs. Nature 425:257–263. https://doi.org/10.1038/nature01958

Palatnik JF, Wollmann H, Schommer C et al (2007) Sequence and Expression differences underlie functional specialization of Arabidopsis MicroRNAs miR159 and miR319. Dev Cell 13:115–125. https://doi.org/10.1016/j.devcel.2007.04.012

Park MY, Wu G, Gonzalez-Sulser A et al (2005) Nuclear processing and export of microRNAs in Arabidopsis. Proc Natl Acad Sci U S A 102:3691–3696. https://doi.org/10.1073/pnas.0405570102

Peng T, Lv Q, Zhang J et al (2011) Differential expression of the microRNAs in superior and inferior spikelets in rice (*Oryza sativa*). J Exp Bot 62:4943–4954. https://doi.org/10.1093/jxb/err205
Pujol G, Baskin TI, Casamayor A et al (2000) The *Arabidopsis thaliana* PPX/PP4 phosphatases: molecular cloning and structural organization of the genes and immunolocalization of the proteins to plastids. Plant Mol Biol 44:499–511
Qu F, Ye X, Morris TJ (2008) Arabidopsis DRB4, AGO1, AGO7, and RDR6 participate in a DCL4-initiated antiviral RNA silencing pathway negatively regulated by DCL1. Proc Natl Acad Sci U S A 105:14732–14737. https://doi.org/10.1073/pnas.0805760105
Raghuram B, Sheikh AH, Rustagi Y, Sinha AK (2015) microRNA biogenesis factor DRB1 is a phosphorylation target of mitogen activated protein kinase MPK3 in both rice and Arabidopsis. FEBS J 282:521–536. https://doi.org/10.1111/febs.13159
Rajagopalan R, Vaucheret H, Trejo J, Bartel DP (2006) A diverse and evolutionarily fluid set of microRNAs in *Arabidopsis thaliana*. Genes Dev 20:3407–3425. https://doi.org/10.1101/gad.1476406
Ramachandran V, Chen X (2008) Degradation of microRNAs by a family of exoribonucleases in Arabidopsis. Science 321:1490–1492. https://doi.org/10.1126/science.1163728
Reinhart BJ, Weinstein EG, Rhoades MW et al (2002) microRNAs in plants. Genes Dev 16:1616–1626. https://doi.org/10.1101/gad.1004402
Reis RS, Hart-Smith G, Eamens AL et al (2015a) Gene regulation by translational inhibition is determined by Dicer partnering proteins. Nat Plants 1:1–6. https://doi.org/10.1038/nplants.2014.27
Reis RS, Hart-Smith G, Eamens AL et al (2015b) microRNA Regulatory Mechanisms Play Different Roles in Arabidopsis. J Proteome Res 14:4743–4751. https://doi.org/10.1021/acs.jproteome.5b00616
Ren G, Chen X, Yu B (2012a) Uridylation of miRNAs by HEN1 SUPPRESSOR1 in Arabidopsis. Curr Biol 22:695–700. https://doi.org/10.1016/j.cub.2012.02.052
Ren G, Xie M, Dou Y et al (2012b) Regulation of miRNA abundance by RNA binding protein TOUGH in Arabidopsis. Proc Natl Acad Sci U S A 109:12817–12821. https://doi.org/10.1073/pnas.1204915109
Rogers K, Chen X (2013) Biogenesis, turnover, and mode of action of plant microRNAs. Plant Cell 25:2383–2399. https://doi.org/10.1105/tpc.113.113159
Rosbash M, Seraphin B (1991) Who's on first? The U1 snRNP-5′ splice site interaction and splicing. Trends Biochem Sci 16:187–190
Schwab R, Speth C, Laubinger S, Voinnet O (2013) Enhanced microRNA accumulation through stemloop-adjacent introns. EMBO Rep 14:615–621. https://doi.org/10.1038/embor.2013.58
Shen Y, Ji G, Haas BJ et al (2008) Genome level analysis of rice mRNA 3′-end processing signals and alternative polyadenylation. Nucleic Acids Res 36:3150–3161. https://doi.org/10.1093/nar/gkn158
Shomron N, Levy C (2009) MicroRNA-biogenesis and pre-mRNA splicing crosstalk. J Biomed Biotechnol 2009:594678. https://doi.org/10.1155/2009/594678
Song L, Axtell MJ, Fedoroff NV (2010) RNA secondary structural determinants of miRNA precursor processing in Arabidopsis. Curr Biol 20:37–41. https://doi.org/10.1016/j.cub.2009.10.076
Souret FF, Kastenmayer JP, Green PJ (2004) AtXRN4 degrades mRNA in Arabidopsis and its substrates include selected miRNA targets. Mol Cell 15:173–183. https://doi.org/10.1016/j.molcel.2004.06.006
Stepien A, Knop K, Dolata J et al (2017) Post transcriptional coordination of splicing and miRNA biogenesis in plants. Wiley Interdiscip Rev RNA 8:e1403. https://doi.org/10.1002/wrna.1403
Su C, Li Z, Cheng J et al (2017) The protein phosphatase 4 and SMEK1 complex dephosphorylates HYL1 to promote miRNA biogenesis by antagonizing the MAPK cascade in Arabidopsis. Dev Cell 41:527–539.e5. https://doi.org/10.1016/j.devcel.2017.05.008
Świda-Barteczka A, Krieger-Liszkay A, Bilger W et al (2018) The plastid-nucleus located DNA/RNA binding protein WHIRLY1 regulates microRNA-levels during stress in barley (*Hordeum vulgare* L.). RNA Biol 15:886–891. https://doi.org/10.1080/15476286.2018.1481695

Takeda A, Iwasaki S, Watanabe T et al (2008) The mechanism selecting the guide strand from small RNA duplexes is different among argonaute proteins. Plant Cell Physiol 49:493–500. https://doi.org/10.1093/pcp/pcn043

Talmor-Neiman M, Stav R, Frank W et al (2006) Novel micro-RNAs and intermediates of micro-RNA biogenesis from moss. Plant J 47:25–37. https://doi.org/10.1111/j.1365-313X.2006.02768.x

Tu B, Liu L, Xu C et al (2015) Distinct and cooperative activities of HESO1 and URT1 nucleotidyl transferases in microRNA turnover in Arabidopsis. PLoS Genet 11:e1005119. https://doi.org/10.1371/journal.pgen.1005119

Vaucheret H (2008) Plant ARGONAUTES. Trends Plant Sci 13:350–358. https://doi.org/10.1016/j.tplants.2008.04.007

Vaucheret H, Vazquez F, Crété P, Bartel DP (2004) The action of ARGONAUTE1 in the miRNA pathway and its regulation by the miRNA pathway are crucial for plant development. Genes Dev 18:1187–1197. https://doi.org/10.1101/gad.1201404.called

Voinnet O (2009) Origin, biogenesis, and activity of plant microRNAs. Cell 136:669–687. https://doi.org/10.1016/j.cell.2009.01.046

Wang F, Perry SE (2013) Identification of direct targets of FUSCA3, a key regulator of Arabidopsis seed development. Plant Physiol 161:1251–1264. https://doi.org/10.1104/pp.112.212282

Wang L, Song X, Gu L et al (2013) NOT2 proteins promote polymerase II-dependent transcription and interact with multiple microRNA biogenesis factors in Arabidopsis. Plant Cell 25:715–727. https://doi.org/10.1105/tpc.112.105882

Wang X, Zhang S, Dou Y et al (2015) Synergistic and independent actions of multiple terminal nucleotidyl transferases in the 3′ tailing of small RNAs in Arabidopsis. PLoS Genet 11:e1005091. https://doi.org/10.1371/journal.pgen.1005091

Wang X, Wang Y, Dou Y, et al (2018a) Degradation of unmethylated miRNA/miRNA*s by a DEDDy-type 3′ to 5′ exoribonuclease Atrimmer 2 in Arabidopsis. Proc Natl Acad Sci U S A 201721917. https://doi.org/10.1073/pnas.1721917115

Wang Z, Ma Z, Castillo-González C et al (2018b) SWI2/SNF2 ATPase CHR2 remodels pri-miRNAs via Serrate to impede miRNA production. Nature 557:516–521. https://doi.org/10.1038/s41586-018-0135-x

Wang S, Quan L, Li S et al (2019) The PROTEIN PHOSPHATASE4 complex promotes transcription and processing of primary microRNAs in Arabidopsis. Plant Cell 31:486–501. https://doi.org/10.1105/tpc.18.00556

Werner S, Wollmann H, Schneeberger K, Weigel D (2010) Structure determinants for accurate processing of miR172a in *Arabidopsis thaliana*. Curr Biol 20:42–48. https://doi.org/10.1016/j.cub.2009.10.073

Wu G, Poethig RS (2006) Temporal regulation of shoot development in *Arabidopsis thaliana* by miR156 and its target SPL3. Development 133:3539 LP–3547. https://doi.org/10.1242/dev.02521

Wu H-J, Wang Z-M, Wang M, Wang X-J (2013a) Widespread long noncoding RNAs as endogenous target mimics for microRNAs in plants. Plant Physiol 161:1875–1884. https://doi.org/10.1104/pp.113.215962

Wu X, Shi Y, Li J et al (2013b) A role for the RNA-binding protein MOS2 in microRNA maturation in Arabidopsis. Cell Res 23:645–657. https://doi.org/10.1038/cr.2013.23

Wyman SK, Knouf EC, Parkin RK et al (2011) Post-transcriptional generation of miRNA variants by multiple nucleotidyl transferases contributes to miRNA transcriptome complexity. Genome Res 21:1450–1461. https://doi.org/10.1101/gr.118059.110

Xie Z (2005) Expression of Arabidopsis MIRNA genes. Plant Physiol 138:2145–2154. https://doi.org/10.1104/pp.105.062943

Xie Z, Kasschau KD, Carrington JC (2003) Negative feedback regulation of Dicer-Like1 in Arabidopsis by microRNA-guided mRNA degradation. Curr Biol 13:784–789

Yamasaki H, Hayashi M, Fukazawa M et al (2009) SQUAMOSA promoter binding Protein-Like7 is a central regulator for copper homeostasis in Arabidopsis. Plant Cell 21:347–361. https://doi.org/10.1105/tpc.108.060137

Yan J, Wang P, Wang B et al (2017) The SnRK2 kinases modulate miRNA accumulation in Arabidopsis. PLoS Genet 13:e1006753. https://doi.org/10.1371/journal.pgen.1006753

Yan J, Zhang H, Zheng Y, Ding Y (2015) Comparative expression profiling of miRNAs between the cytoplasmic male sterile line MeixiangA and its maintainer line MeixiangB during rice anther development. Planta 241:109–123. https://doi.org/10.1007/s00425-014-2167-2

Yang Z, Ebright YW, Yu B, Chen X (2006) HEN1 recognizes 21–24 nt small RNA duplexes and deposits a methyl group onto the 2′ OH of the 3′ terminal nucleotide. Nucleic Acids Res 34:667–675. https://doi.org/10.1093/nar/gkj474

Yang L, Wu G, Poethig RS (2012) Mutations in the GW-repeat protein SUO reveal a developmental function for microRNA-mediated translational repression in Arabidopsis. Proc Natl Acad Sci U S A 109:315–320. https://doi.org/10.1073/pnas.1114673109

Yant L, Mathieu J, Dinh TT et al (2010) Orchestration of the floral transition and floral development in Arabidopsis by the bifunctional transcription factor APETALA2. Plant Cell 22:2156 LP–2170. https://doi.org/10.1105/tpc.110.075606

Yu B, Yang Z, Li J et al (2005) Methylation as a crucial step in plant microRNA biogenesis. Science 307:932–935. https://doi.org/10.1126/science.1107130

Yu Y, Jia T, Chen X (2017) The "how" and "where" of plant microRNAs. New Phytol 216:1002–1017. https://doi.org/10.1111/nph.14834

Zhang L, Chia J-M, Kumari S et al (2009) A genome-wide characterization of microRNA genes in maize. PLoS Genet 5:e1000716. https://doi.org/10.1371/journal.pgen.1000716

Zhang S, Xie M, Ren G, Yu B (2013) CDC5, a DNA binding protein, positively regulates posttranscriptional processing and/or transcription of primary microRNA transcripts. Proc Natl Acad Sci 110:17588–17593. https://doi.org/10.1073/pnas.1310644110

Zhang J, Huang M, Liang J et al (2016) Genome-wide mining for microRNAs and their targets in *Betula luminifera* using high-throughput sequencing and degradome analyses. Tree Genet Genomes 12:99. https://doi.org/10.1007/s11295-016-1047-2

Zhang Z, Hu F, Sung MW et al (2017) RISC-interacting clearing 3′–5′ exoribonucleases (RICEs) degrade uridylated cleavage fragments to maintain functional RISC in *Arabidopsis thaliana*. Elife 6:e24466. https://doi.org/10.7554/eLife.24466

Zhao Y, Yu Y, Zhai J et al (2012) The Arabidopsis nucleotidyl transferase HESO1 uridylates unmethylated small RNAs to trigger their degradation. Curr Biol 22:689–694. https://doi.org/10.1016/j.cub.2012.02.051

Zhu H, Hu F, Wang R et al (2011) Arabidopsis Argonaute10 specifically sequesters miR166/165 to regulate shoot apical meristem development. Cell 145:242–256. https://doi.org/10.1016/j.cell.2011.03.024

Chapter 2
Technologies to Address Plant microRNA Functions

Sachin Teotia, Deepali Singh and Guiliang Tang

Abstract microRNAs (miRNAs) are small non-coding RNAs which play critical roles in growth, development and responses to abiotic and biotic stresses in plants and animals. Functional characterization of these miRNAs is important in understanding their roles and applications of the gained knowledge in crop improvement programs. Various approaches have been adopted to study functions of miRNAs in plants. These approaches involve overexpressing miRNAs for gain-of-function studies or their decoys/mimics in the form of target mimics (TMs), short tandem target mimics (STTMs) and sponges (SPs), for loss-of-function analyses. These decoys sequester and/or degrade mature target miRNAs, preventing the accumulation of miRNAs in sufficient amount to suppress their respective target genes. To suppress the expression of MIRNA genes or target genes, controlling agronomic traits, RNA knockdown techniques, such as RNAi or CRISPR/Cas13a can be used. Furthermore, gene-editing tools, such as CRISPR/Cas-based systems, can complement RNAi-based approaches to edit and/or alter expression of genes encoding miRNA or their targets. All constructs, targeting miRNAs or their target genes, can be expressed using constitutive, ubiquitous, tissue-specific or inducible promoters to achieve silencing in a desired manner, spatially or temporally. Altogether, modulating the expression of a miRNA, or of its target gene(s), can help manipulating their functions to engineer desirable traits in plants for crop improvement.

Keywords miRNA · Target mimic · STTM · Artificial miRNA · CRISPR

Sachin Teotia and Deepali Singh contributed equally to this work.

S. Teotia (✉)
Department of Biotechnology, Sharda University, Greater Noida 201306, India
e-mail: sachin25@gmail.com

D. Singh
School of Biotechnology, Gautam Buddha University, Greater Noida 201312, India
e-mail: deepali@gbu.ac.in

G. Tang (✉)
Department of Biological Sciences, Michigan Technological University,
Houghton, MI 49931, USA
e-mail: gtang1@mtu.edu

C. Miguel et al. (eds.), *Plant microRNAs*, Concepts and Strategies in Plant Sciences, https://doi.org/10.1007/978-3-030-35772-6_2

2.1 Introduction

microRNAs (miRNAs) play critical roles in growth, development and responses to abiotic and biotic stresses in plants (Brosnan and Voinnet 2009; Ghildiyal and Zamore 2009; Matzke et al. 2009; Simon and Meyers 2010). miRNAs are important in negatively regulating one or more target genes, including transcription factors (TFs), at the post-transcriptional level (Jones-Rhoades et al. 2006). To date, hundreds of miRNAs have been identified from a variety of plant species (Griffiths-Jones et al. 2008). The availability of sequence information of whole genomes, transcriptomes, and miRNAomes of many plant species has opened the door for functional studies of miRNAs (Nobuta et al. 2010). miRNAs have been primarily discovered by sequencing, followed by their identification through bioinformatic predictions. This is followed by cloning and functional validations through loss or gain-of-function mutants. Such studies were also facilitated by genetic-screening based on activation-tagging or dominant-mutants (Aukerman and Sakai 2003; Palatnik et al. 2003). Traditional approaches to study functions of miRNAs (such as chemical/radiation/insertional mutagenesis) are largely ineffective due to the small size of miRNAs and the fact that many miRNA families are composed of multiple members with potentially redundant functions. Indirect approaches to investigate miRNA functions relies upon the generation and analysis of transgenic lines overexpressing either the target miRNAs or the miRNA-resistant target genes containing silent mutations (Baker et al. 2005; Mallory et al. 2005). However, this approach could only partially reveal the functions of a particular miRNA in vivo, as a miRNA usually regulates multiple target genes with either redundant or distinct functions (Bartel 2009). Thus, an ideal approach to study functions of miRNAs is to simultaneously inactivate all members of the miRNA family. Target mimics (TMs), short tandem target mimics (STTMs) and molecular sponges (SPs), along with artificial miRNAs (amiRNAs) have been commonly used for loss-of-function studies of miRNAs (Teotia et al. 2016). Genome editing tools have further complemented the RNA-based techniques to regulate miRNA expression. Here, we discuss the strategies used for functional interrogation of miRNAs in plants (Fig. 2.1; Table 2.1).

2.2 Gain-of-Function Strategies

The initial approaches to characterize miRNAs relied upon overexpressing them using the *35S* promoter to analyze gain-of-function effects. These approaches involved overexpressing: pri-miRNAs in rice (miR166) (Ding et al. 2018); pre-miRNAs in Arabidopsis (miR394), switchgrass (miR156), wheat (miR1139) and Medicago (miR166) (Tian et al. 2018; Fu et al. 2012; Boualem et al. 2008; Liu et al. 2018); or full-length cDNA of MIRNA genes, such as miR319 in rice and switchgrass (Yang et al. 2013; Liu et al. 2019). In another study, using Arabidopsis miR168a backbone, Tang lab created two-hit artificial miRNA vector system, which can also overexpress endogenous miRNAs. Endogenous miRNA

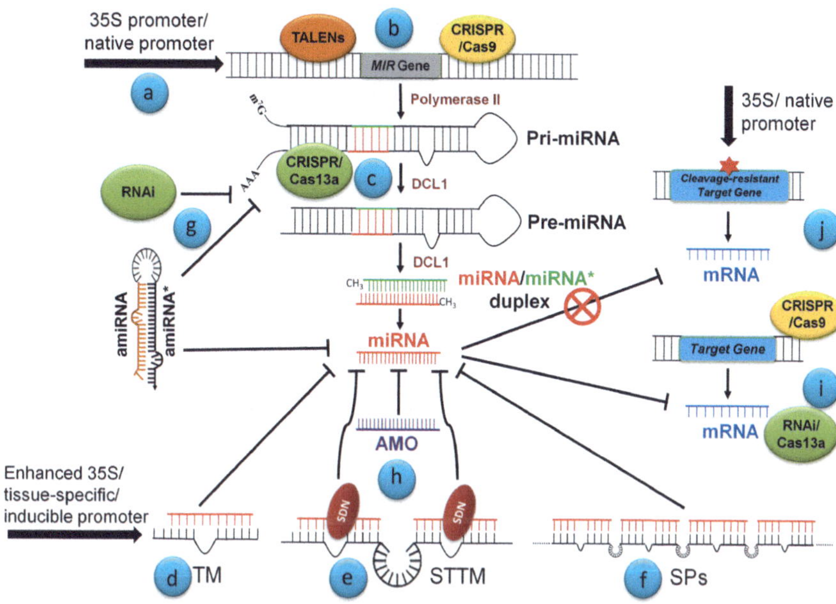

Fig. 2.1 Technologies to alter expression of miRNAs in plants. During miRNA biogenesis, a mature miRNA is generated by the cellular Dicer-like (DCL) enzymes from the primary stem-looped precursors. **a** Overexpression of any miRNA can be achieved by expressing *MIRNA genes* using constitutive or native promoters. Silencing of such miRNAs can be achieved through the editing of *MIRNA genes* through CRISPR/Cas9 system or TALENs (**b**) or of MIRNA transcripts through CRISPR/Cas13a system (**c**). Mature miRNAs can be deactivated using TMs (**d**), STTMs (**e**), and SPs (**f**) through sequestration and/or cleavage. amiRNAs or hairpin RNAi constructs (**g**) can be used to target any pri-/pre-miRNA or mature miRNA. **h** AMOs can downregulate the expression of target miRNAs. **i** To modulate any miRNA-regulated pathway, the miRNA target gene DNA can also be edited through CRISPR/Cas9 or the target gene mRNA can be silenced through RNAi or CRISPR/Cas13a system. **j** Expression of miRNA cleavage-resistant target genes can be done to analyze the function of miRNA-gene network. These target genes have mutations at the cleavage site and thus escape cleavage by their respective miRNAs. amiRNA, Artificial microRNA; AMO, Anti-microRNA oligonucleotides; CRISPR/Cas, Clusters of regularly interspaced short palindromic repeats/CRISPR-associated proteins; SDN, Small RNA degrading nuclease; SP, Molecular Sponges; STTM, Short tandem target mimic; TM, Target mimic; TALENs, Transcription activator-like effector nucleases

are cloned into two-hit amiRNA vectors by designing the endogenous miRNA* sequences in complementarity with mismatches only at the positions corresponding to the P1 and P12 of the miRNA strand. miR165/166 in Arabidopsis was successfully over-expressed using such a strategy (Ji et al. 2011). The constructs overexpressing miRNAs can be stably transformed into plants but can also be delivered through *Agrobacterium*-mediated infiltration for transient expression (Zhang et al. 2017a).

2.3 Loss-of-Function Strategies

2.3.1 *Targeting Mature miRNAs*

2.3.1.1 Target Mimics (TMs)

In plants, most miRNA targets are cleaved and show almost perfect complementarity with the miRNAs around the cleavage site. An extremely useful approach involves knocking down miRNAs by target decoys/mimics, which possess significant complementary to the endogenous miRNAs targeted to be knocked down. The design of these mimics are based upon the discovery of target mimicry such as shown by the non-protein-coding gene *IPS1 (INDUCED BY PHOSPHATE STARVATION1)* from *Arabidopsis thaliana* (Franco-Zorrilla et al. 2007) or endogenous target mimicry (eTMs) shown by numerous long non-coding RNA (lncRNA) or circular non-coding RNAs (circRNAs) that are differentially expressed usually in response to various stresses in plants (Karakülah et al. 2016; Liu et al. 2017). Similarly, retrotransposon-derived transcripts have been reported to act as decoys for miR171 in rice (Cho and Paszkowski 2017).

IPS1 or other endogenous mimics escape cleavage by miRNAs on the basis of the mismatched sequences forming a loop at the cleavage site. Such mimics act as natural sponges by sequestering miRNAs and thus fine-tune miRNA accumulation in response to various stresses. IPS1-based TM has been adapted to block the functions of several other miRNAs in Arabidopsis (Franco-Zorrilla et al. 2007; Wu et al. 2009). The TM approach was applied to study miRNA functions in Arabidopsis on a large scale (Todesco et al. 2010). The endogenous miR399 target site of *IPS1* was modified to express artificial TMs for the 73 Arabidopsis MIRNA gene families to create several MIM lines. Using this approach, the expression of many miRNAs was successfully knocked-down, many of which showed visible phenotypes, such as in the case of miR156/157, miR159, miR160, miR164, miR165/166, miR167,

Table 2.1 A summary of different strategies, along with their examples, to study miRNA functions in plants

Strategy	Targeted miRNA(s)	Plant	References
Gain-of-function studies			
Overexpressing pri/pre-miRNAs	miR166	Rice	Ding et al. (2018)
	miR394	Arabidopsis	Tian et al. (2018)
	miR156	Switchgrass	Fu et al. (2012)
	miR1139	Wheat	Boualem et al. (2008)
	miR166	Medicago	Liu et al. (2018)
Overexpressing mature miRNAs	miR165/166	Arabidopsis	Ji et al. (2011)
Loss-of-function studies			
Target mimics (TMs)	miR-399	Arabidopsis	Franco-Zorrilla et al. (2007)

(continued)

Table 2.1 (continued)

Strategy	Targeted miRNA(s)	Plant	References
	miR169, miR171, miR161, miR159, miR319, miR156, miR157, miR170, miR171, miR172, miR158, miR160, miR162, miR163, miR164, miR165, miR166, miR167, miR168, miR173, miR390, miR391, miR393, miR394, miR395, miR396, miR397, miR398, miR400, miR401, miR402, miR403, miR404, miR405, miR406, miR408, miR413, miR414, miR415, miR416, miR417, miR418, miR419, miR420, miR426, miR427, miR447, miR771, miR772, miR773, miR774, miR775, miR776, miR777, miR778, miR779, miR780, miR781, miR782, miR783, miR823, miR824, miR828, miR834, miR838, miR841, miR847, miR858, miR859	Arabidopsis	Todesco et al. (2010)
	miR773	Arabidopsis	Salvador-Guirao et al. (2018)
	miR396	Arabidopsis	Soto-Suárez et al. (2017)
Short tandem target mimics (STTMs)	miR165/166, miR156/157, miR160	Arabidopsis	Yan et al. (2012)
	miR171, miR390	Wild tomato (*S. habrochaites); Arabidopsis*	Hou et al. (2019)
	miR172, miR319; miR165/166	Tobacco and tomato	Sha et al. (2014)

(continued)

Table 2.1 (continued)

Strategy	Targeted miRNA(s)	Plant	References
	35 miRNA families including: miR398, miR172, miR156, miR166, miR159, miR160, miR171, miR441, miR1428	Rice	Zhang et al. (2017b)
	miR156/157, miR160, miR165/166, miR167, miR171, miR319, miR159, miR319/159, miR172 among others	Arabidopsis, tomato, rice, and maize	Peng et al. (2018), Teotia and Tang (2017)
	milR-1, milR-7 AF-milR-4, AF-milR-7	*Metarhizium robertsii* *Aspergillus flavus*	Wang et al. (2019)
	miR160, miR165/166, double mutant	Arabidopsis	Yang et al. (2019)
	miR482/2118 family	Tomato	Canto-Pastor et al. (2019)
	miR171	Tomato	Kravchik et al. (2019)
	miR1916	Tomato	Chen et al. (2019)
	miR390	Poplar	He et al. (2018)
	miR482b	Tomato	Jiang et al. (2018)
	miR166	Rice	Zhang et al. (2017b), Teotia et al. (2017)
	miR159	Rice	Zhao et al. (2017)
	miR396	Tomato	Cao et al. (2016)
	miR165/166, miR159	Tobacco	Zhao et al. (2016b)
	miR160	Tomato	Damodharan et al. (2016)
	miR160	Soybean	Nizampatnam et al. (2015)
	miR165/166	Arabidopsis and tomato	Jia et al. (2015)
	miR472a	Poplar	Su et al. (2018)
	miR396	Medicago	Proust et al. (2018)
	miR1916	Tomato	Chen et al. (2019)
	miR165/166	Cotton	Gu et al. (2014)
	miR1514a	Bean	Sosa-Valencia et al. (2017)

(continued)

Table 2.1 (continued)

Strategy	Targeted miRNA(s)	Plant	References
	miR9863	Barley	Liu et al. (2014b)
	miR9678	Wheat	Guo et al. (2018)
Molecular Sponges (SPs)	miR159, miR165/166, miR319	Arabidopsis	Reichel and Millar (2015)
Anti-microRNA oligonucleotides (AMOs)	miR156, miR160, miR167, miR169, miR169, miR171, miR390, miR820, miR164	Rice	He et al. (2016)
Hairpin RNAi	miR163, miR171a	Arabidopsis	Vaistij et al. (2010)
Artificial miRNAs	miR159, miR319, miR164	Arabidopsis	Eamens et al. (2011)
CRISPR-Cas system	miR1514, miR1509	Soybean	Jacobs et al. (2015)
	miR156	Rice	Li et al. (2016)
	OsMIR408, OsMIR528, miR815a/b/c, miR820a/b/c	Rice	Zhou et al. (2017)
	MIR169a, MIR827a	Arabidopsis	Zhao et al. (2016a)
	OsMIR397b miR159a and b	Rice Arabidopsis	Zhou et al. (2019) Lowder et al. (2015)
	miR160a	Rice	Bi et al. (2019)
TALENs	miR159b, miR390, miR394, miR398b, miR408	Rice	Bi et al. (2019)
Expressing cleavage-resistant target genes	miR160-resistant *ARF17* (*rARF17*)	Arabidopsis	Mallory et al. (2005)
	miR172-resistant *TOE3* (*rTOE3*)	Arabidopsis	Jung et al. (2014)
	miR156-resistant *SPL* (*rSPL3*)	Arabidopsis	Wu and Poethig (2006)
	miR169-resistant *NF-YA* (*rNF-YA*)	Arabidopsis	Xu et al. (2014)
	miR172- resistant *SMZ* (*rSMZ*)	Arabidopsis	Mathieu et al. (2009)
	microRNA166-resistant *OsHB3* (*rOSHB3*)	Rice	Itoh et al. (2008)
Delivering TMs and STTMs or overexpressing miRNA in plants through viral vectors	Many miRNAs silenced by TMs and STTMs through VbMS	Tobacco, tomato	Sha et al. (2014)
	miR165/166 (TM)	Tobacco, tomato	Liao et al. (2015)

(continued)

Table 2.1 (continued)

Strategy	Targeted miRNA(s)	Plant	References
	miR159 (TM)	Arabidopsis	Du et al. (2014)
	miR159a, miR3134a	Wheat	Jiao et al. (2015)
	miR156, miR319, miR164	Arabidopsis	Yan et al. (2014)
	miR165/166, miR3134a	Wheat	Yang et al. (2018)
	miR156, miR166 miR156	Wheat *Nicotiana benthamiana*	Jian et al. (2017) Ju et al. (2017)
	miR165/166, miR159 miR156b, miR165a	*Nicotiana benthamiana*	Zhao et al. (2016b) Tang et al. (2010)

miR170/171, miR172, miR319, miR393, and miR394. Using TMs, other groups studied the roles of other miRNAs such as of miR377 and miR396 in imparting resistance to infection by fungal pathogens in Arabidopsis (Salvador-Guirao et al. 2018; Soto-Suárez et al. 2017).

The TM approach offers a big advantage as a single non-cleavable target site expressed via vector, can reduce the levels of all miRNA members of a family, available for real endogenous targets. However, this approach does not allow suppression of expression of an individual MIRNA gene of a MIRNA family. In those cases, a CRISPR-based approach to edit an individual MIRNA gene, or an amiRNA to target individual pre-miRNA, can be used for loss-of-function studies. Alternatively, these individual miRNAs can be overexpressed to study their gain-of-function effects. TMs can have variable efficiency to downregulate the expression of miRNAs. A dual-luciferase-based miRNA sensor system to quantitatively evaluate plant miRNA complementarity requirements at both the mRNA and protein levels can be used to test the efficacy of chosen miRNA target sites for designing mimics (Liu et al. 2014a). A recent study showed that minor mutations (nucleotide substitutions) in flanking sequences of miRNA binding sites in *IPS1* backbone can strongly enhance or reduce (depending upon the target miRNA) TM-miRNA interaction (Wong et al. 2018).

2.3.1.2 Short Tandem Target Mimics (STTMs)

A modification of TM was designed, named as short tandem target mimic (STTM) (Yan et al. 2012). STTM is composed of two short TM sequences mimicking small RNA target sites, in tandem, separated by a tested linker (of at least 48 nt). Such a structure has been shown to degrade small RNAs by small RNA degrading nucleases (Yan et al. 2012). The efficacy of STTMs was demonstrated through the strong and specific developmental defects obtained by targeting three miRNAs: miR165/166, miR156/157, miR160 and endogenous trans-acting siRNAs (tasiRNAs), D7(+) and D8(+), in Arabidopsis (Yan et al. 2012). Since then, STTMs have been successfully applied to silence miRNAs in model plants and few staple crops, including tomato

(Cao et al. 2016; Jia et al. 2015; Jiang et al. 2018; Hou et al. 2019; Canto-Pastor et al. 2019; Kravchik et al. 2019; Chen et al. 2019), rice (Zhang et al. 2017b; Zhao et al. 2017, 2019; Peng et al. 2018), wheat (Guo et al. 2018), tobacco (Sha et al. 2014), Medicago (Proust et al. 2018), soybean (Wong et al. 2014; Nizampatnam et al. 2015; Bao et al. 2018), poplar (Su et al. 2018), cotton (Gu et al. 2014), common bean (Sosa-Valencia et al. 2017), and barley (Liu et al. 2014b). Using STTMs, an extensive functional study of miRNAs led to the development of a resource for functional studies and crop improvement in case of rice (Zhang et al. 2017b), and Arabidopsis, rice, tomato and maize (Peng et al. 2018). STTM was also used to degrade several fungal miRNA-like RNAs in two different fungal species, *Metarhizium robertsii* and *Aspergillus flavus* (Wang et al. 2019).

STTMs have been introduced into plant cells through stable transformation (Yan et al. 2012), Agrobacterium-mediated transient expression (Zhang et al. 2017a), and viral vectors (Sha et al. 2014; Jian et al. 2017). STTMs have been designed for different conditions or combinations: (i) for the members of a miRNA family sharing identical mature miRNA sequences, (ii) for the members of a miRNA family having varied sequences, and (iii) for two different miRNA families with distinct or partially conserved mature miRNA sequences, Fig. 2.1 of Peng et al. 2018. STTMs have been expressed under enhanced 35S promoter (Yan et al. 2012; Teotia et al. 2017; Teotia and Tang 2017), and inducible and seed-specific promoters (Peng et al. 2018; Zhao et al. 2019).

2.3.1.3 Molecular Sponges (SPs)

Similar to using mimics in plants, animal systems use 'miRNA sponges (SPs)' to be effective at inhibiting miRNA action (Ebert et al. 2007). These sponges have a mismatch at the miRNA cleavage site and when expressed at high levels inside the cells, compete with the real targets to bind to the miRNAs. To increase the efficacy of a sponge for its target miRNA, multiple binding sites are inserted in a sponge. These sponges are able to target entire miRNA families to avoid the problems of functional redundancy.

Reichel and Millar (2015) effectively applied the SPs technology in plants, using multiple miRNA binding sites (up to 15, separated by a 4 nt spacer) for inhibiting miR159 and miR165/166 families in Arabidopsis. In contrast to 3 nucleotide (nt) mismatches in STTMs and TMs at the miRNA cleavage site, SPs have only 2 'central mismatch' (cm). It was found that multiple miRNA binding sites with mismatches at the cleavage site (cmSPs) could strongly repress miRNA activity but the efficacy of the repression varied on the targeted miRNA family (Reichel and Millar 2015). In order to compare the efficacy of TMs, STTMs, and SPs for silencing miRNAs in plants, two highly conserved plant miRNA families (miR159 and miR165/166) were targeted by these approaches. The results showed that MIM159 was better than STTM159 or SP159 in silencing miR159, but STTM165 and SP165 were better than

MIM165 to silence miR165/166 in Arabidopsis (Reichel and Millar 2015). Therefore, silencing efficacy depends upon the chosen tool and the miRNA(s), intended to be targeted.

2.3.1.4 Anti-microRNA Oligonucleotides (AMOs)

Anti-miRNA oligonucleotide (AMO) technology was also used, initially in animals and subsequently in plants, to suppress miRNA activity. AMOs are chemically modified antisense oligonucleotides which are complementary to mature miRNAs. Based upon complementarity, AMOs bind to mature miRNAs and block them from binding to their real targets. Thus, modified synthetic AMOs are useful in unraveling the function of miRNAs and their targets. AMOs can be designed and modified to improve their efficiency by attracting target miRNAs more effectively and prevent them to be bound to target mRNAs (Lima et al. 2018). AMOs, fully complementary to the mature sequence of the target miRNA and modified with 2′-O-methyl have been reported to be delivered in plants. Sucrose-mediated delivery of 2′-O-Me AMOs to rice protoplasts and intact leaves lead to efficient inhibition of miRNAs from the same family with concomitant de-repression of their target genes (He et al. 2016). This repression was sequence-specific and did not affect the expression of other miRNAs. This method is a good alternative to transient expression of target mimics for quick assessment of the function of miRNAs in plants.

2.3.2 Targeting pri/pre-miRNAs

2.3.2.1 RNA Interference and Artificial miRNAs

RNA interference (RNAi) has also been used to suppress the accumulation of miRNAs. RNAi constructs targeting the pri-miRNA transcripts and the promoter regions of miR163 and miR171a were successfully used to knockdown the expression of those miRNAs (Vaistij et al. 2010). This work suggested that hairpin RNAi constructs targeting the region overlapping transcriptional start sites should maximize the chances of miRNA silencing. This suppression may arise due to silencing of the MIRNA gene promoters, transcriptionally, or of the pri-miRNA transcripts, post-transcriptionally.

Another approach of silencing miRNAs is through using artificial miRNA (amiRNAs). In plants, amiRNAs (generally 21 nt) are being designed by replacing miRNA/miRNA* with amiRNA/amiRNA* in an existing miRNA precursor backbone. These amiRNAs have been successfully used to silence endogenous genes or trans-genes (Eamens et al. 2009; Alvarez et al. 2006; Qu et al. 2007; Schwab et al. 2006). Through targeting a conserved region of a mature miRNA of a MIRNA gene family, an amiRNA can induce silencing against all MIRNA gene family members. Unlike, MIMs and STTMs, this approach offers an advantage of targeting an individual member of a MIRNA gene family by designing an amiRNA targeting an unconserved stem-loop region of the precursor transcript (Eamens et al. 2011). Using this approach, a sequence of mature miR159a/b/c was targeted using a common amiRNA, while miR164b was targeted using a unique sequence binding to regions in pri/pre-MIRNA164b. Individual targeting of miRNA164b, did not affect the expression of other miRNAs of miR164 family (Eamens et al. 2011). amiRNAs offer another advantage of showing fewer off-target effects, compared to traditional RNAi (Schwab et al. 2006). amiRNAs use only a 21-nt sequence to target a transcript, whereas multiple siRNAs arising from a hairpin double-stranded RNAi construct, have a potential of off-target binding at unintended places in the transcriptome (Xu et al. 2006).

2.3.2.2 CRISPR/Cas System and TALENs

Clustered Regularly Interspaced Short Palindromic Repeats/CRISPR-associated protein-9 nuclease (CRISPR/Cas9) and CRISPR/Cpf1 systems, guided by RNA, named as guide RNA (gRNA), have been optimized for plant genome editing (Osakabe et al. 2015, 2016; Wang et al. 2018). CRISPR/Cas9-induced non-homologous end joining (NHEJ) can introduce indels (insertions-deletions) at pre-miRNA sequences, or the miRNA processing sites of MIRNA genes, or at the mRNA binding sites, which may impede miRNA biogenesis or its pairing with target mRNAs (Chang et al. 2016; Zhou et al. 2017). Furthermore, homology-directed repair (HDR) can lead to full deletion or knock-in of MIRNA genes or their promoter sequences (Zhao et al. 2016a).

Usually, a codon-optimized Cas9 is expressed under a 35S or other promoters, along with gRNAs, which are expressed under U3 or U6 promoters. Together with a selection marker, Cas9 and gRNA(s) are introduced into plant cells through biolistic method or Agrobacterium-mediated transformation/infiltration (Liang et al. 2017). Using CRISPR-Cas9 system, specific mutations in the miR156 recognition site of the rice *ipa1* gene (ideal plant architecture 1) were introduced, which improved several traits related to plant architecture (Li et al. 2016). Mono - and biallelic mutations in several MIRNA genes of T0 rice lines were achieved using CRISPR/Cas9, resulting in loss of function for target miRNAs (Zhou et al. 2017). In another study, four MIR167 genes in *Arabidopsis thaliana* were completely deleted using this system (Yao et al. 2019).

In a different approach, by fusing dCas9 (a mutant of Cas9 without endonuclease activity) with a transcriptional repressor, (pco-dCas9-3X(SRDX)), silencing of miR159a and miR159b was achieved by designing gRNAs targeting miRNA promoters (Lowder et al. 2015).

CRISPR/Cas13a uses a novel nuclease type (class II type VI-A endoribonuclease) also guided by gRNA for targeting and cleaving single-stranded RNA (ssRNA or mRNA) (Ali et al. 2018; Aman et al. 2018). Similar to amiRNAs, this system can be used to target pri-/pre-miRNA transcripts of MIRNA genes.

Very recently, it has been shown that TALENs (transcription activator-like effector nucleases) and CRISPR/Cas9 are equally effective in modifying miRNA precursor structure, disrupting miRNA processing, and generating miRNA null mutant plants (Bi et al. 2019) (Table 1).However, the short length of MIRNA genes reduces the number of possible gRNA/Cas9 targeting the MIRNA gene (Jacobs et al. 2015). Furthermore, redundancy of miRNA functions across the family makes it difficult to edit all MIRNA genes simultaneously. This creates a major limitation for studying loss-of-function miRNA mutants, using this approach. Nevertheless, multiplexing of gRNAs can be done to target multiple genes in a single transformation. Off-target effects are another concern with CRISPR/Cas system. Therefore, it is important to analyze several independent gene-edited lines and select those showing a common phenotype.

2.4 Expressing Cleavage-Resistant Target Genes

Another approach to study the functions of miRNAs is through overexpressing their target genes. This mimics the condition of downregulating miRNA expression, which upregulates their targets. Since over-expressed targets also get cleaved by miRNAs, cleavage resistant-targets were created by creating synonymous mutations at the miRNA cleavage site. Such strategy enabled engineered target genes to resist cleavage from their respective miRNAs. There are numerous examples where such approach has been used, such as creating transgenic plants expressing miR160-resistant ARF17 (*rARF17*) (Mallory et al. 2005), miR172-resistant *rTOE3* (Jung et al. 2014), miR156-resistant *rSPL3* (Wu and Poethig 2006), miR169-resistant *rNF-YA* (Xu et al. 2014), miR172-resistant *rSMZ* (Mathieu et al. 2009) in *Arabidopsis*, and miR166-resistant *rOSHB3* in rice (Itoh et al. 2008) (Table 2.1). Programmable target base editing, without the need for a foreign DNA donor or double-stranded DNA cleavage, was shown to be done using CRISPR-Cas9 nickase-cytidine deaminase fusion, converting cytosine to thymine in rice, wheat and maize (Zong et al. 2017). Similarly, an adenosine deaminase fused to the nickase CRISPR/Cas9 converted A-T to G-C in rice and wheat plants (Li et al. 2018). Such precise base editing can be done at the miRNA cleavage site of the target genes to make them cleavage resistant by their targeting miRNAs.

2.5 Delivering TMs and STTMs or Overexpressing miRNAs in Plants Through Viral Vectors

Viral vectors have an advantage of transiently expressing genes and are useful in high-throughput analysis of miRNA functions in a short time. Sha et al. (2014) tested the virus-based microRNA silencing (VbMS)-mediated inhibition of miR172, miR319, and miR165/166 activity through *Tobacco rattle virus* (TRV)-based expression of STTMs targeting miRNAs in tobacco and tomato. They successfully demonstrated that VbMS through TRV-based expression of TMs can be used to block the function of miRNAs in plants, though only 20–30% of plants infected with TRV variants expressing either TM or STTM exhibited the expected phenotypes (Sha et al. 2014). In another study, TMs of miR156, miR319, and miR164 were introduced into TRV vector system and then inoculated into *Arabidopsis*. The resulting phenotypes were consistent with previous reports, and the expression of targets of the miRNAs was also increased, indicating inhibition of miRNAs activity (Yan et al. 2014).

Cucumber mosaic virus (CMV) strain LS based vector, with TM sequences inserted in the RNA3 sequence immediately downstream of the coat protein (CP) gene, demonstrated efficient inhibition of activity of artificial and endogenous miRNAs in Arabidopsis (Du et al. 2014). LS-CMV vector strain was further improved by incorporating the virus-derived sequences into T-DNAs, allowing the use of CMV-derived TM vectors using Agro-infection. Tobacco plants Agro-infected with LS-MIM165/166 showed severe phenotype in comparison to control plants. Similarly, 28 dpi tomato plants, Agro-infected with LS-MIM165/166, showed stunting and remarkable deformation of upper compound leaves (Liao et al. 2015), indicating the loss of miRNA activity.

Zhao et al. (2016) demonstrated that the *Potato virus X* (PVX)-based TM expression causes strong miRNA silencing in *Nicotiana benthamiana*. The PVX-based expression of STTMs against miR165/166 and miR159 caused strong miRNA silencing in *Nicotiana benthamiana* and the plants showed the corresponding phenotypes in all infected plants (Zhao et al. 2016b).

Barley stripe mosaic virus (BSMV)-based expression system was used to express miRNA TMs to suppress miR159a and miR3134a in wheat plants. The relative expression levels of mature miR159a and miR3134a decreased with increasing transcript levels of their target genes (Jiao et al. 2015). BSMV vector has also been used for overexpression of miR156 and miR166, as well as knockdown of endogenous miR156 in wheat (Jian et al. 2017).

Chinese wheat mosaic virus (CWMV) can infect both monocots and dicot at an optimal temperature of 17 °C for systemic infection. CWMV-induced silencing system was used to suppress miR165/166 and miR3134a through expression of miRNA TMs. It was also observed that CWMV-induced silencing system was more efficient compared to BSMV or *Foxtail mosaic virus* (FoMV) in wheat (Yang et al. 2018).

Furthermore, an overexpression system for endogenous mature miRNAs and siRNAs has been developed which is delivered in plants by a viral satellite DNA vector (Ju et al. 2017). This viral system successfully overexpressed endogenous or artificial miRNAs, siRNAs and trans-acting siRNAs in *Nicotiana benthamiana*. This system uses Arabidopsis *MIR390a* backbone in *Tomato yellow leaf curl China virus* (TYLCCNV) vector. Functional analysis of endogenous miRNAs such as miR156, and siRNAs, such as *athTAS3a* 5′D8(+) was carried out using this system. In another similar study, a modified cabbage leaf-curl geminivirus (CaLCuV) vector was used to express artificial and endogenous miRNAs in plants. CaLCuV vector has been used to express precursors of Arabidopsis miR156b and miR165a in *N. benthamiana* plants (Tang et al. 2010).

2.6 Studying Interactions of miRNAs

STTMs have also been used to simultaneously downregulate multiple unrelated miRNAs by using increased number of tandem repeats targeting different miRNAs (Fei et al. 2015). STTM sequences were designed with only one construct for suppression of all miRNA families together (miR1507, miR2109, or miR2118). Such combined suppression can help understand the interactions of miRNAs belonging to diverse families. Similarly, the interaction of miR159 and miR319 was studied by creating STTM159/319 (Peng et al. 2018).

Another approach to study interactions of miRNAs is to make genetic crosses of two STTM lines targeting more than one miRNA, such as STTM165/166 and STTM160 in Arabidopsis (Yang et al. 2019). Using this approach, the interplay and regulatory circuit between miR160 and miR165/166 and its effect on leaf development and drought tolerance in Arabidopsis was studied. A double STTM of miR160 and miR165/166 in Arabidopsis was generated by crossing STTM160 with STTM165/166. The double STTM plants exhibited a series of compromised phenotypes in leaf development and drought tolerance in comparison to phenotypic alterations in the single STTM lines. Further analyses suggested that the expression levels of auxin and ABA signaling genes in the double STTM were compromised compared to the plants carrying single STTMs (Yang et al. 2019).

2.7 Conclusions and Future Perspectives

Boosted by the advancements of high throughput sequencing, more and more miRNAs are being discovered, not only for the model species but also for the species with unsequenced genomes and transcriptomes. Discovery of miRNAs provides an excellent platform for their functional interrogation, which is very important as miRNAs play key roles in numerous physiological and developmental processes of plants. Functional characterization of miRNAs can be done by adopting gain- and loss-of-function strategies. Overexpression or knockdown of miRNAs can lead to concomitant downregulation or upregulation of their target genes, respectively. Expression of *MIRNA genes* can be modulated using CaMV 35S, tissue-specific or inducible promoters. Target mimics in the form of TMs, STTMs, and SPs are effective in knocking down miRNA expression. These mimics can also be expressed using multiple promoters, as discussed above. Transient expression of these decoys using viral vectors and Agrobacterium-infiltration can be done for quick functional analysis of miRNAs. Exogenous delivery of these decoys through nanoparticles can be another quick way of functional analysis of these miRNAs but awaits further standardization. Using all of these approaches, functions of miRNAs can be manipulated for crop improvement, for better yield and stress resistance.

References

Ali Z, Mahas A, Mahfouz M (2018) CRISPR/Cas13 as a Tool for RNA Interference. Trends Plant Sci 23(5):374–378

Alvarez JP, Pekker I, Goldshmidt A, Blum E, Amsellem Z, Eshed Y (2006) Endogenous and synthetic microRNAs stimulate simultaneous, efficient and localized regulation of multiple targets in diverse species. Plant Cell 18:1134–1151

Aman R, Ali Z, Butt H, Mahas A, Aljedaani F, Khan MZ, Ding S, Mahfouz M (2018) RNA virus interference via CRISPR/Cas13a system in plants. Genome Biol 19(1):1

Aukerman MJ, Sakai H (2003) Regulation of flowering time and floral organ identity by a microRNA and its APETALA2-like target genes. Plant Cell 15:2730–2741

Baker CC, Sieber P, Wellmer F, Meyerowitz EM (2005) The early extra petals1 mutant uncovers a role for microRNA miR164c in regulating petal number in Arabidopsis. Curr Biol 15:303–315

Bao D, Ganbaatar O, Cui X, Yu R, Bao W, Falk BW, Wuriyanghan H (2018) Down regulation of genes coding for core RNAi components and disease resistance proteins via corresponding microRNAs might be correlated with successful *Soybean mosaic virus* infection in soybean. Mol Plant Pathol 19:948–960

Bartel DP (2009) microRNAs: target recognition and regulatory functions. Cell 136:215–233

Bi H, Fei Q, Li R et al (2019) Disruption of miRNA sequences by TALENs and CRISPR/Cas9 induces varied lengths of miRNA production. Plant Biotechnol J. https://doi.org/10.1111/pbi.13315

Boualem A, Laporte P, Jovanovic M, Laffont C, Plet J, Combier JP, Niebel A, Crespi M, Frugier F (2008) microRNA166 controls root and nodule development in *Medicago truncatula*. Plant J 54(5):876–887

Brosnan CA, Voinnet O (2009) The long and the short of noncoding RNAs. Curr Opin Cell Biol 21:416–425

Canto-Pastor A, Santos BAMC, Valli AA, Summers W, Schornack S, Baulcombe DC (2019) Enhanced resistance to bacterial and oomycete pathogens by short tandem target mimic RNAs in tomato. Proc Natl Acad Sci U S A 116(7):2755–2760

Cao D, Wang J, Ju Z, Liu Q, Li S, Tian H, Fu D, Zhu H, Luo Y, Zhu B (2016) Regulations on growth and development in tomato cotyledon, flower and fruit via destruction of miR396 with short tandem target mimic. Plant Sci 247:1–12

Chang H, Yi B, Ma R, Zhang X, Zhao H, Xi Y (2016) CRISPR/cas9, a novel genomic tool to knock down microRNA in vitro and in vivo. Sci Rep 6:22312

Chen L, Meng J, He XL, Zhang M, Luan YS (2019) *Solanum lycopersicum* microRNA1916 targets multiple target genes and negatively regulates the immune response in tomato. Plant Cell Environ 42(4):1393–1407

Cho J, Paszkowski J (2017) Regulation of rice root development by a retrotransposon acting as a microRNA sponge. eLife 6:e30038

Damodharan S, Zhao D, Arazi T (2016) A common miRNA160-based mechanism regulates ovary patterning, floral organ abscission and lamina outgrowth in tomato. Plant J 86(6):458–471

Ding Y, Gong S, Wang Y, Wang F, Bao H, Sun J, Cai C, Yi K, Chen Z, Zhu C (2018) microRNA166 modulates cadmium tolerance and accumulation in rice. Plant Physiol 177(4):1691–1703

Du ZY, Chen A, Chen W, Westwood JH, Baulcombe DC, Carr JP (2014) Using a viral vector to reveal the role of microRNA159 in disease symptom induction by a severe strain of *Cucumber mosaic virus*. Plant Physiol 164:1378–1388

Eamens AL, Smith NA, Curtin SJ, Wang MB, Waterhouse PM (2009) The *Arabidopsis thaliana* double stranded RNA binding protein DRB1 directs guide strand selection from microRNA duplexes. RNA 15:2219–2235

Eamens AL, Agius C, Smith NA, Waterhouse PM, Wang MB (2011) Efficient silencing of endogenous microRNAs using artificial microRNAs in *Arabidopsis thaliana*. Mol Plant 4:157–170

Ebert MS, Neilson JR, Sharp PA (2007) MicroRNA sponges: competitive inhibitors of small RNAs in mammalian cells. Nat Methods 4:721–726

Fei Q, Li P, Teng C, Meyers BC (2015) Secondary siRNAs from Medicago NB-LRRs modulated via miRNA-target interactions and their abundances. Plant J 83(3):451–465

Franco-Zorrilla JM1, Valli A, Todesco M, Mateos I, Puga MI, Rubio-Somoza I, Leyva A, Weigel D, García JA, Paz-Ares J (2007) Target mimicry provides a new mechanism for regulation of microRNA activity. Nat Genet 39(8):1033–1037

Fu C, Sunkar R, Zhou C, Shen H, Zhang JY, Matts J, Wolf J, Mann DG, Stewart CN Jr, Tang Y, Wang ZY (2012) Overexpression of miR156 in switchgrass (*Panicum virgatum* L.) results in various morphological alterations and leads to improved biomass production. Plant Biotechnol J 10(4):443–452

Ghildiyal M, Zamore PD (2009) Small silencing RNAs: an expanding universe. Nat Rev Genet 10:94–108

Griffiths-Jones S, Saini HK, van Dongen S, Enright AJ (2008) miRBase: Tools for microRNA genomics. Nucleic Acids Res 36(database issue):D154–D158

Gu, Z, Huang C, Li F, Zhou X (2014) A versatile system for functional analysis of genes and microRNAs in cotton. Plant Biotechnol J 12:638–649

Guo G, Liu X, Sun F, Cao J, Huo N, Wuda B, Xin M, Hu Z, Du J, Xia R, Rossi V, Peng H, Ni Z, Sun Q, Yao Y (2018) Wheat miR9678 affects seed germination by generating phased siRNAs and modulating abscisic acid/gibberellins signaling. Plant Cell 30:796–814

He L, Xie M, Huang J, Zhang T, Shi S, Tang T (2016) Efficient and specific inhibition of plant microRNA function by anti-microRNA oligonucleotides (AMOs) in vitro and in vivo. Plant Cell Rep 35(4):933–945

He F, Xu C, Fu X, Shen Y, Guo L, Leng M, Luo K (2018) The microRNA390/TRANS-ACTING SHORT INTERFERING RNA3 module mediates lateral root growth under salt stress via the auxin pathway. Plant Physiol 177(2):775–791

Hou Y, Jiang F, Zheng X, Wu Z (2019) Identification and analysis of oxygen responsive microRNAs in the root of wild tomato (*S. habrochaites*). BMC Plant Biol 12 19(1):100

Itoh J, Hibara K, Sato Y, Nagato Y (2008) Developmental role and auxin responsiveness of Class III homeodomain leucine zipper gene family members in rice. Plant Physiol 147(4):1960–1975

Jacobs TB, LaFayette PR, Schmitz RJ, Parrott WA (2015) Targeted genome modifications in soybean with CRISPR/Cas9. BMC Biotechnol 15:1–10

Ji L, Liu X, Yan J, Wang W, Yumul RE, Kim YJ, Dinh TT, Liu J, Cui X, Zheng B, Agarwal M, Liu C, Cao X, Tang G, Chen X (2011) ARGONAUTE10 and ARGONAUTE1 regulate the termination of floral stem cells through two microRNAs in Arabidopsis. PLoS Genet 7:e1001358

Jia X, Ding N, Fan W, Yan J, Gu Y, Tang X, Li R, Tang G (2015) Functional plasticity of miR165/166 in plant development revealed by small tandem target mimic. Plant Sci 233:11–21

Jian C, Han R, Chi Q, Wang S, Ma M, Liu X, Zhao H (2017) Virus-Based microRNA silencing and overexpressing in common wheat (*Triticum aestivum* L.). Front Plant Sci 10(8):500

Jiang N, Meng J, Cui J, Sun G, Luan Y (2018) Function identification of miR482b, a negative regulator during tomato resistance to *Phytophthora infestans*. Hortic Res 5:9

Jiao J, Wang Y, Selvaraj JN, Xing F, Liu Y (2015) Barley Stripe Mosaic Virus (BSMV) induced microRNA silencing in common wheat (*Triticum aestivum* L.). PLoS One 8 10(5):e0126621

Jones-Rhoades MW, Bartel DP, Bartel B (2006) microRNAs and their regulatory roles in plants. Annu Rev Plant Biol 57:19–53

Ju Z, Cao D, Gao C, Zuo J, Zhai B, Li S, Zhu H, Fu D, Luo Y, Zhu B (2017) A viral satellite DNA vector TYLCCNV) for functional analysis of miRNAs and siRNAs in plants. Plant Physiol 173:1940–1952

Jung JH, Lee S, Yun J, Lee M, Park CM (2014) The miR172 target TOE3 represses AGAMOUS expression during Arabidopsis floral patterning. Plant Sci 215–216:29–38

Karakülah G, Yücebilgili Kurtoğlu K, Unver T (2016) PeTMbase: a database of plant endogenous target mimics (eTMs). PLoS ONE 11:e0167698

Kravchik M, Stav R, Belausov E, Arazi T (2019) Functional characterization of microRNA171 family in tomato. Plants (Basel) 4:8(1). pii:E10

Li M, Li X, Zhou Z, Wu P, Fang M, Pan X, Lin Q, Luo W, Wu G, Li H (2016) Reassessment of the four yield-related genes Gn1a, DEP1, GS3 and IPA1 in rice using a CRISPR/Cas9 system. Front Plant Sci 7:377

Li C, Zong Y, Wang Y, Jin S, Zhang D, Song Q, Zhang R, Gao C (2018) Expanded base editing in rice and wheat using a Cas9-adenosine deaminase fusion. Genome Biol 29;19(1):59

Liang Z, Chen K, Li T, Zhang Y, Wang Y, Zhao Q, Liu J, Zhang H, Liu C, Ran Y, Gao C (2017) Efficient DNA-free genome editing of bread wheat using CRISPR/Cas9 ribonucleoprotein complexes. Nat Commun 8:14261

Liao Q, Tu Y, Carr JP, Du Z (2015) An improved cucumber mosaic virus-based vector for efficient decoying of plant microRNAs. Sci Rep 17(5):13178

Lima JF, Cerqueira L, Figueiredo C, Oliveira C, Azevedo NF (2018) Anti-miRNA oligonucleotides: a comprehensive guide for design. RNA Biol 4 15(3):338–352

Liu Q, Wang F, Axtell MJ (2014a) Analysis of complementarity requirements for plant microRNA targeting using a *Nicotiana benthamiana* quantitative transient assay. Plant Cell 26(2):741–753

Liu J, Cheng X, Liu D, Xu W, Wise R, Shen QH (2014b) The miR9863 family regulates distinct Mla alleles in barley to attenuate NLR receptor-triggered disease resistance and cell-death signaling. PLoS Genet 10:e1004755

Liu T, Zhang L, Chen G, Shi T (2017) Identifying and characterizing the circular RNAs during the lifespan of Arabidopsis leaves. Front Plant Sci 8:1278

Liu Z, Wang X, Chen X, Shi G, Bai Q, Xiao K (2018) TaMIR1139: a wheat miRNA responsive to Pi-starvation, acts a critical mediator in modulating plant tolerance to Pi deprivation. Plant Cell Rep 37(9):1293–1309

Liu Y, Li D, Yan J, Wang K, Luo H, Zhang W (2019) MiR319-mediated ethylene biosynthesis, signaling and salt stress response in switchgrass. Plant Biotechnol J. https://doi.org/10.1111/pbi.13154

Lowder LG, Zhang D, Baltes NJ, Paul JW, Tang X, Zheng X, Voytas DF, Hsieh TF, Zhang Y, Qi Y (2015) A CRISPR/Cas9 toolbox for multiplexed plant genome editing and transcriptional regulation. Plant Physiol 169:971–985

Mallory AC, Bartel DP, Bartel B (2005) MicroRNA-directed regulation of Arabidopsis AUXIN RESPONSE FACTOR17 is essential for proper development and modulates expression of early auxin response genes. Plant Cell 17:1360–1375

Mathieu J, Yant LJ, Murdter F, Kuttner F, Schmid M (2009) Repression of flowering by the miR172 target SMZ. PLoS Biol 7:e1000148

Matzke M, Kanno T, Daxinger L, Huettel B, Matzke AJ (2009) RNA-mediated chromatin-based silencing in plants. Curr Opin Cell Biol 21:367–376

Nizampatnam NR, Schreier SJ, Damodaran S, Adhikari S, Subramanian S (2015) microRNA160 dictates stage-specific auxin and cytokinin sensitivities and directs soybean nodule development. Plant J. 84(1):140–153

Nobuta K, McCormick K, Nakano M, Meyers BC (2010) Bioinformatics analysis of small RNAs in plants using next generation sequencing technologies. Methods Mol Biol 592:89–106

Osakabe Y, Osakabe K (2015) Genome editing with engineered nucleases in plants. Plant Cell Physiol 56:389–400

Osakabe Y, Watanabe T, Sugano SS, Ueta R, Ishihara R, Shinozaki K, Osakabe K (2016) Optimization of CRISPR/Cas9 genome editing to modify abiotic stress responses in plants. Sci Rep 6:26685

Palatnik JF, Allen E, Wu X, Schommer C, Schwab R, Carrington JC, Weigel D (2003) Control of leaf morphogenesis by microRNAs. Nature 425:257–263

Peng T, Qiao M, Liu H, Teotia S, Zhang Z et al (2018) A resource for inactivation of microRNAs using short tandem target mimic technology in model and crop plants. Mol Plant 11:1400–1417

Proust H, Bazin J, Sorin C, Hartmann C, Crespi M, Lelandais-Brière C (2018) Stable inactivation of microRNAs in *Medicago truncatula* roots. Methods Mol Biol 1822:123–132

Qu J, Ye J, Fang R (2007) Artificial microRNA-mediated virus resistance in plants. J Virol 81:6690–6699

Reichel M, Millar AA (2015) Specificity of plant microRNA target MIMICs: cross targeting of miR159 and miR319. J Plant Physiol 180:45–48

Salvador-Guirao R, Baldrich P, Weigel D, Rubio-Somoza I, San Segundo B (2018) The microRNA miR773 is involved in the Arabidopsis immune response to fungal pathogens. Mol Plant Microbe Interact 31(2):249–259

Schwab R, Ossowski S, Riester M, Warthmann N, Weigel D (2006) Highly specific gene silencing by artificial microRNAs in Arabidopsis. Plant Cell 18:1121–1133

Sha A, Zhao J, Yin K, Tang Y, Wang Y, Wei X, Hong Y, Liu Y (2014) Virus-based microRNA silencing in plants. Plant Physiol 164(1):36–47

Simon SA, Meyers BC (2010) small RNA-mediated epigenetic modifications in plants. Curr Opin Plant Biol 14:148–155

Sosa-Valencia G, Palomar M, Covarrubias AA, Reyes JL (2017) The legume miR1514a modulates a NAC transcription factor transcript to trigger phasiRNA formation in response to drought. J Exp Bot 68:2013–2026

Soto-Suárez M, Baldrich P, Weigel D, Rubio-Somoza I, San Segundo B (2017) The Arabidopsis miR396 mediates pathogen-associated molecular pattern-triggered immune responses against fungal pathogens. Sci Rep 7:44898

Su Y, Li HG, Wang Y, Li S, Wang HL, Yu L, He F, Yang Y, Feng CH, Shuai P, Liu C, Yin W, Xia X (2018) Poplar miR472a targeting NBS-LRRs is involved in effective defence against the necrotrophic fungus *Cytospora chrysosperma*. J Exp Bot 26;69(22):5519–5530

Tang Y, Wang F, Zhao J, Xie K, Hong Y, Liu Y (2010) Virus-based microRNA expression for gene functional analysis in plants. Plant Physiol 153(2):632–641

Teotia S, Singh D, Tang X, Tang G (2016) Essential RNA-based technologies and their applications in plant functional genomics. Trends Biotechnol 34(2):106–123

Teotia S, Tang G (2017) Silencing of stress-regulated miRNAs in plants by Short Tandem Target Mimic (STTM) approach. Methods Mol Biol 1631:337–348

Teotia S, Zhang D, Tang G (2017) Knockdown of rice microRNA166 by Short Tandem Target Mimic (STTM). Methods Mol Biol 1654:337–349

Tian X, Song L, Wang Y, Jin W, Tong F, Wu F (2018) miR394 Acts as a negative regulator of Arabidopsis resistance to *B. cinerea* infection by targeting LCR. Front Plant Sci 3(9):903

Todesco M, Rubio-Somoza I, Paz-Ares J, Weigel D (2010) A collection of target mimics for comprehensive analysis of microRNA function in Arabidopsis thaliana. PLoS Genet 6(7):e1001031. https://doi.org/10.1371/journal.pgen.1001031

Vaistij FE, Elias L, George GL, Jones L (2010) Suppression of microRNA accumulation via RNA interference in *Arabidopsis thaliana*. Plant Mol Biol 73:391–397

Wang P, Zhang J, Sun L, Ma Y, Xu J, Liang S, Deng J, Tan J, Zhang Q, Tu L, Daniell H, Jin S, Zhang X (2018) High efficient multisites genome editing in allotetraploid cotton (*Gossypium hirsutum*) using CRISPR/Cas9 system. Plant Biotechnol J 16(1):137–150

Wang Y, Wang Z, Yang W, Xie X, Cheng H, Qin L, Tang G, Huang B (2019) Degradation of fungal microRNAs triggered by short tandem target mimics is via the small-RNA-degrading nuclease. Appl Environ Microbiol; 18;85(9). pii:e03132–18

Wong J, Gao L, Yang Y, Zhai JX, Arikit S, Yu Y, Duan SY, Chan V, Xiong Q, Yan J, Li S, Liu R, Wang Y, Tang G, Meyers BC, Chen X, Ma W (2014) Roles of small RNAs in soybean defense against *Phytophthora sojae* infection. Plant J 79:928–940

Wong G, Alonso-Peral M, Li B, Li J, Millar AA (2018) microRNA *MIMIC* binding sites: minor flanking nucleotide alterations can strongly impact *MIMIC* silencing efficacy in Arabidopsis. Plant Direct 23;2(10):e00088

Wu G, Poethig RS (2006) Temporal regulation of shoot development in *Arabidopsis thaliana* by miR156 and its target SPL3. Development 133:3539–3547

Wu G, Park MY, Conway SR, Wang JW, Weigel D, Poethig RS (2009) The sequential action of miR156 and miR172 regulates developmental timing in Arabidopsis. Cell 138:750–759

Xu P, Zhang Y, Kang L, Roossinck MJ, Mysore KS (2006) Computational estimation and experimental verification of off-target silencing during posttranscriptional gene silencing in plants. Plant Physiol 142:429–440

Xu MY, Zhang L, Li WW, Hu XL, Wang MB, Fan YL, Zhang CY, Wang L (2014) Stress induced early flowering is mediated by miR169 in *Arabidopsis thaliana*. J Exp Bot 65:89–101

Yan J, Gu Y, Jia X, Kang W, Pan S, Tang X, Chen X, Tang G (2012) Effective small RNA destruction by the expression of a short tandem target mimic in Arabidopsis. Plant Cell 24:415–427

Yan F, Guo W, Wu G, Lu Y, Peng J, Zheng H, Lin L, Chen J (2014) A virus-based miRNA suppression (VbMS) system for miRNA loss-of-function analysis in plants. Biotechnol J 9(5):702–708

Yang C, Li D, Mao D, Liu X, Ji C, Li X, Zhao X, Cheng Z, Chen C, Zhu L (2013) Overexpression of microRNA319 impacts leaf morphogenesis and leads to enhanced cold tolerance in rice (*Oryza sativa* L.). Plant Cell Environ 36(12):2207–2218

Yang J, Zhang TY, Liao QS, He L, Li J, Zhang HM, Chen X, Li J, Yang J, Li JB, Chen JP (2018) Chinese wheat Mosaic virus-induced gene silencing in monocots and dicots at low temperature. Front Plant Sci 14(9):1627

Yang T, Wang Y, Teotia S, Wang Z, Shi C, Sun H, Gu Y, Zhang Z, Tang G (2019) The interaction between miR160 and miR165/166 in the control of leaf development and drought tolerance in Arabidopsis. Sci Rep 26;9(1):2832

Yao X, Chen J, Zhou J, Yu H, Ge C, Zhang M, Gao X, Dai X, Yang Z-N, Zhao Y (2019) An essential role for miRNA167 in maternal control of embryonic and seed development. Plant Physiol https://doi.org/10.1104/pp.19.00127

Zhang Q, Li Y, Zhang Y, Wu C, Wang S, Hao L, Wang S, Li T (2017a) Md-miR156ab and Md-miR395 target WRKY transcription factors to influence apple resistance to leaf spot disease. Front Plant Sci 8:526

Zhang H, Zhang J, Yan J, Gou F, Mao Y, Tang G, Botella JR (2017b) Short tandem target mimic rice lines uncover functions of miRNAs in regulating important agronomic traits. Proc Natl Acad Sci U S A 114:5277–5282

Zhao Y, Zhang C, Liu W, Gao W, Liu C, Song G, Li W-X, Mao L, Chen B, Xu Y, Li X, Xie C (2016a) An alternative strategy for targeted gene replacement in plants using a dual-sgRNA/Cas9 design. Sci Rep 6:23890

Zhao J, Liu Q, Hu P, Jia Q, Liu N, Yin K, Cheng Y, Yan F, Chen J, Liu Y (2016b) An efficient Potato virus X-based microRNA silencing in *Nicotiana benthamiana*. Sci Rep 3(6):20573

Zhao Y, Wen H, Teotia S, Du Y, Zhang J, Li J, Sun H, Tang G, Peng T, Zhao Q (2017) Suppression of microRNA159 impacts multiple agronomic traits in rice (*Oryza sativa* L.). BMC Plant Biol 21;17(1):215

Zhao YF, Peng T, Sun HZ, Teotia S, Wen HL, Du YX, Zhang J, Li JZ, Tang GL, Xue HW, Zhao QZ (2019) miR1432-OsACOT (Acyl-CoA thioesterase) module determines grain yield via enhancing grain filling rate in rice. Plant Biotechnol J 17(4):712–723

Zhou J, Deng K, Cheng Y, Zhong Z, Tian L, Tang X, Tang A, Zheng X, Zhang T, Qi Y, Zhang Y (2017) CRISPR/Cas9 based genome editing reveals new insights into microRNA function and regulation in rice. Front Plant Sci 8:1598

Zhou J, Zhong Z, Chen H, Li Q, Zheng X, Qi Y, Zhang Y (2019) Knocking out microRNA genes in rice with CRISPR-Cas9. In: Qi Y (ed) Plant genome editing with CRISPR systems. Methods in molecular biology, vol 1917. Humana Press, New York, NY

Zong Y, Wang Y, Li C, Zhang R, Chen K, Ran Y, Qiu JL, Wang D, Gao C (2017) Precise base editing in rice, wheat and maize with a Cas9-cytidine deaminase fusion. Nat Biotechnol 35(5):438–440

Chapter 3
Strategies and Resources for the Identification of microRNAs in Non-model Plants

Bruno Vasques Costa and Inês Chaves

Abstract High-throughput sequencing of small RNAs is becoming increasingly affordable and accessible. The amount of data and information that has been generated over the last decade is enormous and continues to increase. Now the major constraint is the analysis and the organization of the data in order to extract useful biological information. Consequently, several software packages have been developed to analyze small RNA datasets and dedicated databases have been created to organize these data and information. However, these tools have limitations that need to be addressed. In this chapter, we present the most commonly used strategies and discuss their advantages and limitations.

Keywords small RNA · Software · Databases · microRNA prediction · Target prediction

3.1 Introduction

Plant small RNAs (sRNAs) are involved in genome stability and in the regulation of gene expression at transcriptional and post-transcriptional level (Treiber et al. 2019) during growth and development (Dugas and Bartel 2004; Guleria et al. 2011), as well as in response to abiotic stress and pathogens (Sunkar 2010; Liu and Vance 2010; Ren et al. 2012). MicroRNAs (miRNAs) are the best characterized and well-studied class of plant sRNAs.

B. V. Costa · I. Chaves (✉)
iBET, Instituto de Biologia Experimental e Tecnológica, Oeiras, Portugal
e-mail: ichaves@itqb.unl.pt

B. V. Costa
Faculdade de Ciências, BioISI—Biosystems & Integrative Sciences Institute, Universidade de Lisboa, Lisbon, Portugal

I. Chaves
Instituto de Tecnologia Química e Biológica, Universidade Nova de Lisboa (ITQB NOVA), Oeiras, Portugal

C. Miguel et al. (eds.), *Plant microRNAs*, Concepts and Strategies in Plant Sciences, https://doi.org/10.1007/978-3-030-35772-6_3

The most used method to analyze miRNAs is through deep sequencing of the small RNA fraction isolated from plant samples. Conserved miRNAs are identified and annotated by a blast-like search against a database, being the miRBase the most commonly used reference database for this purpose (Morin et al. 2008). However, the identification of non-conserved species-specific miRNAs, requires a more sophisticated computational approach, by means of the prediction of the stem-loop secondary structure of the potential pre-miRNA molecule. Once the miRNAs are identified, the expression of each miRNA is calculated using computational analysis tools, to determine the number of reads present.

The regulatory role of miRNAs in different pathways or processes is achieved by targeting mRNAs for cleavage or repressing translation, therefore inactivating the target mRNAs. The identification of the miRNA targets is obtained by the query of RNA strands with near-complete complementarity.

The sequencing of the miRNA-cleaved RNA strands, called Degradome-Seq, provides a comprehensive means of analyzing patterns of RNA degradation enabling the identification of microRNA cleavage sites. The preparation of degradome-seq libraries using a modified 5′ end sequencing protocol is followed by high-throughput next-generation sequencing by parallel analysis of RNA ends (PARE). Sequencing of the 5′ ends allows the identification of over-represented 5′ ends matching target sites to known miRNA sequences (Lin et al. 2019).

The analysis of miRNA regulatory networks in different plant species subjected to different stresses or in developmental stages is helpful to understand differential patterns of miRNA gene regulation and to correlate the obtained results among species.

Plant miRNAs have a biogenesis pathway different from animals using different enzymes and other proteins. In plants, miRNAs are produced by the nuclear RNase III Dicer-like1 (DCL1), which cuts both the pri-miRNA and the pre-miRNA in the nucleus. The miRNA-miRNA* complex is exported to the cytoplasm and incorporated into the RISC (RNA-induced silencing complex) whereas in animals the pri-miRNA is cut by the nuclear RNase III enzyme Drosha and the (pre-miRNA) hairpin is exported to the cytoplasm by Exp5 to be processed into a single miRNA/miRNA* duplex by Dicer (Millar and Waterhouse 2005).

Therefore, the programs used to identify miRNAs in plants are based on the different characteristics of this type of sRNAs, and therefore are not generic but specific to plant species. Moreover, the secondary structure of miRNA precursors in plants, as well as the identification of their targets, differs from animal species. In plants, the recognition of the target is achieved by near-perfect complementarity between both RNA strands, whereas in animals this complementarity is only required for the first 6–8 bp typically located in the 5′ end of the miRNA (Plotnikova et al. 2019). In this way, the algorithms used to identify and predict plant miRNAs have to be designed in order to contemplate these specificities.

The identification of the diverse classes of sRNAs in non-model plant species without a sequenced genome available faces a number of additional challenges. These challenges are related to the need to incorporate the several available information sources, e.g., searching either for homologous sequences in available transcriptomic and genomic databases for the given species or close relatives, or for identical sRNAs, usually miRNAs or tasiRNAs, previously identified and annotated in other plant species. Additional complexities need to be considered, like the search for species-specific novel miRNAs, requiring the computational identification of their larger precursors and their characteristic hairpin secondary structure.

Given the fact that the data in databases is updated over time, the results of previously performed sRNA analysis are very likely to become obsolete if not revised, recomputed, and updated periodically. Moreover, public sRNA databases are prone to contain erroneous annotation results given the high frequency of false positives generated by the prediction of novel miRNAs, where the features for the pre-miRNA hairpin secondary structure, are not observed, and a manual curation of that prediction is therefore preferred (Taylor et al. 2017).

In recent years, the scientific community has discussed and congregated the fundamental principles to ensure the quality of data by implementing a set of good practices that have been acknowledged as essential for the publication of open data. These principles state that the data should be FAIR, which means Findable, Accessible Interoperable, and Re-usable (Stall et al. 2019). Regarding FAIR principles, a dataset should be exportable in a standard format to make it interoperable with other systems and in a size that makes it convenient to be moved throughout systems, while retaining the significant properties that characterize the data. Furthermore, data should be well documented with ample metadata and provide clear identification of provenance information. In the end, FAIR principles will help ensure the quality of the data and should be at the core of any data repository that aims at providing reusable data (Xie et al. 2015).

The provision of provenance information along-side the database entries is paramount for researchers to know the settings in which the data were generated. Additionally, a versioning system should be integrated to keep track of changes generated by updates. The provenance information should detail a summary of the experiment that generated the data, and the conditions in which the data was produced.

3.2 Available Tools and Resources

Understanding the molecular regulation of a specific biological process under analysis usually requires the characterization of the full sRNA repertoire. This is achieved by the application of a sequence of bioinformatics analysis steps in order to: (i) pre-process the sRNA sequences by controlling their quality and by filtering unwanted sequences like tRNAs or rRNAs; (ii) identify known or conserved miRNAs; (iii) predict novel miRNAs and their precursors, as well as tasiRNAs; and (iv) compile the obtained results in tables and figures for a comprehensive analysis report (Fig. 3.1). In the end, an integrated overview of the results obtained should facilitate the interpretation of the processes going on, by giving an overview of the active regulatory network.

At first glance, a quick internet search might give the impression that there is an abundance of tools to work with miRNA; in June 2019, there were more than 9

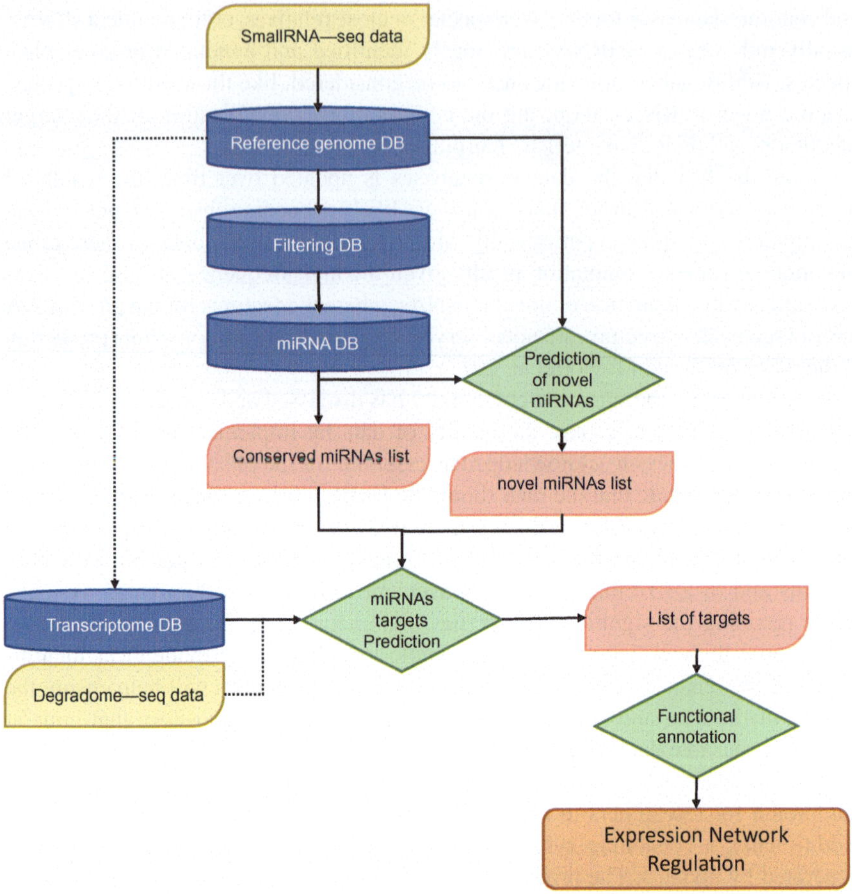

Fig. 3.1 Overview of the common workflow of bioinformatics analysis of sRNAs

million results for the term 'miRNA tools'. However, many of the tools are offline, not maintained, or designed for animals, specifically for vertebrates. The remaining few that are designed for plants, generally only allow the use of a selected amount of model species and there is often a lack of documentation both explaining how to get the tool working or describing their operations. Additionally, many of the existing tools were designed for analysis of sRNA in organisms with fully sequenced genomes and therefore these species represent the majority of the miRNAs deposited in public databases.

A list of several databases available for plant miRNAs were compiled in Table 3.1. The importance of having curated repositories of biological data for genetics and genomics research has been of concern in the molecular biology community (Batemen 2007). However, due to the lack of maintenance and curation of existing and

Table 3.1 List of working databases with plant miRNAs

Name	Description	Last update	Link	References
miRBase	Published miRNA sequences and annotation. Each entry represents a predicted hairpin portion of a miRNA transcript, with information on the location and sequence of the mature miRNA sequence	2018	http://www.mirbase.org/	Kozomara and Griffiths-Jones (2011)
miRNEST 2.0	Integrative collection of animal, plant and virus microRNA data	2015	http://rhesus.amu.edu.pl/mirnest/copy/	Szczesniak and Makalowska (2014)
miREX^2	Comprehensive platform for comparative analysis of primary microRNA expression data. RT–qPCR-based gene expression profiles are stored in a universal and expandable database scheme and wrapped by an intuitive user-friendly interface.	2015	http://www.combio.pl/mirex2	Zielezinski et al. (2015)
Tomato functional genomics database	Tomato sRNA collected from samples of different tissues and development stages	2013	http://ted.bti.cornell.edu	http://ted.bti.cornell.edu/cgi-bin/TFGD/sRNA/home.cgi

(continued)

Table 3.1 (continued)

Name	Description	Last update	Link	References
Plant sRNA portal	Dynamic database that allows users to submit raw data and have a pipeline generating the annotation automatically	2019	https://srna-portal.biodata.pt	Costa et al. (2018)
miRTarBase	Contains miRNA-target interactions (MTIs) experimentally validated by reporter assays, western blot, or microarray experiments with miRNA overexpression or knockdown	2017	http://mirtarbase.mbc.nctu.edu.tw	Chou et al. (2018)
TarBase v8	A database of experimentally supported miRNA:gene interactions	2017	http://carolina.imis.athena-innovation.gr/diana_tools/web/index.php?r=tarbasev8%2Findex	Karagkouni et al. (2018), Chou et al. (2018)

novel data, databases are always at risk to quickly become out-of-date, and corrupted, and therefore meaningless.

The concern about curation and maintenance of databases and tools available have been object of attention by the journal Nucleic Acids Research, in its Database issue, which looked back at the last 12 years, selecting a 'golden set' which have consistently served as authoritative, comprehensive, and convenient data resources widely used by the entire research community (Galperin et al. 2017).

The first and most comprehensive database for miRNAs is the miRBase (Kozomara and Griffiths-Jones 2011). The latest update of miRBase was released in 2018, four years after its previous update. In the current version (v22.1) 10,350 mature plant miRNAs are annotated, from 82 species. Species that have the full genome sequenced have a considerably higher volume of deposited miRNAs (600 *Piceaabies*, 298 *Physcomitrella patens*, 428 *Arabidopsis thaliana*, 322 *Malus domestica*, 738 *Oryza sativa*, 401 from *Populustrichocarpa*). However, the deposited plant miRNAs in miRBase are a small fraction (Kozomara and Griffiths-Jones 2011) of the miRNAs annotated for plants. Some of published and annotated miRNAs in plants, even for sequenced species are not deposited there, for instance *Eucalyptus globulus* or *Eucalyptus grandis* (Pappas et al. 2015).

miRNEST provides animal, plant and virus miRNA data from high-throughput predictions as well as from external databases, predicted plant miRNA targets and experimental target support, and integrates data from 15 external databases, which includes, e.g., sequences, polymorphism, expression, promoters, mirtrons, miRNA gene structures, degradome data. In its current version (2015), this database contains miRNAs from 522 animal and plant species and 22 viruses.

MiRTar and miRTarBase are databases supporting data on miRNA-target interactions. These databases are mostly useful to widely studied plant species, such as *A. thaliana*. Interestingly, a curation ranking is provided according to the validation experiments available.

Plant sRNA portal provides integration of miRNA data from miRBase with data from other sources. Data submitted to the database are automatically annotated by a pipeline based on the settings defined by the user. Although data are submitted by the user, the submission process incorporates provenance information about the analysis to ensure compliance with the FAIR principles. The quality of the datasets is ensured through a ranking system applied automatically. Since the database is dynamic, the incorporation of new data can complete previous information gaps, and increase the overall quality of the datasets.

The analysis of RNA-seq data in general, and sRNA analysis, in particular, is currently a very active field of bioinformatics, with a large number of computational tools being developed for this purpose (Conesa et al. 2016; Sablok et al. 2013; Yang and Li 2011; An et al. 2014). These tools comprise both commercial and open-source solutions and are implemented and delivered in various forms, from web-based to developer-hosted solutions (Table 3.2).

There are different methods of miRNA identification that can be divided according to the algorithm strategy with varying degrees of accuracy. The adopted strategy will

Table 3.2 List of working tools for the identification of plant miRNAs

Name	Description	Status/type	Homepage
miRPursuit (Chaves et al. 2017)	Multi-sample analysis pipeline Comparative and de novo miRNA TA-SI sRNAs identification Uses RNA sequence data	Downloadable Last release in 2017 Command-line	https://github.com/forestbiotech-lab/miRPursuit/releases
miRPlant (An et al. 2014)	Comparative and de novo miRNA identification Uses RNA sequence data Limited to model species	Downloadable Last release in 2018 Java app/command-line	http://www.australianprostatecentre.org/research/software/mirplant
sRNA Workbench (Stocks et al. 2018)	Comparative and de novo miRNA TA-SI sRNAs identification Uses RNA sequence data	Downloadable Last release in 2018 Java app/command-line	http://srna-workbench.cmp.uea.ac.uk/
HHMMiR (Kadri et al. 2009)	Machine learning Hidden Markov model *denovo* miRNA identification Uses genome data	Downloadable Last release in 2009 Java/perl app	https://www.csb.pitt.edu/Faculty/benos/ ?page_id=51

certainly depend on the experimental design or the data available when conducting a study, and can be based on:

- Comparative prediction, based on the fact that miRNA sequences are highly conserved among species. Therefore, if the sequence exists in one species there is a high probability that it also exists in a closely related species.
- De novo prediction, done by combing through the genome, using a sliding window and assessing the possibility that a set of sequences are sRNAs. These sequences are evaluated through the calculation of the free energy of the RNA secondary structure of the pre-miRNA.
- Hybrid prediction methods couple elements from both methods above.

Comparative methods generally provide a higher degree of accuracy since the base sequence has already been validated in some way while de novo methods can generate many false negatives and require further validation. Comparative prediction of miRNAs is done as a first approach to identify miRNAs in high throughput sequence data (HTSD). An additional analysis approach for HTSD is to run *de novo* prediction methods for the sequences that were not identified by comparative methods. When

HTSD is not available *de novo* prediction methods can be used on a genome to try to identify novel miRNAs in that species.

With the growing amount of data that is being deposited in databases, it is becoming feasible to start using machine learning methods as a hybrid approach. Machine learning methods require labeled data to train a model, that can then be used as a classifier of miRNA sequences based on the secondary structure of the precursor or the genome context. Based on this prior knowledge of miRNA identified sequences, machine learning methods will increasingly become a more reliable approach (Gomes et al. 2013).

sRNA workbench (Stocks et al. 2018) and miRPursuit pipeline (Costa et al. 2017), run an end-to-end analysis of high-throughput sRNA sequence data in model and non-model plants, annotating conserved miRNA, and predicting novel miRNAs and tasiRNAs. MiRPursuit pipeline produces an expression matrix of the libraries from the input dataset (Chaves et al. 2017). This pipeline is being improved to enable the annotation of metadata about the analysis, with a thorough description of the provenance information and the improvement of the data exportation tools.

MirPlant uses a comparative approach as well as a *de novo* prediction method but is limited to certain model species. Additional species can be added if present in miRBase but it requires some advanced bioinformatics experience to setup.

HHMMiR (Kadri et al. 2009), adopts a new approach relying on a Hierarchical Hidden Markov Model (HHMM) to predict whether a genomic region is likely to produce a miRNA precursor that would generate a miRNA sequence. The prediction is based on a model that was generated, using publicly available databases.

3.3 Further Considerations

The criteria for the prediction of targets for novel miRNAs is not usually considered by the users. The available software for prediction of miRNA targets (Folkes et al. 2012; Dai and Zhao 2011; Addo-Quaye et al. 2009) gives long lists of targets that need to be analyzed and manually curated. The understanding of the global role of any miRNA in the cell and its biological impact requires that an integrative analysis is performed considering gene ontologies (Ashburner et al. 2000), Kegg Pathways (Kanehisa and Goto 2000), Brenda-Enzyme DB (Schomburg et al. 2004), among others. Therefore, a powerful approach for the prediction of high-confidence targets should consider all the available information and allow to infer the potential interactome of the annotated sRNAs through approaches such as graph drawing.

Machine learning algorithms could also be considered for generating and automatically screening the precursors of sRNAs. This would provide an indispensable pre-screening that would simplify the work of curators by eliminating poorly classified precursors while learning from the decisions the curator made as a way of a supervised training classification.

References

Addo-Quaye C, Miller W, Axtell MJ (2009) CleaveLand: a pipeline for using degradome data to find cleaved small RNA targets. Bioinformatics 25(1):130–131

An J, Lai J, Sajjanhar A, Lehman ML, Nelson CC (2014) miRPlant: an integrated tool for identification of plant miRNA from RNA sequencing data. BMC Bioinform 15:275

Ashburner M, Ball CA, Blake JA et al (2000) Gene ontology: tool for the unification of biology. The gene ontology consortium. Nat Genet 25(1):25–29

Batemen A (2007) Editorial. Nucl Acids Res 35(1):D1–D2

Chaves I, Costa BV, Rodrigues AS, Bohn A, Miguel CM (2017) miRPursuit—a pipeline for automated analyses of small RNAs in model and nonmodel plants. FEBS Lett 591(15):2261–2268

Chou C-H, Shrestha S, Yang C-D et al (2018) miRTarBase update 2018: a resource for experimentally validated microRNA-target interactions. Nucl Acids Res 46(D1):D296–D302

Conesa A, Madrigal P, Tarazona S et al (2016) A survey of best practices for RNA-seq data analysis. Genome Biol 17(1):13

Costa BV, Chaves I, Rodrigues AS, Bohn A, Miguel CM (2017) miRPursuit. GitHub. http://github.com/forestbiotech-lab/miRPursuit

Costa BV, Chaves I, Miguel CM, Silva MJ (2018). Building the plant miRNA service. F1000 Research Limited

Dai X, Zhao PX (2011) psRNATarget: a plant small RNA target analysis server. Nucl Acids Res 39(web server issue):W155–W159

de Pappas MCR, Pappas GJ, Grattapaglia D (2015) Genome-wide discovery and validation of Eucalyptus small RNAs reveals variable patterns of conservation and diversity across species of *Myrtaceae*. BMC Genom 16:1113

Dugas DV, Bartel B (2004) microRNA regulation of gene expression in plants. Curr Opin Plant Biol 7(5):512–520

Folkes L, Moxon S, Woolfenden HC et al (2012) PAREsnip: a tool for rapid genome-wide discovery of small RNA/target interactions evidenced through degradome sequencing. Nucl Acids Res 40(13):e103

Galperin MY, Fernández-Suárez XM, Rigden DJ (2017) The 24th annual nucleic acids research database issue: a look back and upcoming changes. Nucl Acids Res 45(D1):D1–D11

Gomes CPC, Cho J-H, Hood L, Franco OL, Pereira RW, Wang K (2013) A review of computational tools in microRNA discovery. Front Genet 4:81

Guleria P, Mahajan M, Bhardwaj J, Yadav SK (2011) Plant small RNAs: biogenesis, mode of action and their roles in abiotic stresses. Genomics Proteomics Bioinform Beijing Genomics Inst 9(6):183–199

Kadri S, Hinman V, Benos PV (2009) HHMMiR: efficient de novo prediction of microRNAs using hierarchical hidden Markov models. BMC Bioinform 10(Suppl 1):S35

Kanehisa M, Goto S (2000) KEGG: kyoto encyclopedia of genes and genomes. Nucl Acids Res 28(1):27–30

Karagkouni D, Paraskevopoulou MD, Chatzopoulos S et al (2018) DIANA-TarBase v8: a decade-long collection of experimentally supported miRNA-gene interactions. Nucl Acids Res 46(D1):D239–D245

Kozomara A, Griffiths-Jones S (2011) miRBase: integrating microRNA annotation and deep-sequencing data. Nucl Acids Res 39(database issue):D152–D157

Lin S-S, Chen Y, Lu M-YJ (2019) Degradome sequencing in plants. Methods Mol Biol 1932:197–213

Liu J, Vance CP (2010) Crucial roles of sucrose and miRNA399 in systemic signaling of P deficiency—a tale of two team players? Plant Signal Behav 5(12):1–5

Millar AA, Waterhouse PM (2005) Plant and animal microRNAs: similarities and differences. Funct Integr Genomics 5(3):129–135

Morin RD, Aksay G, Dolgosheina E et al (2008) Comparative analysis of the small RNA transcriptomes of *Pinus contorta* and *Oryza sativa*. Genome Res 18(4):571–584

Plotnikova A, Kellner MJ, Schon MA, Mosiolek M, Nodine MD (2019) MicroRNA dynamics and functions during Arabidopsis embryogenesis. Plant Cell 31(12):2929–2946

Ren Y, Chen L, Zhang Y, Kang X, Zhang Z, Wang Y (2012) Identification of novel and conserved. Funct Integr Genomics 12:1–13

Sablok G, Milev I, Minkov G et al (2013) isomiRex: web-based identification of microRNAs, isomiR variations and differential expression using next-generation sequencing datasets. FEBS Lett 587(16):2629–2634

Schomburg I, Chang A, Ebeling C et al (2004) BRENDA, the enzyme database: updates and major new developments. Nucl Acids Res 32(database issue):D431–D433

Stall S, Yarmey L, Cutcher-Gershenfeld J et al (2019) Make scientific data FAIR. Nature 570(7759):27–29

Stocks MB, Mohorianu I, Beckers M et al (2018) The UEA sRNA Workbench (version 4.4): a comprehensive suite of tools for analyzing miRNAs and sRNAs. Bioinformatics 34(19):3382–3384

Sunkar R (2010) microRNAs with macro-effects on plant stress responses. Semin Cell Dev Biol 21(8):805–811

Szczesniak MW, Makalowska I (2014) miRNEST 2.0: a database of plant and animal microRNAs. Nucl Acids Res 42(database issue):D74–D77

Taylor RS, Tarver JE, Foroozani A, Donoghue PCJ (2017) microRNA annotation of plant genomes—do it right or not at all. Bioessays News Rev Mol Cell Devel Biol 39(2):1600113

Treiber T, Treiber N, Meister G (2019) Regulation of microRNA biogenesis and its crosstalk with other cellular pathways. Nat Rev Mol Cell Biol 20(1):5–20

Xie M, Zhang S, Yu B (2015) microRNA biogenesis, degradation and activity in plants. Cell Mol Life Sci 72(1):87–99

Yang X, Li L (2011) miRDeep-P: a computational tool for analyzing the microRNA transcriptome in plants. Bioinformatics 27(18):2614–2615

Zielezinski A, Dolata J, Alaba S et al (2015) mirEX 2.0—an integrated environment for expression profiling of plant microRNAs. BMC Plant Biol 15:144

Chapter 4
microRNAs in the Formation of Epigenetic Memory in Plants: The Case of Norway Spruce Embryos

Igor Yakovlev, Marcos Viejo and Carl Gunnar Fossdal

Abstract MicroRNAs (miRNA, miR) are short non-protein coding RNA molecules that are involved in both the nuclear and the posttranscriptional regulation of gene expression. miRNAs are endogenous mediators of RNA-interference forming part of the epigenetic machinery and influence gene expression post-transcriptionally without affecting the DNA sequence. In Norway spruce, epigenetic memory is established in response to the temperature conditions during embryogenesis and it affects the timing of bud burst and bud set, vitally important adaptive traits for long-lived forest species. Somatic embryogenesis at different epitype inducing (EpI) temperatures closely mimics the natural processes of epigenetic memory formation in seeds, giving rise to epigenetically different clonal plants in a reproducible and predictable manner, with respect to altered bud phenology. Developing Norway spruce embryos possesses a more complex small non-coding RNA (sRNA) structure than that reported for other somatic tissues. A variety of the predicted miRNAs showed distinct EpI temperature-dependent expression patterns. These putative EpI miRNAs target spruce genes with a wide range of functions, including genes known to be involved in epigenetic regulation, which in turn could provide a feedback process leading to the placement of epigenetic marks. Major features of miRNAs are related to their pleiotropic and synergistic actions, whereby a single miRNA can have several potential mRNA targets, and a single mRNA usually also has multiple miRNA binding sites. Moreover, miRNAs binding to a single mRNA often acts in a synergistic fashion. Fine-tuning of the miRNA production likely participates in both developmental regulation and epigenetic memory formation. Further functional and expression studies are necessary in order to elucidate the common miRNA-mediated regulatory mechanisms that underlie memory formation in plants. The use of artificial miRNAs, as well as overexpression and knockout/down of both miRNAs and their targets, will be the best techniques for determining the specific roles of individual miRNAs in memorizing

---

Electronic supplementary material The online version of this chapter (https://doi.org/10.1007/978-3-030-35772-6_4) contains supplementary material, which is available to authorized users.

I. Yakovlev (✉) · M. Viejo · C. G. Fossdal
Division of Forestry and Forest Resources, Norwegian Institute for Bioeconomy Research, Pb 115, 1431 Ås, Norway
e-mail: yai@nibio.no

C. Miguel et al. (eds.), *Plant microRNAs*, Concepts and Strategies in Plant Sciences, https://doi.org/10.1007/978-3-030-35772-6_4

the response to environmental stresses. In the present minireview, we illustrate the current knowledge regarding the epigenetics–miRNA regulatory networks aiming to provide biological insights into epigenetic memory formation in plants with a particular emphasis in spruce.

Keywords Conifers · *Picea* · Epigenetic memory · Epigenetic regulators · miRNAs · Somatic embryos

List of Abbreviations

DEG	Differentially expressed gene
DEM	Differentially expressed defined miRNA
EpI	Epitype inducing
isomiRs	Isoform microRNAs
miRNA	MicroRNA
SE	Somatic embryogenesis
sRNA	Small non-coding RNA

4.1 Introduction

Many environmental stresses, including extreme temperature changes, drought, salinity, pathogen infections, and herbivory attacks, may occur more than once in the lifespan of a plant. It would be beneficial for plants, as sessile organisms that cannot move instantly and adapt behaviorally, to create reversible epigenetic response mechanisms that take advantage of the information that is acquired from their environment and to pass it onto their progeny. The likelihood that the progeny will experience a similar environment is rather high, given that seeds are often dispersed locally. Thus, passing on such information to the progeny would have an adaptive value, and epigenetic inheritance is the prime candidate for its transmission (Grossniklaus et al. 2013). Epigenetic regulation can be mediated through a dynamic interplay between sRNAs, DNA methylation, histone modifications, histone variants, and chromatin architecture, which together modulate transcriptional silencing, activation and the accessibility of DNA in a variety of ways (Heo and Sung 2011; Simon and Meyers 2011; Lee 2012; Mirouze 2012; Bond and Baulcombe 2013). Indeed, plants have an extensive repertoire of epigenetic regulation mechanisms based on DNA methylation, histone modification, and non-coding RNAs. This includes a unique epigenetic pathway, RNA-directed DNA methylation, whereby DNA methylation and chromatin modifications are guided in a sequence-specific manner by sRNAs (Law and Jacobsen 2010).

Multiple studies in the last few years have established that plants assign sRNAs as mediators of epigenetic modifications for transcriptional or post-transcriptional regulation of gene expression in a sequence-specific manner to respond to environmental cues (Sunkar et al. 2012; Zhang 2015; Fu et al. 2017). Such sRNAs can be regulated developmentally or induced by biotic or abiotic stresses (Khraiwesh et al. 2012), providing great versatility to epigenetic gene regulation and the potential for transmitting acquired information to the progeny (Norouzitallab et al. 2019). Experiments in several species indicate that the effects of stress or other environmental conditions affect the epigenome and might be inherited and persistent for several generations even without any subsequent stress exposure (Molinier et al. 2006; Whittle et al. 2009; Verhoeven et al. 2010). It is not surprising then that the transgenerational epigenetic inheritance of environmentally induced changes has attracted so much attention (Heard and Martienssen 2014; Bilichak and Kovalchuk 2016; Blake and Watson 2016).

The epigenetic memory is defined as a heritable change in gene expression or behavior that is induced by a previous stimulus (D'Urso and Brickner 2014). The stimulus can be developmental, biotic or environmental. The underlying molecular mechanisms behind the epigenetic memory are currently under active investigation, mostly based on model organisms with short generation times (Thellier and Lüttge 2013; D'Urso and Brickner 2014; Iwasaki and Paszkowski 2014). Memorizing specific responses, laid down by epigenetic mechanisms, could provide significant strategic benefits to plants, since the most successful reply could be tuned or reenacted as a consequence of a modified environmental condition and this would be retained in future cell lineages and potentially inherited and altered by selection in future generations (Bräutigam et al. 2013; Baulcombe and Dean 2014; Iwasaki and Paszkowski 2014; Kinoshita and Seki 2014). Much remains to be known about the enigmatic repertoire of epigenetic mechanisms that operate in forest trees but earlier studies firmly confirmed the presence of epitype inducing (EpI) temperature-dependent plant phenotypes (Yakovlev et al. 2012; Liu et al. 2015) and significant transcriptomic changes in such epitypes (Yakovlev et al. 2016).

Molecular models of epigenetic memory generally include the translation of external and/or internal cues into sequence-specific signals, which are transformed into precise epigenetic modifications that in their turn should be maintained through mitotic and/or meiotic cell divisions. Then the modifications should be sensed and implemented in transcriptional changes and pathway shifts. When necessary, the modifications may be removed (Thellier and Lüttge 2013). Environmental cues might activate as yet unidentified pathways that elicit epigenetic changes both in genes and transposable elements, generating heritable epigenetic variations in natural populations (Saze 2008). The inheritance of epigenetic memory is one of the main issues for many research and applied programs in forestry. These variations can be considered as mitotically or meiotically heritable epigenetic memory (Iwasaki and Paszkowski 2014).

MicroRNAs (miRNA, miR) are short non-protein coding RNA molecules that are involved both in the nuclear and in the posttranscriptional regulation of gene expression, repressing the translation of their mRNA target without affecting the

DNA sequence of the targeted gene. miRNAs are endogenous mediators of RNA-interference forming part of the epigenetic machinery (He and Hannon 2004). Recently, accumulating studies have shown that a subgroup of miRNAs is regulated epigenetically. Although epigenetics and miRNAs have been frequently linked in many reviews (Castel and Martienssen 2013; Gruber and Zavolan 2013; Baulcombe and Dean 2014; Stuwe et al. 2014; Sow et al. 2018), few reviews have focused upon the relationship between epigenetics and miRNA in memorizing stress responses (Avramova 2015). In the present minireview, we illustrate the current knowledge regarding the epigenetics–miRNA regulatory networks aiming to provide biological insights into epigenetic memory in plants with a particular emphasis in spruce.

4.2 Environmentally Induced Epigenetic Memory

Plant memory is often characterized by heightened molecular responses upon exposure to subsequent stress, which can be composed of an enhanced response, a more efficient response, or a more rapid response (Crisp et al. 2016). The stimulus can be developmental, biotic or environmental. We define the epigenetic memory as mitotically and/or meiotically heritable changes in gene expression due to environmental stimuli that affect adaptive traits that cannot be explained by the changes in the primary DNA sequence. Such type of memory is identified mostly due to specific time of its formation and specific features related to perennial plants. There are distinct life periods when the epigenome is especially sensitive to modifications, as during early development, transmission to reproductive development and at the end of the life cycle. Nevertheless, the embryogenesis period is the most sensitive to environmental factors due to the high rates of DNA synthesis and the establishment of the epigenome (Eckardt 2006; Patkin and Sofronov 2013). Temperature differences during embryogenesis in Norway spruce cause visible phenotypic differences, mostly in phenology (Rohde and Junttila 2008).

Phenology (annual life cycle events) in trees is considered to be highly heritable and be under strong genetic selection. Thus, it came as a great surprise when an epigenetic memory of the temperature sum during embryogenesis of Norway spruce (*Picea abies*) was shown to consistently and reproducibly affect phenology in the resulting trees in a manner often ascribed to ecotypes and the gradual phenotypic differences observed in clines (Refs. in Yakovlev et al 2012; Carneros et al. 2017). It has been firmly established that environmental conditions during embryo formation affect the performance of newly formed trees for multiple adaptive traits, such as timing of bud set, cold acclimation, spring dehardening, and flushing. In common garden experiments, timing is delayed if seeds are produced at warmer than normal temperatures and vice versa, it is advanced if seeds are produced at colder than normal temperature. On the other hand, the conditions during the pre-zygotic stage and fertilization did not affect the phenology (Johnsen et al. 2005b). The effect of temperature also interacts with the daylengths effects, and plants subjected to high temperature and long days in contrast to low temperature and short days expressed similar characteristics to cold-subjected individuals from the previously mentioned study: autumn bud set and spring flushing took place earlier and they were more frost-hardy in the autumn (Johnsen et al. 2005a). The observed phenomenon was confirmed by SE experiments when genetically uniform somatic embryos were grown at different temperatures during sub-culturing and maturation, and the regenerated clones were tested in a common garden (Kvaalen and Johnsen 2008). It has been concluded that the environmental conditions during embryogenesis result in the formation of epitypes, and the epigenetic memory is mitotically propagated resulting in significant and long-lasting phenotypic changes in seasonal timing of bud phenology and other adaptive traits. Epitype testing experiments in Hoxmark (Norway) and SE, mimicking the process ongoing during zygotic development, became very effective experimental systems for studying the molecular basis of epigenetic memory formation, including transcriptional and microRNA changes among epitypes (Yakovlev et al. 2010, 2014, 2016; Yakovlev and Fossdal 2017).

Exposure of clonal, somatic Norway spruce embryos to EpI temperatures during embryogenesis shifts the timing of bud set in the trees (epitypes) corresponding to a 4–6 °C latitudinal ecotype difference. Also delayed bud set, bud burst, and de-acclimation towards frost were influenced by a high embryogenic temperature sum (Kvaalen and Johnsen 2008; Carneros et al. 2017). This indicates that a strikingly large part of the variability in bud phenology between natural ecotypes can be explained by the memory of EpI temperature conditions during embryogenesis. Thus, the clinal variation patterns in bud set and other phenological traits cannot be due only to classic Mendelian selection, but must rather be a combination of this and the epigenetic memory.

This epigenetic memory phenomenon is phenotypically best described in Norway spruce, but it has also been described in other gymnosperms (*Picea*, *Pinus*, and *Larix* species) and more recently also in angiosperm trees such as black poplar (*Populus nigra*). Maternal temperature during seed maturation in full-sib families of black poplar considerably impacts on seed germination and the phenological responses of even two-year-old seedlings in common garden experiment, suggesting the existence of a memory of maternal temperature during seed maturation. Warmer (+10 °C) maternal temperatures decreased the seed germination success. The seedlings from the warmer maternal environment also displayed later bud burst and earlier bud set. The seedlings resulting from the colder maternal environment (−10 °C) grew taller than those from the warmer environment during the first, but not second, growing season (Dewan et al. 2018).

Environmental conditions during germination and early growth can affect phenology in a similar way. In a nursery experiment with Norway spruce and European larch, Gömöry et al. (2015) found that provenances sown in a warm nursery delayed budburst consistently compared to those grown in a cold nursery, wherever they were later transplanted (Gömöry et al. 2017).

Another finding described the response of flowering phenology to soil temperature in the perennial plant mouse-ear chickweed (*Cerastium fontanum*). Plants memorize the soil temperature of their origin. In the common garden, plants responded to spring warming in a counter-gradient fashion; plants originating from warmer microsites flowered relatively later than those originating from colder microsites (Valdés et al. 2018). However, in this study authors explained the observed results exclusively through genomic adaptation and did not consider epigenetic influence.

Therefore, the concept of epigenetic memory is coming from hypothesis into a developing theory, trying to explain molecular mechanisms underlying environmental-dependent phenotypic changes (D'Urso and Brickner 2014; Iwasaki and Paszkowski 2014). Nonetheless, more researches at the phenotypic and molecular levels are needed for a better understanding of the mechanisms behind the observed effects of the maternal environment.

4.3 Identification of Conserved and Novel miRNAs in Spruce

Conserved and novel miRNAs have been identified in angiosperm and gymnosperm species (Montes et al. 2014), including multiple conifers—pines (Lu et al. 2007; Oh et al. 2008; Wan et al. 2012b; Quinn et al. 2014), spruces (Yakovlev et al. 2010; Nystedt et al. 2013) and larches (Zhang et al. 2012, 2013), but their diversity, abundance, and regulatory functions are still largely to be studied.

Based on spruce genome project small RNA-seq datasets, Nystedt et al. (2013) reported 2719 de novo miRNA gene annotations in spruce using UEA sRNA workbench tools. Later, Xia et al (2015) reanalyzed the small RNA sequencing data from 22 different tissue samples (~352 million reads) in a spruce genome paper (Nystedt et al. 2013). Their relatively more stringent pipeline for de novo annotation generated a set of 585 miRNA loci. Nearly half of these miRNA loci were novel and without any known homologs in the miRBase. Subsequent comparison of the mature sequences with ginkgo sRNAs indicated that approximately one-third of the new loci might be spruce-specific. To focus on the genetic pathways initiating and establishing the epigenetic memory response in Norway spruce, miRNAs were obtained using deep sRNA sequencing on embryos from contrasting EpI temperatures (Yakovlev and Fossdal 2017).

In silico analysis of miRNAs in Norway spruce's embryos allowed the identification of around 2000 miRNA candidates. Search and annotation of conserved miRNAs was based on the miRBase v21 database allowing up to two mismatches. A total of 636 conserved miRNAs were defined and these belonged to 51 miRNA families. These could originate from at least 99 defined precursors identified for the majority of miRNA families, but not for all. This was considered sufficient for supporting the internal origin of the defined conserved miRNAs. Additionally, 1316 novel miRNAs were defined that had no homology to miRBase v21 annotations. They could belong to 630 families and could originate from 740 predicted precursors (Yakovlev and Fossdal 2017). This result reconfirms earlier studies that reported the presence of a large number of miRNAs in Norway spruce (Nystedt et al. 2013) and in other gymnosperm species (Wan et al. 2012a, b; Wang et al. 2015), and firmly rejects the earlier hypothesis claiming that the 24 nt miRNA and siRNA class are missing (Dolgosheina et al. 2008) and the notion that the DCL4-mediated miRNA biogenesis is absent within gymnosperm plants.

The majority of the identified miRNAs in spruce embryos were 21 nt (41%) and 22 nt (34%) in length, all other length classes counting less than 10%. More often the miRNAs at the initial positions contain uridine (U—37%) and adenine (A—25%) and less C and G (16 and 22%, respectively).

Large numbers of miRNA isoforms (isomiRs) were found for nearly all conserved miRNA families. On average, there were 15 isomiRs per family but varied from 1 to 102. The largest quantity of isomiRs was found for the highly conserved family miRNA166, with 102 miRNAs. More than 40 miRNAs were found across the identified miRNA families—of miR156, miR159, miR946, miR950, miR951, miR1311, and miR3701. Eleven families had more than 10 but less than 40member miRNAs and these included miR319, miR390, miR396, miR397, miR482, miR947, miR1312, miR1316, miR3705, miR3710, and miR3712. The remaining 39 families had less than 10 isomiRs (Yakovlev and Fossdal 2017).

Various mechanisms are associated with the diversification of miRNA sequences, including imprecise DCL processing or post-transcriptional modifications, like trimming and tailing (Li et al. 2014). For the majority of isomiRs, the specific genomic fragments with precursors were not found among genomic data. Most probably they originated from common precursors and were post-transcriptionally modified. The presence of isomiRs could be also explained by the possibility of miRNAs to originate from several miRNA genes. In this case, any SNP changes in miRNA genes would cause the creation of specific isomiRs. The generation of isomiRs from the same miRNA locus may extend its functional influence. As miRNA isoforms vary in size and sequence from the canonical miRNA, alternative targets may be regulated and differential AGO loading could arise, resulting in diverse regulatory outcomes (Neilsen et al. 2012; Ameres and Zamore 2013; Li et al. 2014). Large amount of isoforms for conserved miRNAs are present in non-model species (e.g. Mica et al. 2009; Lin and Lai 2013; Liu et al. 2014), but these isomiRs are not well characterized and, in most cases, their origins and functions if any remain unknown (Neilsen et al. 2012). It has previously been demonstrated that temperature treatments altered the expression of a specific subset of mature miRNAs and displayed differential expression of multiple miRNA isoforms (Baev et al. 2014).

Star miRNAs (*miRNA) could originate from the complementary (passenger) strand of the precursor (miRNA/*miRNA duplex) and generally, it was thought to be degraded after the formation of mature miRNAs. Recently several studies have demonstrated that many *miRNAs have abundances similar to or higher than their corresponding miRNAs in specific biological processes, tissues and times, and their biological functions are not an occasional event but universal in plant species (Liu et al. 2017). High levels of *miRNAs have been reported in spruce embryos (Yakovlev and Fossdal 2017). In many cases both predicted miRNAs (guide and *miRNA) from the two strands were expressed and occasionally in an opposite manner. sRNAs in that study were originated from whole embryos containing various tissues and cell types, and since different tissues and cell types may preferentially express either the

−5p or the −3p variant (or vice versa), it is impossible in the present material to define with sufficient certainty which variant is the main miRNA and which is the star form. Future cell type specific studies should be performed to clear up this matter.

4.4 miRNAs and Their Targets Profiling in Spruce Embryos During Formation of an Epigenetic Memory

Preliminary analysis of partial complementarity pairing of predicted miRNAs with predicted gene models in spruce embryos revealed that 2050 miRNAs could be matched to 6058 annotated gene models, as defined in Norway spruce v1 coding sequences (Nystedt et al. 2013). Most of them had no annotations and matches to the database and annotations were defined for around 1400 gene models which belong to families of a wide range of biological functions (Yakovlev and Fossdal 2017).

Usually, miRNAs regulate posttranscriptional repression of mRNAs via two different mechanisms: the miRNAs induce mRNA translational repression or the miRNAs induce mRNA cleavage mediated by ARGONAUTE protein (Bartel 2009). Since full complementarity between a miRNA and its target is not always required, it is difficult to establish clear correlations between miRNA transcript profiles and their putative targets transcript profiles, as in many cases the same gene transcript may be regulated by several miRNAs and, conversely, the same miRNA may target several gene transcripts. High redundancy of miRNA–mRNA interplay and the multiplicity of miRNA genes and miRNA binding sites in the UTR of target genes may play a synergistic or additive function in the regulation of such targets (Bartel 2009). Hence, the role of miRNA in gene expression would most likely be that of a fine-tuning process rather than an ON/OFF switch. One gene may be targeted by any up and downregulated miRNA at the same time in order to attain the optimum concentration required for a specific function (Herranz and Cohen 2010). This is the case for the mediation of cell fate decisions, where miRNAs act in synergy with other transcription regulators to establish gene regulatory networks (Herranz and Cohen 2010).

The highest number of predicted target gene models, which could be regulated by miRNAs, were found among multiple repeats containing proteins gene families such as Leucine-rich repeat (LRR) genes, protein kinase domain, NB-ARC (nucleotide-binding adaptor R-gene shared) domain, ATPase family associated with various cellular activities (AAA), Toll-Interleukin receptor (TIR) domain, Clathrin heavy chains/VPS (vacuolar protein sorting-associated), tetra- (TPR), and pentatricopeptide (PPR) repeat genes among others. TIR and NBS-LRR genes are considered as one of the first lines of defense against pathogen infection (Dangl and Jones 2001; Meyers et al. 2005). However, the spruce in vitro cultures were axenic, free from any pathogens, so we could expect to not find defense-related gene expression. Even so, a large amount of differentially expressed LRR-containing genes showing clear dependence on epitype inducing temperature was observed. TIR, NBS, and LRR domain-containing proteins may fulfill more general roles in signal transduction from external environment (both biotic and abiotic) and conversion into molecular responses of diverse nature. TPR proteins can promote the formation of highly specific multiprotein complexes and can support the binding of different ligands (Zeytuni and Zarivach 2012). A typical PPR protein can bind one or several organellar transcripts, and influence their expression by altering RNA sequence, turnover, processing, or translation (Barkan and Small 2014). Clathrin/VPS domain proteins are another large group of proteins involved in the vesicular sorting and trafficking pathways and essential for body plan development, defense, and response to the environment (Chen et al. 2011). They often contain penta- and tetratricopeptide repeat domains, which could be the targets for regulation by miRNAs. In response to temperature differences, these may help to finely tuning intracellular traffic or the delivery of signaling molecules, but it is hard to envisage otherwise how they may directly contribute to temperature-dependent formation of epigenetic memory in the spruce embryos.

The expression analysis of the identified spruce miRNAs revealed 654 differentially expressed miRNAs (DEMs) with high expression level, which could be grouped into 12 clusters, based on their transcript profiles at different EpI temperatures (Yakovlev and Fossdal 2017). Their expression patterns then were compared with the expression patterns of the putative differentially expressed target gene models at different EpI temperatures to refine potential cognate miRNA–mRNA target pairs. The miRNA–mRNA target pairs with the highest negative correlation (below −0.6) were considered as prospective miRNAs regulated gene models. Correlations between target transcripts and miRNAs at all temperatures could help to build a robust definition of functional pairs, providing further insight into temperature-dependent processes leading to the formation of epigenetic memory in developing embryos.

In total, 1921 miRNA–mRNAs (DEM–DEG) pairs were defined, consisting of 470 miRNAs and 1139 target genes, including 930 annotated gene models from around 212 gene families with diverse biological functions and 209 gene models without match at the NCBI databases. Similar to the whole set of miRNAs, the largest number of DEM–DEG pairs were found in gene families coding for tandem repeat domain (TRD) containing proteins. Among the gene families are the 166 LRR gene models, which could be targeted by 278 miRNAs; 90 NB-ARC targeted by 169 miRNAs, and 52 TIRs targeted by 138 miRNAs. TIR, NBS-LRR proteins could be targeted by both conserved (miR482, miR946, miR950, miR1311-1316, miR3710, etc.) and novel miRNAs, like Pab-miRnY45_str, Pab-miRn00543, Pab-miRn00468, Pab-miRn00930, Pab-miRn00202_3p, Pab-miRn00386, Pab-miRn00492, Pab-miRn00930, Pab-miRn01804_5p, Pab-miRnB5, etc. Clathrin/VPS domain proteins often contain PPR and TPR repeat domains, and these could be the targets for regulation by miRNAs based on duplex sequence similarity. 30 Clathrin/VPS and 128 PPR/TPR gene models could be targeted by 202 miRNAs, including Pab-miRn00676, which might regulate the translation of 43 genes (Yakovlev and Fossdal 2017).

DEMs were also shown to target several transcription factors such as Squamosa promoter-binding (SPB) protein (target of miR156), plant transcription factor NOZZLE-like (miR159), Myb-like (miR159, miR482), Homeobox domain bZIP transcription factor (miR166), HD-ZIP protein (miR165/miR166), CCAAT-binding transcription factor (miR169), and AP2-like transcription factors (miR172, miR950), involved in the developmental timing and transition from juvenility. Besides transcription factors, other targets included F-box protein (miR396), laccase (miR397) or plant U-box (PUB) proteins (miR946). In Norway spruce embryos, around 317 DEGs could be regulated by several miRNAs and 290 DEMs could target more than one gene.

As a result, the pleiotropic and synergistic actions of miRNAs during embryo development can be confirmed, whereby a single miRNA can have several potential mRNA targets. Moreover, a single mRNA usually also has multiple miRNA binding sites.

4.5 Differentially Expressed miRNAs Targeting Epigenetic Regulators

miRNAs are especially important in controlling the expression of epigenetic regulators. The reciprocal actions of the miRNAs and the epigenetic pathways appear to form a miRNA-epigenetic feedback loop and have an extensive influence on gene expression regulation. However, the crosstalk between the epigenetic regulators is not fully understood (Yao et al. 2019).

A large number of epigenetic regulators in spruce have previously been described (Yakovlev et al. 2016). The general analysis of all the miRNAs and their predicted epigenetic regulators targets revealed multiple matches between miRNAs and epigenetic regulators. In total, around 84 conserved miRNAs from 32 families and 280 novel miRNAs had matches and could target and putatively post-transcriptionally regulate 683 gene models spanning all pathways of epigenetic regulation (Yakovlev and Fossdal 2017).

Among them, 204 DEM–DEG pairs were defined, showing opposite transcription patterns and considered as probably involved in post-transcriptional silencing of epigenetic regulators during epitype formation in Norway spruce embryos. This network consists of 22 conserved miRNAs from 12 families and 98 novel miRNAs putatively regulating the expression of 197 of predicted targets (Table 4.1 and Table S1). miRNAs were mostly participating in the regulation of genes related to methylation modifications, both in DNA and histones. Epigenetic regulators gene models targeted by miRNAs include several gene families, including DNA methyltransferases, histone deacetylases, and polycomb group genes, like WD domain, G-beta repeat; SNF2 family N-terminal domain; DEAD/DEAH box helicase; BRCA1 C Terminus (BRCT) domain; KH (K Homology) RNA-binding domain; PHD-finger; Core histone H2A/H2B/H3/H4; SET domain; BTB/POZ (BR-C, ttk and bab/Pox virus, and Zinc finger) domain; E1-E2 ATPase; ThiF family of Ubiquitin-activating enzyme (E1 enzyme) among others.

Table 4.1 Predicted miRNAs targeting epigenetic regulators, with the distribution of the target gene models by the type of epigenetic modification

	Number of DEMs	Number of target DEGs
DNA methylation	3	3
Histone methylation	34	37
Histone acetylation	7	7
Histone (protein) phosphorylation	69	133
Histone ubiquitination (sumoylation)	9	9
Chromatin remodeling	7	7
sRNA pathways	4	4
Thermosensing	1	1
Total	120	197

Conserved miRNAs were found also to target genes involved in miRNA and other sRNAs biogenesis pathways. Several miRNAs from miR156, miR162, and miR482 could target *DCL1-like* gene transcripts, although transcription patterns did not confirm any regulatory effect. The novel miRNAs Pab-miRn0030.3 was shown to regulate *ARGONAUTE7* (*AGO7*), Pab-miRn0009.3_3p to regulate 3-hydroxy-3-methylglutaryl coenzyme A reductase (*HMG1*), while Pab-miRn0305.2_3p and Pab-miR482.29 were found to regulate different *SUPPRESSOR OF GENE SILENCING 3* (*SGS3*) genes.

In addition, presented data has confirmed the existence of tight regulatory feedback loops within the miRNA and siRNA pathways in both gymnosperms and angiosperms (Henderson and Jacobsen 2007; Niu et al. 2015). Recently, the opposite has also been shown, where miRNAs expression could be regulated by specific genes in response to changes in the extracellular microenvironment and considered as one of the major mechanisms for epigenetic modifications of the cell. Ion channels/transporters have been reported to transduce extracellular signals into miRNA transcript level changes, which, in turn, regulate target genes, and a potential link between cells and their microenvironment has been proposed (Jiang et al. 2012).

This complicated network of feedback regulation between miRNAs and epigenetic pathways appears to form an epigenetics–miRNA regulatory circuit, and organizes the whole gene expression profile. However, the finding that the putative targets of miRNAs include all types of epigenetic modifiers indicates that miRNAs are central players involved in the formation of the epigenetic memory or at least in regulating the expression of the epigenetic machinery. In light of their important functions in the epigenetic memory formation, future validation work on these miRNAs and their targets are required.

4.6 Conclusion

There is a recent advocacy for the role of miRNAs as key epigenetic regulators of memory formation to respond to environmental cues (Zhang 2015; Crisp et al. 2016). Despite the great progress that has been made for the identification of both protein-coding genes and sRNAs responsive to environmental stresses, these studies are still in their infancy. The majority of plant miRNA-related research has focused on the identification of miRNAs from different plant species and just a few studies aimed at the functional analysis of these miRNAs.

In the described analysis of sRNAs repertoire during formation of the epigenetic memory in Norway spruce, a number of conserved and novel miRNAs were defined. A variety of the predicted miRNAs showed distinct EpI temperature-dependent expression patterns. They could play crucial roles in regulating embryo development and epigenetic memory formation. These putative EpI miRNAs target spruce genes with a wide range of functions, including genes known to be involved in epigenetic regulation, which in turn could provide a feedback process leading to the formation of epigenetic marks. Fine-tuning of the miRNA production likely participates in both developmental regulation and epigenetic memory formation in Norway spruce.

Despite the established links between miRNAs and their target genes, further function and expression studies are necessary in order to elucidate the common miRNA-mediated regulatory mechanisms that underlie memory formation in plants. The use of artificial miRNAs, as well as overexpression and knockout/down of both miRNAs and their targets, will be the best techniques for determining the specific roles of individual miRNAs in memorizing the response to environmental stresses.

References

Ameres SL, Zamore PD (2013) Diversifying microRNA sequence and function. Nat Rev Mol Cell Biol 14(8):475–488. https://doi.org/10.1038/nrm3611

Avramova Z (2015) Transcriptional 'memory' of a stress: transient chromatin and memory (epigenetic) marks at stress-response genes. Plant J 83(1):149–159. https://doi.org/10.1111/tpj.12832

Baev V, Milev I, Naydenov M, Vachev T, Apostolova E, Mehterov N et al (2014) Insight into small RNA abundance and expression in high- and low-temperature stress response using deep sequencing in Arabidopsis. Plant Physiol Biochem 84:105–114. https://doi.org/10.1016/j.plaphy.2014.09.007

Barkan A, Small I (2014) Pentatricopeptide repeat proteins in plants. Annu Rev Plant Biol 65(1):415–442. https://doi.org/10.1146/annurev-arplant-050213-040159

Bartel DP (2009) MicroRNAs: target recognition and regulatory functions. Cell 136(2):215–233

Baulcombe DC, Dean C (2014) Epigenetic regulation in plant responses to the environment. Cold Spring Harb Perspect Biol 6:a019471. https://doi.org/10.1101/cshperspect.a019471

Bilichak A, Kovalchuk I (2016) Transgenerational response to stress in plants and its application for breeding. J Exp Bot. https://doi.org/10.1093/jxb/erw066

Blake GET, Watson ED (2016) Unravelling the complex mechanisms of transgenerational epigenetic inheritance. Curr Opin Chem Biol 33:101–107. https://doi.org/10.1016/j.cbpa.2016.06.008

Bond DM, Baulcombe DC (2013) Small RNAs and heritable epigenetic variation in plants. Trends Cell Biol. http://dx.doi.org/10.1016/j.tcb.2013.08.001

Bräutigam K, Vining KJ, Lafon-Placette C, Fossdal CG, Mirouze M, Marcos JG et al (2013) Epigenetic regulation of adaptive responses of forest tree species to the environment. Ecol Evol 3(2):399–415. https://doi.org/10.1002/ece3.461

Carneros E, Yakovlev I, Viejo M, Olsen JE, Fossdal CG (2017) The epigenetic memory of temperature during embryogenesis modifies the expression of bud burst-related genes in Norway spruce epitypes. Planta 246:553–566. https://doi.org/10.1007/s00425-017-2713-9

Castel SE, Martienssen RA (2013) RNA interference in the nucleus: roles for small RNAs in transcription, epigenetics and beyond. Nat Rev Genet 14(2):100–112

Chen X, Irani NG, Friml J (2011) Clathrin-mediated endocytosis: the gateway into plant cells. Curr Opin Plant Biol 14(6):674–682. https://doi.org/10.1016/j.pbi.2011.08.006

Crisp PA, Ganguly D, Eichten SR, Borevitz JO, Pogson BJ (2016) Reconsidering plant memory: intersections between stress recovery, RNA turnover, and epigenetics. Sci Adv 2(2). https://doi.org/10.1126/sciadv.1501340

D'Urso A, Brickner JH (2014) Mechanisms of epigenetic memory. Trends Genet 30(6):230–236. https://doi.org/10.1016/j.tig.2014.04.004

Dangl JL, Jones JDG (2001) Plant pathogens and integrated defence responses to infection. Nature 411(6839):826–833

Dewan S, Vander Mijnsbrugge K, De Frenne P, Steenackers M, Michiels B, Verheyen K (2018) Maternal temperature during seed maturation affects seed germination and timing of bud set in seedlings of European black poplar. For Ecol Manage 410:126–135. https://doi.org/10.1016/j.foreco.2018.01.002

Dolgosheina EV, Morin RD, Aksay G, Sahinalp SC, Magrini V, Mardis ER et al (2008) Conifers have a unique small RNA silencing signature. RNA 14(8):1508–1515. https://doi.org/10.1261/rna.1052008

Eckardt NA (2006) Genetic and epigenetic regulation of embryogenesis. Plant Cell 18(4):781–784. https://doi.org/10.1105/tpc.106.042440

Fu R, Zhang M, Zhao Y, He X, Ding C, Wang S et al (2017) Identification of salt tolerance-related microRNAs and their targets in maize (*Zea mays* L.) using high-throughput sequencing and degradome analysis. Front Plant Sci 8(864). https://doi.org/10.3389/fpls.2017.00864

Gömöry D, Foffová E, Longauer R, Krajmerová D (2015) Memory effects associated with early-growth environment in Norway spruce and European larch. Eur J Forest Res 134(1):89–97. https://doi.org/10.1007/s10342-014-0835-1

Gömöry D, Hrivnák M, Krajmerová D, Longauer R (2017) Epigenetic memory effects in forest trees: a victory of "Michurinian biology"? Central Eur Forest J 63(4):173–179. https://doi.org/10.1515/forj-2017-0024

Grossniklaus U, Kelly WG, Ferguson-Smith AC, Pembrey M, Lindquist S (2013) Transgenerational epigenetic inheritance: how important is it? Nat Rev Genet 14:228. https://doi.org/10.1038/nrg3435

Gruber AJ, Zavolan M (2013) Modulation of epigenetic regulators and cell fate decisions by miRNAs. Epigenomics 5(6):671–683

He L, Hannon GJ (2004) MicroRNAs: small RNAs with a big role in gene regulation. Nat Rev Genet 5(7):522–531. https://doi.org/10.1038/nrg1379

Heard E, Martienssen Robert A (2014) Transgenerational epigenetic inheritance: myths and mechanisms. Cell 157(1):95–109. https://doi.org/10.1016/j.cell.2014.02.045

Henderson IR, Jacobsen SE (2007) Epigenetic inheritance in plants. Nature 447. https://doi.org/10.1038/nature05917

Heo JB, Sung S (2011) Vernalization-mediated epigenetic silencing by a long intronic noncoding RNA. Science 331(6013):76–79. https://doi.org/10.1126/science.1197349

Herranz H, Cohen SM (2010) MicroRNAs and gene regulatory networks: managing the impact of noise in biological systems. Genes Dev 24(13):1339–1344. https://doi.org/10.1101/gad.1937010

Iwasaki M, Paszkowski J (2014) Epigenetic memory in plants. EMBO J 33(18):1987–1998. https://doi.org/10.15252/embj.201488883

Jiang X, Zhang J, Chan H (2012) Ion channels/transporters as epigenetic regulators? A microRNA perspective. Sci China Life Sci 55(9):753–760. https://doi.org/10.1007/s11427-012-4369-9

Johnsen Ø, Daehlen OG, Østreng G, Skrøppa T (2005a) Daylength and temperature during seed production interactively affect adaptive performance of *Picea abies* progenies. New Phytol 168(3):589–596. https://doi.org/10.1111/j.1469-8137.2005.01538.x

Johnsen Ø, Fossdal CG, Nagy N, Molmann J, Dælen OG, Skrøppa T (2005b) Climatic adaptation in *Picea abies* progenies is affected by the temperature during zygotic embryogenesis and seed maturation. Plant Cell Environ 28(9):1090–1102. https://doi.org/10.1111/j.1365-3040.2005.01356.x

Khraiwesh B, Zhu J-K, Zhu J (2012) Role of miRNAs and siRNAs in biotic and abiotic stress responses of plants. Biochimica et Biophysica Acta (BBA) Gene Regul Mech 1819(2):137–148. https://doi.org/10.1016/j.bbagrm.2011.05.001

Kinoshita T, Seki M (2014) Epigenetic memory for stress response and adaptation in plants. Plant Cell Physiol 55:1859–1863. https://doi.org/10.1093/pcp/pcu125

Kvaalen H, Johnsen O (2008) Timing of bud set in *Picea abies* is regulated by a memory of temperature during zygotic and somatic embryogenesis. New Phytol 177(1):49–59. https://doi.org/10.1111/j.1469-8137.2007.02222.x

Law JA, Jacobsen SE (2010) Establishing, maintaining and modifying DNA methylation patterns in plants and animals. Nat Rev Genet 11(3):204–220

Lee JT (2012) Epigenetic regulation by long noncoding RNAs. Science 338(6113):1435–1439. https://doi.org/10.1126/science.1231776

Li J, Reichel M, Li Y, Millar AA (2014) The functional scope of plant microRNA-mediated silencing. Trends Plant Sci 19(12):750–756. https://doi.org/10.1016/j.tplants.2014.08.006

Lin Y, Lai Z (2013) Comparative analysis reveals dynamic changes in mirnas and their targets and expression during somatic embryogenesis in Longan (*Dimocarpus longan* Lour.). PLoS ONE 8(4):e60337. https://doi.org/10.1371/journal.pone.0060337

Liu J, Feng L, Li J, He Z (2015) Genetic and epigenetic control of plant heat responses. Front Plant Sci 6:267. https://doi.org/10.3389/fpls.2015.00267

Liu W-W, Meng J, Cui J, Luan Y-S (2017) Characterization and function of microRNA(*)s in plants. Front Plant Sci 8:2200. https://doi.org/10.3389/fpls.2017.02200

Liu W, Yu W, Hou L, Wang X, Zheng F, Wang W et al (2014) Analysis of miRNAs and their targets during adventitious shoot organogenesis of *Acacia crassicarpa*. PLoS ONE 9(4):e93438. https://doi.org/10.1371/journal.pone.0093438

Lu S, Sun Y-H, Amerson H, Chiang VL (2007) MicroRNAs in loblolly pine (*Pinus taeda* L.) and their association with fusiform rust gall development. Plant J 51(6):1077–1098

Meyers BC, Kaushik S, Nandety RS (2005) Evolving disease resistance genes. Curr Opin Plant Biol 8(2):129–134. https://doi.org/10.1016/j.pbi.2005.01.002

Mica E, Piccolo V, Delledonne M, Ferrarini A, Pezzotti M, Casati C et al (2009) High throughput approaches reveal splicing of primary microRNA transcripts and tissue specific expression of mature microRNAs in Vitis vinifera. BMC Genom 10(1):558

Mirouze M (2012) The small RNA-based Odyssey of epigenetic information in plants: from cells to species. DNA Cell Biol 31(12):1650–1656. https://doi.org/10.1089/dna.2012.1681

Molinier J, Ries G, Zipfel C, Hohn B (2006) Transgeneration memory of stress in plants. Nature 442:1046. https://doi.org/10.1038/nature05022

Montes RAC, de Fátima Rosas-Cárdenas F, De Paoli E, Accerbi M, Rymarquis LA, Mahalingam G et al (2014) Sample sequencing of vascular plants demonstrates widespread conservation and divergence of microRNAs. Nat Commun 5. https://doi.org/10.1038/ncomms4722

Neilsen CT, Goodall GJ, Bracken CP (2012) IsomiRs—the overlooked repertoire in the dynamic microRNAome. Trends Genet 28(11):544–549. https://doi.org/10.1016/j.tig.2012.07.005

Niu S-H, Liu C, Yuan H-W, Li P, Li Y, Li W (2015) Identification and expression profiles of sRNAs and their biogenesis and action-related genes in male and female cones of *Pinus tabuliformis*. BMC Genom 16(1):1–13. https://doi.org/10.1186/s12864-015-1885-6

Norouzitallab P, Baruah K, Vanrompay D, Bossier P (2019) Can epigenetics translate environmental cues into phenotypes? Sci Total Environ 647:1281–1293. https://doi.org/10.1016/j.scitotenv.2018.08.063

Nystedt B, Street NR, Wetterbom A, Zuccolo A, Lin Y-C, Scofield DG et al (2013) The Norway spruce genome sequence and conifer genome evolution. Nature 497:579–584. https://doi.org/10.1038/nature12211

Oh TJ, Wartell RM, Cairney J, Pullman GS (2008) Evidence for stage-specific modulation of specific microRNAs (miRNAs) and miRNA processing components in zygotic embryo and female gametophyte of loblolly pine (*Pinus taeda*). New Phytol 179(1):67–80

Patkin EL, Sofronov GA (2013) Population epigenetics, ecotoxicology, and human diseases. Russ J Genetics Appl Res 3(5):338–351. https://doi.org/10.1134/s2079059713050079

Quinn C, Iriyama R, Fernando D (2014) Expression patterns of conserved microRNAs in the male gametophyte of loblolly pine (*Pinus taeda*). Plant Reprod 27(2):69–78. https://doi.org/10.1007/s00497-014-0241-3

Rohde A, Junttila O (2008) Remembrances of an embryo: long-term effects on phenology traits in spruce. New Phytol 177(1):2–5. https://doi.org/10.1111/j.1469-8137.2007.02319.x

Saze H (2008) Epigenetic memory transmission through mitosis and meiosis in plants. Semin Cell Dev Biol 19(6):527–536. https://doi.org/10.1016/j.semcdb.2008.07.017

Simon SA, Meyers BC (2011) Small RNA-mediated epigenetic modifications in plants. Curr Opin Plant Biol 14(2):148–155. https://doi.org/10.1016/j.pbi.2010.11.007

Sow MD, Allona I, Ambroise C, Conde D, Fichot R, Gribkova S et al (2018) Epigenetics in forest trees: state of the art and potential implications for breeding and management in a context of climate change. In: Gallusci P, Bucher E, Mirouze M (eds) Plant epigenetics coming of age for breeding applications. Academic Press, New York, pp 387–453

Stuwe E, Tóth KF, Aravin AA (2014) Small but sturdy: small RNAs in cellular memory and epigenetics. Genes Dev 28(5):423–431. https://doi.org/10.1101/gad.236414.113

Sunkar R, Li Y-F, Jagadeeswaran G (2012) Functions of microRNAs in plant stress responses. Trends Plant Sci 17(4):196–203. https://doi.org/10.1016/j.tplants.2012.01.010

Thellier M, Lüttge U (2013) Plant memory: a tentative model. Plant Biol 15(1):1–12. https://doi.org/10.1111/j.1438-8677.2012.00674.x

Valdés A, Marteinsdottir B, Ehrlén J (2018) A natural heating experiment: phenotypic and genotypic responses of plant phenology to geothermal soil warming. Global Change Biol. https://doi.org/10.1111/gcb.14525

Verhoeven KJF, Jansen JJ, van Dijk PJ, Biere A (2010) Stress-induced DNA methylation changes and their heritability in asexual dandelions. New Phytol 185(4):1108–1118. https://doi.org/10.1111/j.1469-8137.2009.03121.x

Wan L-C, Wang F, Guo X, Lu S, Qiu Z, Zhao Y et al (2012a) Identification and characterization of small non-coding RNAs from Chinese fir by high throughput sequencing. BMC Plant Biol 12(1):146

Wan L-C, Zhang H, Lu S, Zhang L, Qiu Z, Zhao Y et al (2012b) Transcriptome-wide identification and characterization of miRNAs from Pinus densata. BMC Genom 13:132

Wang L, Zhao J, Zhang M, Li W, Luo K, Lu Z et al (2015) Identification and characterization of microRNA expression in *Ginkgo biloba* L. leaves. Tree Genetics Genomes 11(4):1–13. https://doi.org/10.1007/s11295-015-0897-3

Whittle CA, Otto SP, Johnston MO, Krochko JE (2009) Adaptive epigenetic memory of ancestral temperature regime in *Arabidopsis thaliana*. Botany 87(6):650–657. https://doi.org/10.1139/B09-030

Xia R, Xu J, Arikit S, Meyers BC (2015) Extensive families of miRNAs and PHAS Loci in Norway Spruce demonstrate the origins of complex phasiRNA networks in seed plants. Mol Biol Evol 32(11):2905–2918. https://doi.org/10.1093/molbev/msv164

Yakovlev IA, Carneros E, Lee Y, Olsen JE, Fossdal CG (2016) Transcriptional profiling of epigenetic regulators in somatic embryos during temperature induced formation of an epigenetic memory in Norway spruce. Planta 243(5):1237–1249. https://doi.org/10.1007/s00425-016-2484-8

Yakovlev IA, Fossdal CG (2017) In silico analysis of small RNAs suggest roles for novel and conserved miRNAs in the formation of epigenetic memory in somatic embryos of Norway Spruce. Front Physiol 8(674):1–17. https://doi.org/10.3389/fphys.2017.00674

Yakovlev IA, Fossdal CG, Johnsen Ø (2010) MicroRNAs, the epigenetic memory and climatic adaptation in Norway spruce. New Phytol 187(4):1154–1169. https://doi.org/10.1111/j.1469-8137.2010.03341.x

Yakovlev IA, Fossdal CG, Skrøppa T, Olsen JE, Jahren AH, Johnsen Ø (2012) An adaptive epigenetic memory in conifers with important implications for seed production. Seed Sci Res 22(02):63–76. https://doi.org/10.1017/S0960258511000535

Yakovlev IA, Lee Y, Rotter B, Olsen JE, Skrøppa T, Johnsen Ø et al (2014) Temperature-dependent differential transcriptomes during formation of an epigenetic memory in Norway spruce embryogenesis. Tree Genetics Genomes 10:355–366. https://doi.org/10.1007/s11295-013-0691-z

Yao Q, Chen Y, Zhou X (2019) The roles of microRNAs in epigenetic regulation. Curr Opin Chem Biol 51:11–17. https://doi.org/10.1016/j.cbpa.2019.01.024

Zeytuni N, Zarivach R (2012) Structural and functional discussion of the tetra-trico-peptide repeat, a protein interaction module. Structure 20(3):397–405. https://doi.org/10.1016/j.str.2012.01.006

Zhang B (2015) MicroRNA: a new target for improving plant tolerance to abiotic stress. J Exp Bot 66(7):1749–1761. https://doi.org/10.1093/jxb/erv013

Zhang J, Wu T, Li L, Han S, Li X, Zhang S et al (2013) Dynamic expression of small RNA populations in larch (*Larix leptolepis*). Planta 237(1):89–101. https://doi.org/10.1007/s00425-012-1753-4

Zhang J, Zhang S, Han S, Wu T, Li X, Li W et al (2012) Genome-wide identification of microRNAs in larch and stage-specific modulation of 11 conserved microRNAs and their targets during somatic embryogenesis. Planta 236. https://doi.org/10.1007/s00425-012-1643-9

Part II
microRNAs in Plant Development

Chapter 5
microRNA Regulation of Fruit Development

Silvia Farinati, Cristian Forestan, Monica Canton, Serena Varotto and Claudio Bonghi

Abstract A fruit is commonly defined as a mature ovary containing seeds and represents a defining characteristic of the angiosperm phylum. For this reason, being the fruit responsible for seed spreading, it assumes a key role in the adaptive success of angiosperms and a large variety of fruit types have evolved. During fruit development, every step is strictly regulated by complex molecular mechanisms, including post-transcriptional regulation by microRNAs. However, the fragmentary nature of information produced so far prevents a global vision of the system and makes difficult a possible comparison between different fruit types. To overcome this limitation the availability of the genome sequence of many fruit-bearing species, the analysis of their genome structure, gene pathways, and gene function are the first essential steps to increase our understanding of fruit development. Starting from the analysis of the information available in miRNA databases, we have analysed conserved fruit type-specific miRNA families and miRNA expression information available in the literature for identifying potential functions of miRNAs. All this information is discussed by considering evolutionary relationships and structural patterning in different fruit types.

Keywords Dry and fleshy fruit · Fruit patterning · Pit hardening · Fruit types evolution · Rosaceae

---

Silvia Farinati and Cristian Forestan contributed equally to this work.

S. Farinati · C. Forestan · M. Canton · S. Varotto · C. Bonghi (✉)
Department of Agronomy, Food, Natural Resources, Animals and Environment, University of Padova Agripolis, Legnaro, Italy
e-mail: claudio.bonghi@unipd.it

C. Miguel et al. (eds.), *Plant microRNAs*, Concepts and Strategies in Plant Sciences, https://doi.org/10.1007/978-3-030-35772-6_5

5.1 Dry and Fleshy Fruit Systems: An Articulated Scenario with Similarities and Diversities

The fruit is a plant organ harbored uniquely by the angiosperms phylum and commonly defined as a mature ovary containing seeds. For this reason, and for its diversification and specialization, the fruit represents the organ responsible for seed spreading, assuming a key role in the adaptive success of angiosperms in most tropical, subtropical and temperate worldwide zones (Handa et al. 2012). The physiological events driving the fruit growth range from the floral meristem initiation stage up to later stages of fruit ripening and include carpel formation, differentiation/patterning and boundary fruit organ formation (Musseau et al. 2017). In particular, fruit set occurs when the signalling triggered by the pollination and fertilization turns a fully developed ovary into a fast-growing fruit that will soon initiate its ripening process.

Because of the large variety of fruit types and their different dispersal strategies, the angiosperm fruits are principally divided into dry and fleshy categories: for dry fruits a simple 'passive' seed dispersal mechanism is observed, while for fleshy fruits multiple dispersal strategies have been evolved. In detail, the advanced association with higher animals is related to development of fruit features more appealing for their dispersal vectors, like attractiveness through color and flavor (pigments and volatile compounds), softening (cell wall and cuticle degradation), and palatability and nutritional value (aroma, flavor, sugars, antioxidants, vitamins, etc.; Tiffney 1984, 2004). In spite of this articulated scenario with a broad range of fruit types and developmental strategies, model species such as *Arabidopsis thaliana* and *Solanum lycopersicum* L. have provided knowledge on the molecular and biochemical aspects underlying dry and fleshy fruit development, respectively. Several reviews have focused the attention on the main differences that characterize fruit development in these two fruit types (McAtee et al. 2013; Osorio et al. 2013; Seymour et al. 2013; Gómez et al. 2014). In addition, numerous studies compared the apparently different developmental programs in order to identify analogies between the biochemical and physiological processes that occur. While the main steps driving fruit set from fertilization to ripening are common to the different fruit models, it is necessary to better define biochemical and/or physiological characteristics to support possible analogy or diversity. For this aim, gene expression analyses have been performed to characterize fruit development and senescence in *Arabidopsis* (Wagstaff et al. 2009; Carbonell-Bejerano et al. 2010) and in tomato fruit (Eriksson et al. 2004; Alba et al. 2005; Tiwari and Paliyath 2011; Zhang et al. 2013). These analyses highlighted specific events occurring in fleshy fruit without correspondence in dry fruits, such as the breakdown of carbohydrates into sugars, reduction in acids and increase in volatiles responsible for flavor and aroma (Klee and Giovannoni 2011).

For many years *Arabidopsis* has been the plant reference to study physiology and metabolism, genetic and genomic properties, stress responses and all developmental processes, including fruit development (Somerville and Koornneef 2002). By taking this model system as reference for dry fruit, the fruit developmental process can be divided into three main phases: pistil patterning, fruit set, and valve abscission or dehiscence (Ferrándiz et al. 2002). In detail, fruit patterning is established mainly during pistil ontogeny where a complex network of genes with a role in identity of different pistil tissues has been identified (Roeder and Yanofsky 2006; Martínez-Laborda and Vera 2009). The following fruit-set events (from pollination up to ovule fertilization), are triggered mainly by hormonal signals (Dorcey et al. 2009; Ruan et al. 2012). As well known, the fruit of *Arabidopsis* is a silique composed by two valves fused at the level of a specific tissue called valve margin, and its dehiscence, necessary for seed dispersal, is the better-characterized fruit ripening process in *Arabidopsis* (Estornell et al. 2013).

During the initial fruit growth phase, rapid growth in fruit length and width, mainly due to rapid cell division and expansion were driven by hormones, like gibberellins (GAs) and auxin is observed in the tissues surrounding fertilized ovules (Vivian-Smith and Koltunow 1999). Later on, the silique proceeds with the programmed senescence regulated more by biological and genetic factors rather than by the environment. The fact that senescence is a process molecularly regulated is testified by the sharing of specific molecular components between different senescent organs, such as fruits and leaves (Kou et al. 2012).

In comparison to dry fruits, a completely different scenario characterizes the fleshy fruit systems. A large number of fruit types with different anatomical and physiological properties were described in the past, dictated mainly by their fruit structures. Hip, berry, drupe, and pome are the main typical botanical structures symbolizing the fleshy fruits, and for studying some of them a specific model plant has been defined such as tomato for berry, peach (*Prunus persica*) for drupe, and apple (*Malus x domestica*) for pome. These fruits can derive from carpels ('true fruit') or from accessory structures ('false fruit'), offering diverse fruit development patterns; however the main steps driving fruit set from fertilization to ripening are common (Farinati et al. 2017). In spite of the large diversity present in fleshy fruit type, general notions on fleshy fruit were mainly obtained from molecular and physiological investigations on tomato, which represents the model system for berry/fleshy fruit and for the entire horticulture sector. Several authors have reviewed the main molecular mechanisms involved in fruit growth and ripening, using tomato fruit as reference, and comparing available data coming from other fleshy or dry fruit models (Ecker 2013; Osorio et al. 2013; Teyssier et al. 2015; Gallusci et al. 2016; Farinati et al. 2017).

The fleshy fruit growth depends on the fruit type and can be represented by specific mathematical models (Fig. 5.1). For example, a single or a double sigmoid curve describes the berry (tomato and apple) and drupe (peaches) growth. For berry, an expolinear equation based on fruit diameter, best fit to its initial exponential growth phase followed by linear growth (Lakso et al. 1995). A mathematical analysis of drupe growth kinetic identified four main developmental stages, named according to the parameters measured for calculations (S1–S4 for cross diameter, FW1–FW4 for fresh weight, and DW1-DW4 for dry weight). The duration of the different developmental stages varies according to the genotype (Chalmers and Ende 1975). Independently of the parameter, in the second phase (S2, FW2 or DW2) an abrupt ceasing of growth occurs concomitantly with the lignification of endocarp, the event characterizing the development of the drupe.

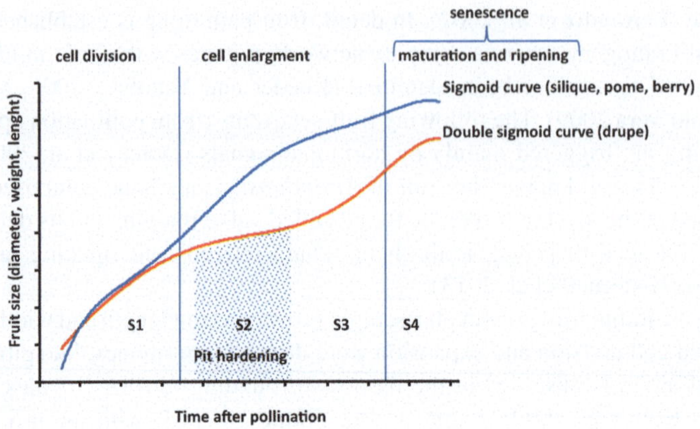

Fig. 5.1 Schematic representation of growth curves of dry and fleshy fruit. Growth of dry and fleshy fruits can be represented by using a sigmoid curve and double sigmoid curve. The sigmoid curve best fits the initial exponential growth phase followed by linear growth that is occurring in silique (dry fruit), pome and berry (fleshy fruit). The double sigmoid curve is the best model for drupe growth in which two exponential growth phases (named S1 and S3) are interrupted by a slow growth phase (S2), in which the lignification of endocarp occurs. In all fruit types, in the early phase of development cell division is predominant, later on the fruit growth is mainly due to cell enlargement and endoreduplication phenomenon. In the last phase of its development, the fruit passes through a series of structural and composition changes, which characterize the maturation and ripening stage. The maturation and ripening stages are part of the senescence program

As previously stated the cultivated tomato represents an appropriate model to analyze tissue morphology because of the large existing phenotypic fruit diversity in this species (Klee and Giovannoni 2011; Pesaresi et al. 2014). Tomato fruit is a bi- or multilocular berry, in which seeds develop attached to the placenta, a parenchymatous tissue that becomes gelatinous and fills the two or more locules formed by the septa that divide the ovary structure, during fruit development and ripening (Bertin 2004; Mintz-Oron et al. 2008). Tomato fruit growth initiates at the fertilization event, defined as stage 0, and then proceeds through several successive major stages (Picken 1984; Gillespie et al. 1993). Immediately after fertilization, a rapid cell division occurs, leading to a progressive increase in pericarp cell number followed by a sharp fall in the rate of cell division (stage 1). During the stage 2, fruit growth relies on cell expansion and leads to a significant increase in weight and is coinciding with endoreduplication process. Usually, this sequence of phases is common in fleshy fruit systems, even if characterized by several different fruit structures (Bergervoet et al. 1996). During the third phase, the fruit enters the mature green (MG) stage (Ho and Hewitt 1986; Giovannoni 2004; Czerednik et al. 2012) in which the liquefaction of locule tissues starts. At this stage, the berry attains its final size (30,000 times the number of cells at anthesis), which varies greatly among cultivars and is very susceptible to environmental influences (Chevalier 2007). After the MG stage, a rapid reorganization of metabolomic profile is observed and this defines the final breaking and ripening (Breaker stage, BR and Red Ripe stage, RR) stages. During the BR stage (analogous to veraison in grape berry fruits) a conversion of chloroplasts into specialized plastids (i.e. chromoplasts) is observed, with a change of color into yellow-orange, followed by a rapid accumulation of carotenoids in association to chlorophyll degradation during the final RR stage (Pesaresi et al. 2014). Tomato fruit is classified as climacteric because of the marked increase of respiration and ethylene production at the onset of ripening to differentiate from non-climacteric fruits, in which these two events are lacking. Regulation of ripening via ethylene is depending on the ability to produce ethylene and the sensitivity to the hormone. Important regulators in ethylene biosynthesis include MADS-domains protein RIPENING INHIBITOR (RIN), COLORLESS NON_RIPENING (CNR), and a NAC domain transcription factor, named NON_RIPENING (NOR). Mutants with altered expression of RIN, CNR and NOR produced fruit with impaired ethylene biosynthesis or perception and therefore unable to ripen (Giovannoni 2004).

Differently from berry, the event characterizing drupe growth type is associated with the endocarp layer and consists of the pit hardening (Fig. 5.1). Endocarp cells can be already recognized at the end of the first phase of exponential growth (S1), because of the phenol compounds accumulating before lignin formation (Masia et al. 1992). After the lignification (S2), the resume of growth (marking the S3 stage) and the following increase in size is partially due to endoreduplication events (Chevalier et al. 2011). Lastly, S4 is the stage where fruits acquire the prerequisite competence to enter the final developmental stage, i.e. ripening (Bonghi et al. 2011). The majority of drupes are climacteric, although some plum cultivars show non-climacteric behavior (also called suppressed climacteric) where these two events are lacking (Abdi et al. 1997).

In addition to information obtained from true fruits, there is information derived from false fruit, such as pome and strawberry (*Fragaria x ananassa* Duch.). In these fruits, the edible part derives from a receptacle, the terminal thickened part of the peduncle (Ireland et al. 2013). In pear (*Pyrus communis*) and apple, the pome development and ripening is a complex process, as demonstrated by deep changes occurring in various metabolic pathways (Wu et al. 2014). Pome growth comprises an initial phase of cell division and enlargement followed by a maturation period starting with no evident morphological alterations (pre-climacteric period) until the onset of climacteric period in which ethylene production and respiration rates are expected to peak (Fonseca et al. 2004; Xia et al. 2012; Ireland et al. 2013). Apples, like tomatoes, can ripen on the tree, while pears need a postharvest period for the progression of ripening (Murayama et al. 2015). The inability of pears to ripen on the tree can be related to the following facts: (1) ethylene is not produced by the fruit until the optimal harvest time is reached, (2) the continuous supply of assimilates to fruit may delay the ripening. The use cold temperature can induce the pears ripening after fruit detachment (Jajo et al. 2015).

In strawberry, 'the true fruit' called achene is present on the surface of the receptacle tip, 'the false fruit part'. Each achene contains a single seed and can be easily removed by manual manipulation (Iannetta et al. 2006; Xia et al. 2015). Strawberry is classified as non-climacteric fruit and achene-derived auxin is the key phytohormone controlling growth and ripening of strawberry receptacle itself (Iannetta et al. 2006). During strawberry development, both the achenes and the receptacle follow separate patterns, but their developmental programs are coordinated. Indeed, a study reporting parallel profiling of primary and secondary metabolism in both structures across several developmental stages, underlined a high level of metabolic synchrony and specialization during development (Csukasi et al. 2012).

All the described events determine the final characteristics and properties of the ripe fruit and, to ensure a correct developmental program, every step is carefully controlled by complex molecular mechanisms under multiple levels of regulation, including post-transcriptional regulation by microRNAs (Seymour et al. 2013; Correa et al. 2018). The biogenesis of plant miRNAs initiates from the transcription of *MIR* genes by RNA polymerase II (PolII), processing of primary miRNA transcripts by DICER-LIKE proteins into mature miRNAs and loading of mature miRNAs into ARGONAUTE proteins to form miRNA-induced silencing complex (miRISC). Epigenetic mechanisms, cis-regulatory elements, and trans-acting regulators are emerging as important components in MIR transcriptional regulation, although little is known about how MIRs are recognized as different from PolII transcribed genes. Transcriptional regulation of MIRs adds a further level of complexity to the sophisticated mechanisms of MIR transcript processing, mature miRNA production and miRNA interaction with target RNA. Many works have demonstrated that miRNA function in a homology-depending manner to suppress the expression of their targets, which are mostly transcription factors regulating gene expression of gene circuits operating during plant development and fruit development and ripening as well (Karlova et al. 2014).

In addition, despite numerous past reports have described the role of gene transcription regulation in every single step during the fruit growth; an open question is the definition of the conservation level of transcriptional control between dry and fleshy fruit-bearing species. In particular, a better characterization of the microRNAs, which have been shown to be able to coordinate the main steps of fruit development and ripening in different plant species, may result in the development of novel strategies for fruit crop management.

5.2 miRNAs Role During Fruit Development and Ripening

miRNAs are considered key regulators of many plant developmental and physiological processes (Liu et al. 2018). They are typically classified as conserved or non-conserved, where conserved miRNAs are those abundant in tissues in which they function and are present in most plant lineages. It is worthy to note that this last characteristic is highly influenced by the number of species with available miRNA annotation. Conserved miRNAs are often found in large families, with each precursor classified by the conserved sequence of the mature miRNA. In addition, they show low sequence variation across plant families, consistent with fundamental roles and conserved targets (Baldrich et al. 2018). They predominantly target genes involved in plant development and stress response. The fact that many genes targeted by miRNAs encode transcription factors gives miRNA a pivotal role in post-transcriptional regulation during plant development and in plant-environment interactions. Important developmental processes, such as cell division/expansion/differentiation, specification of organ identities, phase transition and nutrition homeostasis, regulated by miRNAs are also involved in fruit development. However, despite several molecular notions on the role of single or few miRNAs during fruit development have been reported in recent literature, there is again a limited information, in terms of genome-wide data, regarding the involvement of different miRNA families in multiple developmental fruit stages and their integration with genome-wide transcriptomic analyses.

If these punctual notions clarify specific molecular mechanisms in precise fruit growth phases, they do not allow a complete vision of the process and make difficult a possible comparison between different fruit systems. As previously reported, the use of model fruits is necessary to facilitate the comprehension of a vastness of fruit examples, each of them with peculiar molecular, structural and physiological properties. Moreover, by simplifying this articulated scenario it has been also possible to define the involvement of miRNA and their targets (also defined as miRNA modules; Correa et al. 2018), in fruit growth regulation. From the earliest steps of fruit patterning, at cell and tissue levels, the miRNA spatiotemporal accumulation modulates consequently the expression of specific targets, by restricting their expression domain (Berger et al. 2009; Chen et al. 2009). To support this view, although the two models of the dry and fleshy fruit systems are characterized by very dissimilar developmental stages, interaction between miRNAs modules have a key role in gynoecium patterning, which influences the fruit development as described above. Interestingly, alterations in conserved miRNA modules produce distinct phenotypic changes in gynoecium and fruits of *Arabidopsis* and tomato (Xing et al. 2013; Silva et al. 2014). Understanding what pathways regulated by conserved miRNA modules in different systems are involved in fruit development, and how they differently influence morphological and growth processes, remain important aspects to elucidate in fruit development (Correa et al. 2018).

To approach these questions, we analyzed information available in miRNA databases that have greatly increased thanks to the application of next-generation sequencing (NGS) technologies for the high throughput sequencing of small RNAs (sRNA-Seq). Indeed, decoding the complete miRNA catalogue and their expression patterns provides a basis to study variation at sequence (conserved and lineage-specific miRNAs) and expression level, and their potential function, and could also help to understand the evolutionary relationship between different fruit types.

New sRNA-Seq datasets are deposited in public archives almost daily and currently, more than 900 Arabidopsis sRNA-Seq data sets are available from the Sequence Read Archive (SRA; https://www.ncbi.nlm.nih.gov/sra/ last accessed April 2019) at the National Center for Biotechnology Information (NCBI) database. Surprisingly, of this huge amount of data, only eight datasets have been generated from silique samples (four developmental stages, two biological replicates each). These datasets were recently generated with the aim to create an expression atlas of miRNAs in *Arabidopsis thaliana*, starting from samples from 27 different organ/tissue types, which cover the entire life cycle of Arabidopsis (Xu et al. 2018). The authors showed that most miRNAs are ubiquitously expressed. However, a small set of miRNAs display highly specific expression patterns pointing to different miRNA members within the same family having distinct spatial and temporal expression patterns. By re-analysing these expression data, we identified 72 miRNAs prevalently expressed during siliques development (out of 308 miRNAs included in the atlas), including several family members previously described as regulators of fruit development (Table 5.1). The expression of these miRNAs progressively increases during siliques development (Fig. 5.2a), with only one and five miRNAs exclusively expressed at 2 and 3 days after fertilization (DAF) respectively, and the remaining 66 miRNAs strongly expressed at 5 and 7 DAF. This suggests that the progression from the cell division (2–3 DAF) to the cell expansion (5–7 DAF) phase requires the switching-off of specific gene sets through the activation of their regulative miRNAs.

As previously discussed, a growing body of molecular genetics and transgenic studies support the fundamental role of miRNAs in regulating fruit development in different non-model, crop species, and point at their different impacts depending on the fruit type (Yao et al. 2016). The relevant impact of fruit development on crop productivity and quality together with their contribution to dietary nutrition globally, indeed promoted many research projects investigating miRNAs and their target genes in different species. For example, compared to the only 4 Arabidopsis sRNA-Seq sample types, SRA archives currently include 17, 19 and 30 different sRNA-Seq datasets produced on tomato, peach and apple fruits, respectively.

These sRNA-Seq data sets are of great value in investigating the expression profiles and potential function of miRNAs, however, they were generated from different laboratories worldwide, each of which with different sequencing coverage and quality and represent miRNAs in specific cultivars or developmental stages. This heterogeneity together with the incomplete annotation of microRNAs in non-model species (428 mature miRNA annotated in Arabidopsis, 147, 214 and 322 in tomato, peach and apple, respectively; mirBase22—http://www.mirbase.org/) make difficult to dissect the relationship between miRNA expression/specificity and evolutionary dynamics.

To partially overcome these limitations, consistent miRNA expression datasets for tomato, peach and apple were retrieved from the Next-Gen Sequence Databases produced by the Meyers lab at the Donald Danforth Plant Science Center (https://mpss.danforthcenter.org/). By storing and sharing uniform miRNA expression data from vegetative and fruit samples of these crop species, this database easily allowed to identify fruit-specific miRNAs in each species. Since many miRNAs are encoded by gene families, we explored those families that are present in fruits of at least one analyzed species, to better evaluate species-specificity of miRNA expression (Table 5.1). We found six families (miR164, miR169, miR171, miR172, miR396 and miR399) with conserved expression in all the four species, while miR156, and miR167 represent specific markers of 'true fruits' versus 'false fruits' developmental program/regulation. Interestingly, miR159, miR403, miR477 and miR482 were

Table 5.1 Conserved miRNA families expressed during fruit development in different fruit types. For each miRNA family expressed in fruits of two or more species, the table reports the species-specific miRNA members, the identified targets and the fruit developmental process(es) in which they are involved

miRNA family	*Arabidopsis thaliana*	*Solanum lycopersicum*	*Prunus persica*	*Malus domestica*	Target	Function	References
miR164	ath-miR164a-b	sly-miR164a-b	ppe-miR164a-c	mdm-miR169b-d	NAC transcription factors	Organ boundaries maintenance	Karlova et al. (2013), Hendelman et al. (2013), Correa et al. (2018)
miR169	ath-miR169d-e ath-miR169 g	sly-miR169f	ppe-miR169a-c ppe-miR169f-l	mdm-miR169c mdm-miR169d	NF-YA and CCAAT-binding transcription factors	Flowering, fruit development	Cartolano et al. (2007), Moxon et al. (2008), Karlova et al. (2013), Zeng et al. (2015)
miR171	ath-miR171a-c	sly-miR171e	ppe-miR171a-d ppe-miR171f-g	mdm-miR171c-e mdm-miR171i mdm-miR171 m-n	Scarecrow-like GRAS transcription factor	Fruit set and development	Karlova et al. (2013), Huang et al. (2017), Kravchik et al. (2019)
miR172	ath-miR172c-d	sly-miR172a-b	ppe-miR172c-d	mdm-miR172i-o	AP2-like	Fruit expansion	Wu et al. (2009), Ripoll et al. (2011), Karlova et al. (2013), José Ripoll et al. (2015), Yao et al. (2016)

(continued)

Table 5.1 (continued)

miRNA family	*Arabidopsis thaliana*	*Solanum lycopersicum*	*Prunus persica*	*Malus domestica*	Target	Function	References
miR396	ath-miR396b	sly-miR396a-b	ppe-miR396a-b	mdm-miR396a-b mdm-miR396f-g	GRF transcription factors; DNA methyltransferase 3	Gynoecium and fruit development	Karlova et al. (2013), Liang et al. (2014), Cao et al. (2016), Lee et al. (2017), Correa et al. (2018)
miR399	ath-miR399a-d ath-miR399f	sly-miR399	ppe-miR399c-l	mdm-miR399a-d mdm-miR399i-k	PHOSPHATE 2 (PHO2)	Maintenance of Pi homeostasis; Fruit maturation and quality	Kim et al. (2011), Gao et al. (2015), Wang et al. (2017a, b)
miR156	ath-miR156 h	sly-miR156e	ppe-miR156a-b		SBP/SPL transcription factors	Gynoecium and fruit development	Moxon et al. (2008), Zuo et al., (2012), Xing et al. (2013), Silva et al. (2014), Correa et al. (2018)
miR167	ath-miR167a-b	sly-miR167b	ppe-miR167a-b ppe-miR167d		Auxin response factors (ARFs)	Ovule development and fruit expansion	Wu and Poethig (2006) da Silva et al. (2017)

(continued)

Table 5.1 (continued)

miRNA family	*Arabidopsis thaliana*	*Solanum lycopersicum*	*Prunus persica*	*Malus domestica*	Target	Function	References
miR166	ath-miR166a-g	sly-miR166a-b			HD-ZIP transcription factors	Early fruit development	Moxon et al. (2008), Mohorianu et al. (2011), Karlova et al. (2013)
miR393	ath-miR393a		ppe-miR393a	mdm-miR393 g-h	Auxin F-box protein TIR1	Fruit set	Karlova et al. (2013), Xu et al. (2017)
miR398	ath-miR398b-c		ppe-miR398a	mdm-miR398a	Copper/zinc superoxide dismutases	Stress response, lignification	Guan et al. (2013), Shi et al. (2017a), Zhang et al. (2017)
miR394	ath-miR394a-b		ppe-miR394a-b	mdm-miR394a-b	F-box proteins	Hormones signal pathway	Jones-Rhoades and Bartel (2004), Zuo et al., (2012), Karlova et al. (2013), Wang et al. (2017a)

(continued)

Table 5.1 (continued)

miRNA family	*Arabidopsis thaliana*	*Solanum lycopersicum*	*Prunus persica*	*Malus domestica*	Target	Function	References
miR395	ath-miR395a ath-miR395d-e		ppe-miR395a-b ppe-miR395o	mdm-miR395a-i	High-affinity sulphate transporter 2	Sulphate assimilation and allocation; Anthocyanins and secondary metabolism	Jones-Rhoades and Bartel (2004), Tavares et al. (2013), Karlova et al. (2013), Paim Pinto et al. (2016)
miR160	ath-miR160a-c		ppe-miR160a-b		Auxin response factors (ARFs)	Carpel and fruit development	Moxon et al. (2008), Liu et al. (2010), Karlova et al. (2013), Damodharan et al. (2016)
miR828	ath-miR828		ppe-miR828	mdm-miR828	MYB transcription factors), Ethylene Insensitive 2 (EIN2)	Anthocyanin biosynthesis Ethylene signal transduction	Moxon et al., (2008), Xia et al. (2012), Jia et al. (2015), Qu et al. (2016), Wang et al. (2016)
miR159		sly-miR159	ppe-miR159	mdm-miR159d-f	R2R3 MYB transcription factors	Ovule and early fruit development	Mohorianu et al. (2011), Karlova et al. (2013), da Silva et al. (2017)

(continued)

Table 5.1 (continued)

miRNA family	*Arabidopsis thaliana*	*Solanum lycopersicum*	*Prunus persica*	*Malus domestica*	Target	Function	References
miR482		sly-miR482a-b sly-miR482e	ppe-miR482a-f	mdm-miR482b-c	CC-NBS-LRR, R (Resistance)-like genes; Pectate Lyase	Disease resistance, fruit ripening, ethylene response	Zuo et al. (2012), Karlova et al. (2013), Solofoharivelo et al. (2014), Qu et al. (2016)
miR403		sly-miR403	ppe-miR403		ARGONAUTE 1	Fruit development and maturation	Karlova et al. (2013), Gao et al. (2015), Hou et al. (2017)
miR477			ppe-miR477 ppe-miR477a-b	mdm-miR477a-b	DELLA protein; pentatricopeptide repeat-containing protein	Fatty acid synthesis	Zhang et al. (2012), Solofoharivelo et al. (2014), Qu et al. (2016)

Most relevant publications describing the involvement of each miRNA during fruit development and their identified targets are also included
The following miRNA families found expressed in only one species were not included in the table
Arabidopsis thaliana: miR165, miR170, miR2111, miR2112, miR397, miR408, miR4245, miR5018, miR5020, miR5028, miR5595, miR5639, miR5995, miR773, miR774, miR779, miR781, miR823, miR826, miR827, miR830, miR833, miR837, miR843, miR844, miR845, miR859, miR869, miR870
Solanum lycopersicum: miR10531, miR9475, miR9471, miR6022, miR10533, miR10535, miR10537, miR9470, miR10532, miR6027, miR7981, miR9472, miR6026, miR1919. *Prunus persica*: miR5225, miR8127, miR8131, miR1511, miR6281, miR6267, miR162, miR8122, miR6291, miR530, miR6285, miR8123, miR8124, miR8128, miR8130, miR8132, miR6288, miR6274, miR712. *Malus domestica*: miR10986, miR10985, miR10982, miR11001, miR10990, miR11010, miR11017, miR7128, miR319, miR858, miR10994, miR11006, miR7120, miR11011, miR10989, miR11009, miR10984, miR7124, miR10981, miR11018

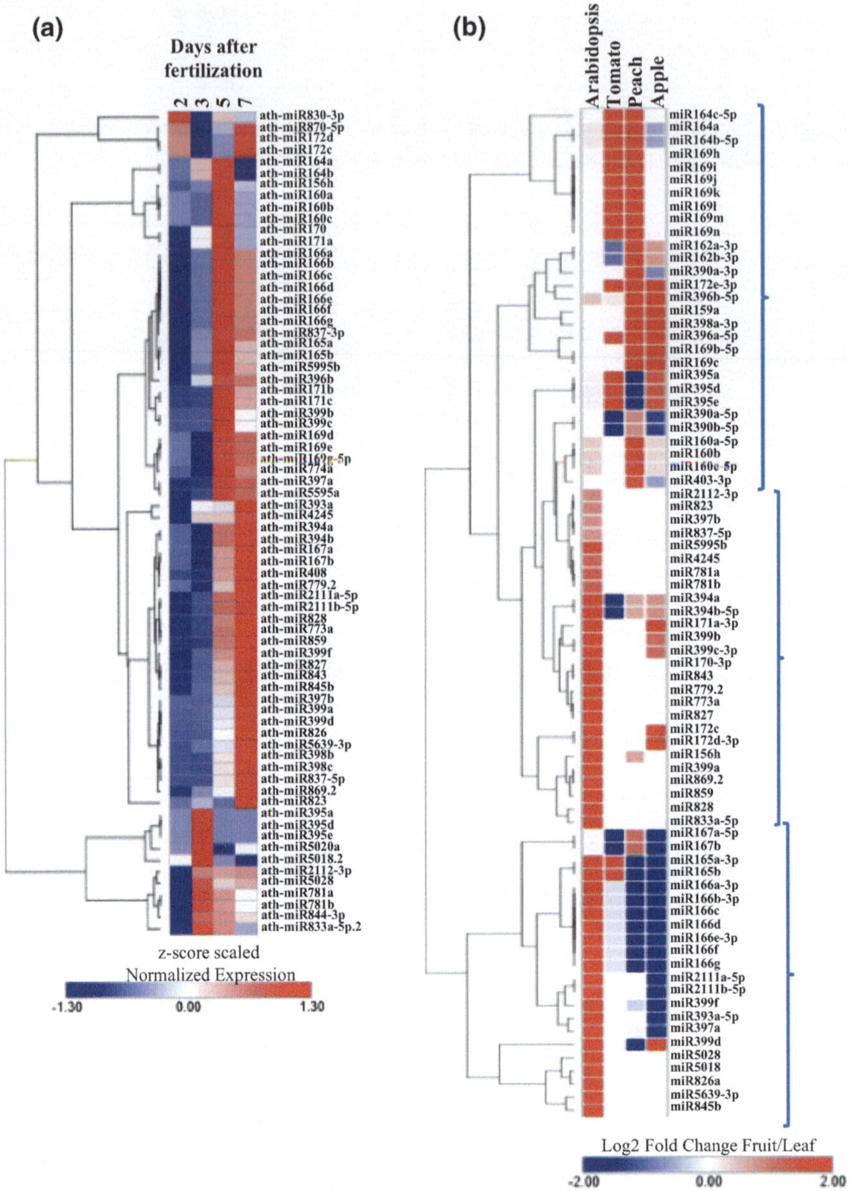
(a)
Days after fertilization
2
3
5
7
ath-miR830-3p
ath-miR870-5p
ath-miR172d
ath-miR172c
ath-miR164a
ath-miR164b
ath-miR156h
ath-miR160a
ath-miR160b
ath-miR160c
ath-miR170
ath-miR171a
ath-miR166a
ath-miR166b
ath-miR166c
ath-miR166d
ath-miR166e
ath-miR166f
ath-miR166g
ath-miR837-3p
ath-miR165a
ath-miR165b
ath-miR5995b
ath-miR396b
ath-miR171b
ath-miR171c
ath-miR399b
ath-miR399c
ath-miR169d
ath-miR169e
ath-miR162a-5p
ath-miR774a
ath-miR397a
ath-miR5595a
ath-miR393a
ath-miR4245
ath-miR394a
ath-miR394b
ath-miR167a
ath-miR167b
ath-miR408
ath-miR779.2
ath-miR2111a-5p
ath-miR2111b-5p
ath-miR828
ath-miR773a
ath-miR859
ath-miR399f
ath-miR827
ath-miR843
ath-miR845b
ath-miR397b
ath-miR399a
ath-miR399d
ath-miR826
ath-miR5639-3p
ath-miR398b
ath-miR398c
ath-miR837-5p
ath-miR869.2
ath-miR823
ath-miR395a
ath-miR395d
ath-miR395e
ath-miR5020a
ath-miR5018.2
ath-miR2112-3p
ath-miR5028
ath-miR781a
ath-miR781b
ath-miR844-3p
ath-miR833a-5p.2
z-score scaled
Normalized Expression
-1.30
0.00
1.30
(b)
Arabidopsis
Tomato
Peach
Apple
miR164c-5p
miR164a
miR164b-5p
miR169h
miR169i
miR169j
miR169k
miR169l
miR169m
miR169n
miR162a-3p
miR162b-3p
miR390a-3p
miR172e-3p
miR396b-5p
miR159a
miR398a-3p
miR396a-5p
miR169b-5p
miR169c
miR395a
miR395d
miR395e
miR390a-5p
miR390b-5p
miR160a-5p
miR160b
miR160c-5p
miR403-3p
miR2112-3p
miR823
miR397b
miR837-5p
miR5995b
miR4245
miR781a
miR781b
miR394a
miR394b-5p
miR171a-3p
miR399b
miR399c-3p
miR170-3p
miR843
miR779.2
miR773a
miR827
miR172c
miR172d-3p
miR156h
miR399a
miR869.2
miR859
miR828
miR833a-5p
miR167a-5p
miR167b
miR165a-3p
miR165b
miR166a-3p
miR166b-3p
miR166c
miR166d
miR166e-3p
miR166f
miR166g
miR2111a-5p
miR2111b-5p
miR399f
miR393a-5p
miR397a
miR399d
miR5028
miR5018
miR826a
miR5639-3p
miR845b
Log2 Fold Change Fruit/Leaf
-2.00
0.00
2.00

◀**Fig. 5.2 Hierarchical clustering of miRNA expression across developmental stages and fruit species. a** Heat map showing the expression profiles of Arabidopsis miRNAs during siliques development (2, 3, 5 and 7 days after fertilization). Normalized expression values (RPM: reads per million) of 72 miRNAs prevalently expressed during silique development were retrieved from Xu et al. (2018). RPM values of the four samples were z-score adjusted before clustering analysis by average linkage and Pearson distance. Heat map was generated using Morpheus software (https://software.broadinstitute.org/morpheus/); red and blue represent high and low expression levels, respectively. **b** Heat map showing the expression of fruit-specific miRNAs in fruits of different species. Arabidopsis miRNA expression values were retrieved from Xu et al. (2018) and mature ath-miRNA sequences were used as queries for similarity searches in the Next-Gen Sequence Databases (https://mpss.danforthcenter.org/) to retrieve expression values of tomato, peach and apple conserved miRNAs. For each species, relative miRNA expression in fruits was calculated as the ratio of the expression in fruits versus leaves RPMs (only 7 DAF Arabidopsis siliques were considered). Finally, log2 fold change values were used for clustering analysis by average linkage and Pearson distance. Heat map was generated using Morpheus software (https://software.broadinstitute.org/morpheus/); red and blue represent high and low expression in fruits compared to leaves, respectively

highlighted as fleshy-fruit specific miRNAs, while several species-specific families were identified. In addition to species-dependent control of fruit development, a large number of specific miRNA families in each species could be explained by the incomplete or divergent annotation of miRNAs in the four different species.

To deepen our investigation, we used a similarity search approach to interrogate each species-specific database by using Arabidopsis mature miRNA sequences as a query. Following this approach, expression profiles of a common dataset of miRNAs were retrieved in the four species. These expression profiles were generated from independent experiments, therefore, for each species, relative miRNA expression in fruits was calculated as the ratio of fruit versus leaves expression values. Hierarchical clustering analysis grouped miRNAs with 1.5 or higher fruit fold increase in at least one species in three main groups (Fig. 5.2b): (i) miRNAs expressed in fleshy fruits (including false fruits); (ii) miRNAs exclusively expressed in siliques (probably not annotated in other species) and (iii) miRNAs overexpressed in siliques and down-regulated (compared to leaves) in fleshy fruits.

The role of single miRNA members that we identified in our analysis were previously described in specific fruit species and in specific growth phases: (i) in Arabidopsis, the miR164-targeted NAC TFs transcripts CUC1 and CUC2 (CUP-SHAPED COTYLEDONS 1 and 2) defines the organ boundary during separation between organ primordia and meristem, controlling the leaf margin serration (Laufs et al. 2004; Nikovics et al. 2006; Hasson et al. 2011; Vialette-Guiraud et al. 2016); (ii) both in Arabidopsis and tomato miRNA156/miRNA157, target transcripts members belonging to SQUAMOSA PROMOTER BINDING PROTEIN_LIKE (SBP/SPL) transcription factor family, regulating the carpel and fruit development (including CNR, which plays pivotal roles in fruit ripening, Giovannoni 2004); (iii) similarly miR396, which is responsible of regulation of several developmental processes, has a key role also in Carpel Margin Meristem (CMM) competence and pluripotency during gynoecium development, through the targeting of GROWTH REGULATING FACTORS members (Karlova et al. 2013; Liang et al. 2014; Cao et al. 2016;

Kim 2017; Correa et al. 2018); (iv) during the fruit expansion process, the role of miR172 is well testified in Arabidopsis, where it limits the growth-repressing activity of *APETALA2-like* genes (Correa et al. 2018); (v) miR828 and miR858 could directly or indirectly control anthocyanin biosynthesis in apples (Xia et al. 2012); (vi) targets of miR477 and mirR482 are DELLA protein GAI1-like, a gene involved in gibberelins action, and pectate lyase, an enzyme participating in cell wall disassembly, respectively (Solofoharivelo et al. 2014). Both miRNAs are expressed only in fleshy fruit where cell expansion is a crucial process, in comparison to dry fruit, for obtaining a mature product with the expected properties; (vii) miR394 targets F-box family proteins which are important participants in the signal transduction pathways of different plant hormones, including ethylene (Qiao et al. 2009). In particular, the stability of the Ethylene Insensitive 2 (EIN2) and 3 (EIN3), two positive ethylene signaling regulators, is modulated by F-box proteins (Qiao et al. 2009). In addition, EIN2 is also targeted by miR828 (Moxon et al. 2008), therefore for the onset of ethylene-dependent ripening events a strong reduction of expression of both mir394 and miR828 is required (Zuo et al., 2012).

All this information drawn a picture of fruit development in which miRNAs can play the same role, targeting orthologue genes in different fruit types, together with additional species-specific roles, evolved during the speciation process by acquiring new targets. A specific example includes a member of the miR156 family in grapevine, for which a new target, the *CONTANS-LIKE* gene instead of *SPL*, the ancestral target for this family, has been identified (Baldrich et al. 2018). In grapevine berry, a *CONTANS-LIKE* gene member, involved in the circadian response, shows a decreasing expression level after veraison (Pilati et al. 2007) in correspondence to the increase of miR156 expression (Pantaleo et al. 2010), suggesting a miRNA control of light signaling circuits during early berry development.

5.3 Conclusions and Future Perspectives

miRNAs play key roles in fruit development, ranging from carpel establishment and patterning to fruit ripening. Interestingly, evolution of miRNA and their targets allow the development of adaptive advantages to plants by using similar regulatory circuits for different developmental processes. In this context, we suggest that the evolution of different fruit types can be possibly influenced by modulating the biogenesis of members of conserved miRNA families. A good example is the evolution of fruit types in Rosaceae family. In this family both dry and fleshy fruits are present, and amongst the latter berry, drupes and pome. Ancestral character reconstruction for fruit types supports independent origins of fleshy fruits from a dry-fruit ancestor, the achenetum. It has been proposed that Whole Genome Duplications (WGDs) and environmental factors, including animals, contributed to the evolution of the many fruit types in Rosaceae (Xiang et al. 2017). WGDs, together with transposable elements (TE; Li et al. 2011), may also contribute to miRNA diversity and their regulatory complexity (Shen et al. 2015; Shi et al. 2017b). In this context, Yao et al. (2016), demonstrated that miR172 has negative influence on fruit growth in crab apples and its negative influence is confirmed by the fact that in domesticated apple, characterized by larger pomes, the expression of miR172p is reduced by the presence of a TE. Interestingly, expression of miR172p was not detected in our comparative analysis, while other members of the family are expressed also in apple fruits (Table 5.1). Worthy to note, miR172 promotes the development of Arabidopsis silique, suggesting a different role in dry and fleshy fruit. These results are consistent considering that silique is a true fruit deriving from ovary tissues, while the pome is a false fruit developing from extra-carpellary tissues deriving mainly from sepals. In Arabidopsis, miRNA172 promotes silique growth through negative regulation of AP2 (its target) that would otherwise hinder the action of AGAMOUS (AG) and FRUITFUL (FUL; Yant et al. 2010), two MADS-box transcription factors that are essential for ovary and silique growth (José Ripoll et al. 2015). Conversely, miRNA172 inhibits apple fruit growth through negative regulation of AP2 that is required for hypanthium development into pome.

Worthy of note is also the impact of miRNAs on fruit patterning and identity during the fruit development. Alteration in fruit patterning and cell identity can result in a final fruit with developmental defects and with less organoleptic properties, with a consequent loss of the economic value. Both in fleshy and dry fruit systems, specific and defined cell/tissue compartments are established, and the role of miRNAs is essential in the organization of a correct and functional final organ structure. For example, in drupe, a precise definition of patterning regulates the growth of different functional fruit 'layers' (epi- endo- and mesocarp), and this is essential for proper fruit development and a correct ripening program. In fact at S2 stage of drupe growth, the mesocarp identity is marked by a low expression level of *Pp-FLESHY*, a bHLH transcription factor belonging to the HECATE3/INDEHISCENT clade, ensured by the increase of expression of a non-conserved miRNA (Botton et al. 2016). In accordance with these results, in a peach mutant (named slow ripening), whose mesocarp is transformed in a 'pseudo-endocarp' for the high accumulation of lignin, a transient increase of *Pp-FLESHY* concomitant with low steady-state expression level of the same miRNA was observed (Botton et al. 2016).

The above-mentioned examples are the mere tip of the iceberg. A better understanding of the mechanisms that control miRNA and target expression and their spatiotemporal regulatory roles is the basis for the application of miRNAs to modulate important fruit traits such as size, shape, seed production, and ripening. In this sense, the availability of the genome sequences of many fruit-bearing crop species, the dissection of their genome structure and gene annotation, together with gene regulation studies, are essential prerequisites. The second major step is the use of technologies able to provide insights on function of genes including the miRNAs. For instance, CRISPR-Cas 9 is a powerful genome-editing technology, which allows for genetic information to be inserted, deleted, or altered at particular loci of the genome. The CRISPR/Cas9 system has many advantages in the genetic manipulation of the plant genome, including ease in assembly, low cost and high efficiency and specificity. CRIPSR/Cas9-based genome editing is currently a breakthrough technology for basic and applied research towards engineering traits of fruit crops.

However, due to the difficulties of genetic transformation and in vitro regeneration in these species, their usually complex genomes, slow growth cycles, or the lack of genomic information, it remains a long and arduous task to apply CRISPR/Cas9 to most fruit crops. Indeed, the use of these technologies might allow subtle changes in miRNA target gene expression which have the potential to modify fruit traits and quality. The rapid development both of new and improved CRISPR/Cas9-based tools delivery systems and the clarification of the mechanisms and regulation of miRNA biogenesis as well as the regulatory roles of miRNAs in fruit developmental plasticity may help overcome some of the barriers that limit genome editing applications in horticultural crops.

References

Abdi N, Holford P, McGlasson W, Mizrahi Y (1997) Ripening behaviour and responses to propylene in four cultivars of Japanese type plums. Postharvest Biol Technol 12:21–34. https://doi.org/10.1016/S0925-5214(97)00041-0

Alba R, Payton P, Fei Z et al (2005) Transcriptome and selected metabolite analyses reveal multiple points of ethylene control during tomato fruit development. Plant Cell 17:2954–2965. https://doi.org/10.1105/tpc.105.036053

Baldrich P, Beric A, Meyers BC (2018) Despacito: the slow evolutionary changes in plant microRNAs. Curr Opin Plant Biol 42:16–22. https://doi.org/10.1016/J.PBI.2018.01.007

Berger Y, Harpaz-Saad S, Brand A et al (2009) The NAC-domain transcription factor GOBLET specifies leaflet boundaries in compound tomato leaves. Development 136:823–832. https://doi.org/10.1242/dev.031625

Bergervoet JHW, Berhoeven HA, Gilissen LJW, Bino RJ (1996) High amounts of nuclear DNA in tomato (*Lycopersicon esculentum* Mill.) pericarp. Plant Sci 116:141–145. https://doi.org/10.1016/0168-9452(96)04383-X

Bertin N (2004) Analysis of the tomato fruit growth response to temperature and plant fruit load in relation to cell division, cell expansion and DNA endoreduplication. Ann Bot 95:439–447. https://doi.org/10.1093/aob/mci042

Bonghi C, Trainotti L, Botton A, et al (2011) A microarray approach to identify genes involved in seed-pericarp cross-talk and development in peach. BMC Plant Biol 11:107. https://doi.org/10.1186/1471-2229-11-107

Botton A, Rasori A, Ziliotto F et al (2016) The peach HECATE3-like gene FLESHY plays a double role during fruit development. Plant Mol Biol 91:97–114. https://doi.org/10.1007/s11103-016-0445-z

Cao D, Wang J, Ju Z et al (2016) Regulations on growth and development in tomato cotyledon, flower and fruit via destruction of miR396 with short tandem target mimic. Plant Sci 247:1–12. https://doi.org/10.1016/j.plantsci.2016.02.012

Carbonell-Bejerano P, Urbez C, Carbonell J et al (2010) A fertilization-independent developmental program triggers partial fruit development and senescence processes in pistils of Arabidopsis. Plant Physiol 154:163–172. https://doi.org/10.1104/pp.110.160044

Cartolano M, Castillo R, Efremova N et al (2007) A conserved microRNA module exerts homeotic control over Petunia hybrida and Antirrhinum majus floral organ identity. Nat Genet 39:901–905. https://doi.org/10.1038/ng2056

Chalmers D, Ende B (1975) A Reappraisal of the growth and development of peach fruit. Funct Plant Biol 2:623. https://doi.org/10.1071/PP9750623

Chen F, Zhang L, An H et al (2009) LWT—food science and technology the nanostructure of hemicellulose of crisp and soft Chinese cherry (*Prunus pseudocerasus* L.) cultivars at different stages of ripeness. LWT Food Sci Technol 42:125–130. https://doi.org/10.1016/j.lwt.2008.03.016

Chevalier C (2007) Cell cycle control and fruit development. Cell cycle control and plant development. Blackwell Publishing Ltd, Oxford, UK, pp 269–293

Chevalier C, Nafati M, Mathieu-Rivet E et al (2011) Elucidating the functional role of endoreduplication in tomato fruit development. Ann Bot 107:1159–1169. https://doi.org/10.1093/aob/mcq257

Correa JP, de O'Silva EM, Nogueira FTS (2018) Molecular Control by non-coding RNAs during fruit development: from gynoecium patterning to fruit ripening. Front Plant Sci 9:1–13. https://doi.org/10.3389/fpls.2018.01760

Csukasi F, Donaire L, Casañal A et al (2012) Two strawberry miR159 family members display developmental-specific expression patterns in the fruit receptacle and cooperatively regulate Fa-GAMYB. New Phytol 195(1):47–57. https://doi.org/10.1111/j.1469-8137.2012.04134.x

Czerednik A, Busscher M, Bielen BAM et al (2012) Regulation of tomato fruit pericarp development by an interplay between CDKB and CDKA1 cell cycle genes. J Exp Bot 63:2605–2617. https://doi.org/10.1093/jxb/err451

da Silva EM, Silva GFF, Bidoia DB et al (2017) micro RNA 159-targeted SlGAMYB transcription factors are required for fruit set in tomato. Plant J 92:95–109. https://doi.org/10.1111/tpj.13637

Damodharan S, Zhao D, Arazi T (2016) A common miRNA160-based mechanism regulates ovary patterning, floral organ abscission and lamina outgrowth in tomato. Plant J 86:458–471. https://doi.org/10.1111/tpj.13127

Dorcey E, Urbez C, Blázquez MA et al (2009) Fertilization-dependent auxin response in ovules triggers fruit development through the modulation of gibberellin metabolism in Arabidopsis. Plant J 58:318–332. https://doi.org/10.1111/j.1365-313X.2008.03781.x

Ecker JR (2013) Epigenetic trigger for tomato ripening. Nat Biotechnol 31:119–120. https://doi.org/10.1038/nbt.2497

Eriksson EM, Bovy A, Manning K et al (2004) Effect of the colorless non-ripening mutation on cell wall biochemistry and gene expression during tomato fruit development and ripening. Plant Physiol 136:4184–4197. https://doi.org/10.1104/pp.104.045765

Estornell LH, Agustí J, Merelo P et al (2013) Elucidating mechanisms underlying organ abscission. Plant Sci 199–200:48–60. https://doi.org/10.1016/j.plantsci.2012.10.008

Farinati S, Rasori A, Varotto S, Bonghi C (2017) Rosaceae fruit development, ripening and post-harvest: an epigenetic perspective. Front Plant Sci 8:1–14. https://doi.org/10.3389/fpls.2017.01247

Ferrándiz C, Pelaz S, Yanofsky MF (2002) Control of carpel and fruit development in Arabidopsis. Annu Rev Biochem 68:321–354. https://doi.org/10.1146/annurev.biochem.68.1.321

Fonseca S, Hackler L, Zvara Á et al (2004) Monitoring gene expression along pear fruit development, ripening and senescence using cDNA microarrays. Plant Sci 167:457–469. https://doi.org/10.1016/j.plantsci.2004.03.033

Gallusci P, Hodgman C, Teyssier E, Seymour GB (2016) DNA methylation and chromatin regulation during fleshy fruit development and ripening. Front Plant Sci 7:1–14. https://doi.org/10.3389/fpls.2016.00807

Gao C, Ju Z, Cao D et al (2015) MicroRNA profiling analysis throughout tomato fruit development and ripening reveals potential regulatory role of RIN on microRNAs accumulation. Plant Biotechnol J 13:370–382. https://doi.org/10.1111/pbi.12297

Gillespie PG, Wagner MC, Hudspeth AJ (1993) Identification of a 120 kd hair-bundle myosin located near stereociliary tips. Neuron 11:581–594

Giovannoni J (2004) Genetic regulation of fruit development and ripening. Plant Cell Online 16:170–181. https://doi.org/10.1105/tpc.019158.Fruit

Gómez MD, Vera-Sirera F, Pérez-Amador MA (2014) Molecular programme of senescence in dry and fleshy fruits. J Exp Bot 65:4515–4526. https://doi.org/10.1093/jxb/eru093

Guan Q, Lu X, Zeng H et al (2013) Heat stress induction of miR398 triggers a regulatory loop that is critical for thermotolerance in Arabidopsis. Plant J 74:840–851. https://doi.org/10.1111/tpj.12169

Handa AK, Tiznado-Hernández M-E, Mattoo AK (2012) Fruit development and ripening: a molecular perspective. Plant Biotechnol Agric 405–424. https://doi.org/10.1016/b978-0-12-381466-1.00026-2

Hasson A, Plessis A, Blein T et al (2011) Evolution and diverse roles of the CUP-SHAPED COTYLEDON genes in Arabidopsis leaf development C W. Plant Cell 23:54–68. https://doi.org/10.1105/tpc.110.081448

Hendelman A, Stav R, Zemach H, Arazi T (2013) The tomato NAC transcription factor SlNAM2 is involved in flower-boundary morphogenesis. J Exp Bot 64:5497–5507. https://doi.org/10.1093/jxb/ert324

Ho LC, Hewitt JD (1986) Fruit development. The tomato crop. Springer, The Netherlands, Dordrecht, pp 201–239

Hou Y, Zhai L, Li X et al (2017) Comparative analysis of fruit ripening-related miRNAs and their targets in blueberry using small RNA and degradome sequencing. Int J Mol Sci 18:2767. https://doi.org/10.3390/ijms18122767

Huang W, Peng S, Xian Z et al (2017) Overexpression of a tomato miR171 target gene SlGRAS24 impacts multiple agronomical traits via regulating gibberellin and auxin homeostasis. Plant Biotechnol J 15:472–488. https://doi.org/10.1111/pbi.12646

Iannetta PPM, Laarhoven L, Medina-escobar N et al (2006) Ethylene and carbon dioxide production by developing strawberries show a correlative pattern that is indicative of ripening climacteric fruit. Physiol Plant 127:247–259. https://doi.org/10.1111/j.1399-3054.2006.00656.x

Ireland HS, Yao J, Tomes S et al (2013) Apple SEPALLATA1/2 -like genes control fruit flesh development and ripening. Plant J 73:1044–1056. https://doi.org/10.1111/tpj.12094

Jajo A, Corso M, Bonghi C et al (2015) Effect of cool storage duration on ripening initiation of "Angelys" pear fruit. Acta Hortic 129–136. https://doi.org/10.17660/actahortic.2015.1079.12

Jia X, Shen J, Liu H et al (2015) Small tandem target mimic-mediated blockage of microRNA858 induces anthocyanin accumulation in tomato. Planta 242:283–293. https://doi.org/10.1007/s00425-015-2305-5

Jones-Rhoades MW, Bartel DP (2004) Computational identification of plant microRNAs and their targets, including a stress-induced miRNA. Mol Cell 14:787–799. https://doi.org/10.1016/J.MOLCEL.2004.05.027

José Ripoll J, Bailey LJ, Mai Q-A et al (2015) microRNA regulation of fruit growth. Nat Plants 1:15036. https://doi.org/10.1038/nplants.2015.36

Karlova R, van Haarst JC, Maliepaard C et al (2013) Identification of microRNA targets in tomato fruit development using high-throughput sequencing and degradome analysis. J Exp Bot 64:1863–1878. https://doi.org/10.1093/jxb/ert049

Karlova R, Chapman N, David K et al (2014) Transcriptional control of fleshy fruit development and ripening. J Exp Bot 65:4527–4541. https://doi.org/10.1093/jxb/eru316

Kim JH (2017) The growth regulating and growth interacting factor duo is essential for specification of carpel margin meristems. https://doi.org/10.1104/pp.17.00960

Kim W, Ahn HJ, Chiou TJ, Ahn JH (2011) The role of the miR399-PHO2 module in the regulation of flowering time in response to different ambient temperatures in *Arabidopsis thaliana*. Mol Cells 32:83–88. https://doi.org/10.1007/s10059-011-1043-1

Klee HJ, Giovannoni JJ (2011) Genetics and control of tomato fruit ripening and quality attributes. Annu Rev Genet 45:41–59. https://doi.org/10.1146/annurev-genet-110410-132507

Kou X, Watkins CB, Gan S (2012) Arabidopsis AtNAP regulates fruit senescence. J Exp Bot 63:695–709. https://doi.org/10.1093/jxb/err313

Kravchik M, Stav R, Belausov E, Arazi T (2019) Functional characterization of microRNA171 family in tomato. Plants 8:10. https://doi.org/10.3390/plants8010010

Lakso AN, Corelli Grappadelli L, Barnard J, Goffinet MC (1995) An expolinear model of the growth pattern of the apple fruit. J Hortic Sci 70:389–394. https://doi.org/10.1080/14620316.1995.11515308

Laufs P, Peaucelle A, Morin H, Traas J (2004) microRNA regulation of the CUC genes is required for boundary size control in Arabidopsis meristems. Development 131:4311–4322. https://doi.org/10.1242/dev.01320

Lee S-J, Lee BH, Jung J-H, et al (2017) The growth regulating and growth interacting factor duo is essential for specification of carpel margin meristems. Plant Physiol 00960. https://doi.org/10.1104/pp.17.00960

Li Y, Li C, Xia J, Jin Y (2011) Domestication of transposable elements into microRNA genes in plants. PLoS ONE 6:e19212. https://doi.org/10.1371/journal.pone.0019212

Liang G, He H, Li Y et al (2014) Molecular Mechanism of microRNA396 mediating pistil development in Arabidopsis. Plant Physiol 164:249–258. https://doi.org/10.1104/pp.113.225144

Liu X, Huang J, Wang Y et al (2010) The role of floral organs in carpels, an Arabidopsis loss-of-function mutation in MicroRNA160a, in organogenesis and the mechanism regulating its expression. Plant J 62:416–428. https://doi.org/10.1111/j.1365-313X.2010.04164.x

Liu J, Abdelfattah A, Norelli J et al (2018) Apple endophytic microbiota of different rootstock/scion combinations suggests a genotype-specific influence. Microbiome 6:1–11. https://doi.org/10.1186/s40168-018-0403-x

Martínez-Laborda A, Vera A (2009) Arabidopsis fruit development. In: Annual plant reviews, pp 172–203

Masia A, Zanchin A, Rascio N, Ramina A (1992) Some biochemical and ultrastructural aspects of peach fruit development. J Am Soc Hortic Sci 117:808–815

McAtee P, Karim S, Schaffer R, David K (2013) A dynamic interplay between phytohormones is required for fruit development, maturation, and ripening. Front Plant Sci 4:1–7. https://doi.org/10.3389/fpls.2013.00079

Mintz-Oron S, Mandel T, Rogachev I et al (2008) Gene expression and metabolism in tomato fruit surface tissues. Plant Physiol 147:823–851. https://doi.org/10.1104/pp.108.116004

Mohorianu I, Schwach F, Jing R et al (2011) Profiling of short RNAs during fleshy fruit development reveals stage-specific sRNAome expression patterns. Plant J 67:232–246. https://doi.org/10.1111/j.1365-313X.2011.04586.x

Moxon S, Jing R, Szittya G et al (2008) Deep sequencing of tomato short RNAs identifies microRNAs targeting genes involved in fruit ripening. Genome Res 18:1602–1609. https://doi.org/10.1101/gr.080127.108

Murayama H, Sai M, Oikawa A, Itai A (2015) Inhibitory factors that affect the ripening of pear fruit on the tree. Hortic J 84:14–20. https://doi.org/10.2503/hortj.mi-015

Musseau C, Just D, Jorly J et al (2017) Identification of two new mechanisms that regulate fruit growth by cell expansion in tomato. Front Plant Sci 8:1–15. https://doi.org/10.3389/fpls.2017.00988

Nikovics K, Blein T, Peaucelle A et al (2006) The balance between the MIR164A and CUC2 genes controls leaf margin serration in Arabidopsis. Plant Cell 18:2929. https://doi.org/10.1105/TPC.106.045617

Osorio S, Scossa F, Fernie AR (2013) Molecular regulation of fruit ripening. Front Plant Sci 4:1–8. https://doi.org/10.3389/fpls.2013.00198

Paim Pinto DL, Brancadoro L, Dal Santo S et al (2016) The Influence of genotype and environment on small RNA profiles in grapevine berry. Front Plant Sci 7:1459. https://doi.org/10.3389/fpls.2016.01459

Pantaleo V, Szittya G, Moxon S et al (2010) Identification of grapevine microRNAs and their targets using high-throughput sequencing and degradome analysis. Plant J 62. https://doi.org/10.1111/j.1365-313x.2010.04208.x

Pesaresi P, Mizzotti C, Colombo M, Masiero S (2014) Genetic regulation and structural changes during tomato fruit development and ripening. Front Plant Sci 5:1–14. https://doi.org/10.3389/fpls.2014.00124

Picken AJF (1984) A review of pollination and fruit set in the tomato (*Lycopersicon esculentum* Mill.). J Hortic Sci 59:1–13. https://doi.org/10.1080/00221589.1984.11515163
Pilati S, Perazzolli M, Malossini A et al (2007) Genome-wide transcriptional analysis of grapevine berry ripening reveals a set of genes similarly modulated during three seasons and the occurrence of an oxidative burst at vèraison. BMC Genom 8:428. https://doi.org/10.1186/1471-2164-8-428
Qiao H, Chang KN, Yazaki J, Ecker JR (2009) Interplay between ethylene, ETP1/ETP2 F-box proteins, and degradation of EIN2 triggers ethylene responses in Arabidopsis. Genes Development 23:512–521. https://doi.org/10.1101/gad.1765709
Qu D, Yan F, Jiang X et al (2016) Identification of microRNAs and their targets associated with fruit-bagging and subsequent sunlight re-exposure in the "Granny Smith" apple exocarp using high-throughput sequencing. Front Plant Sci 7:1–14. https://doi.org/10.3389/fpls.2016.00027
Ripoll JJ, Roeder AHK, Ditta GS, Yanofsky MF (2011) A novel role for the floral homeotic gene APETALA2 during Arabidopsis fruit development. Development 138:5167–5176. https://doi.org/10.1242/DEV.073031
Roeder AHK, Yanofsky MF (2006) Fruit development in Arabidopsis. In: The Arabidopsis book, pp 1–50
Ruan YL, Patrick JW, Bouzayen M et al (2012) Molecular regulation of seed and fruit set. Trends Plant Sci 17:656–665. https://doi.org/10.1016/j.tplants.2012.06.005
Seymour GB, Østergaard L, Chapman NH et al (2013) Fruit development and ripening. Annu Rev Plant Biol 64:219–241. https://doi.org/10.1146/annurev-arplant-050312-120057
Shen E, Zou J, Hubertus Behrens F et al (2015) Identification, evolution, and expression partitioning of miRNAs in allopolyploid Brassica napus. J Exp Bot 66:7241–7253. https://doi.org/10.1093/jxb/erv420
Shi M, Hu X, Wei Y et al (2017a) Genome-wide profiling of small RNAs and degradome revealed conserved regulations of miRNAs on auxin-responsive genes during fruit enlargement in peaches. Int J Mol Sci 18:1–14. https://doi.org/10.3390/ijms18122599
Shi T, Wang K, Yang P (2017b) The evolution of plant microRNAs: insights from a basal eudicot sacred lotus. Plant J 89:442–457. https://doi.org/10.1111/tpj.13394
Silva GFFE, Silva EM, da Silva Azevedo M, et al (2014) microRNA156-targeted SPL/SBP box transcription factors regulate tomato ovary and fruit development. Plant J 78:604–618. https://doi.org/10.1111/tpj.12493
Solofoharivelo MC, van der Walt AP, Stephan D et al (2014) MicroRNAs in fruit trees: discovery, diversity and future research directions. Plant Biol 16:856–865. https://doi.org/10.1111/plb.12153
Somerville C, Koornneef M (2002) A fortunate choice: the history of Arabidopsis as a model plant. Nat Rev 3:3–9
Tavares S, Vesentini D, Fernandes JC et al (2013) Vitis vinifera secondary metabolism as affected by sulfate depletion: diagnosis through phenylpropanoid pathway genes and metabolites. Plant Physiol Biochem 66:118–126. https://doi.org/10.1016/j.plaphy.2013.01.022
Teyssier E, Boureauv L, Chen W et al (2015) Epigenetic regulation during fleshy fruit development and ripening. Appl Plant Genom Biotechnol 133–151. https://doi.org/10.1016/b978-0-08-100068-7.00008-2
Tiffney BH (1984) Seed size, dispersal syndromes, and the rise of the angiosperms: evidence and hypothesis. Missouri Bot Gard 71:551–576
Tiffney BH (2004) Vertebrate dispersal of seed plants through time. Annu Rev Ecol Evol Syst 35:1–29. https://doi.org/10.1146/annurev.ecolsys.34.011802.132535
Tiwari K, Paliyath G (2011) Cloning, expression and functional characterization of the C2 domain from tomato phospholipase Dα. Plant Physiol Biochem 49:18–32. https://doi.org/10.1016/j.plaphy.2010.09.015
Vialette-Guiraud ACM, Chauvet A, Gutierrez-Mazariegos J et al (2016) A conserved role for the NAM/miR164 developmental module reveals a common mechanism underlying carpel margin fusion in monocarpous and syncarpous eurosids. 6:1239. https://doi.org/10.3389/fpls.2015.01239
Vivian-Smith A, Koltunow AM (1999) Genetic analysis of growth-regulator-induced parthenocarpy in Arabidopsis. Plant Physiol 121:437–451

Wagstaff C, Yang TJW, Stead AD et al (2009) A molecular and structural characterization of senescing Arabidopsis siliques and comparison of transcriptional profiles with senescing petals and leaves. Plant J 57:690–705. https://doi.org/10.1111/j.1365-313X.2008.03722.x

Wang Y, Wang Y, Song Z, Zhang H (2016) Repression of MYBL2 by both microRNA858a and HY5 leads to the activation of anthocyanin biosynthetic pathway in Arabidopsis. Mol Plant 9:1395–1405. https://doi.org/10.1016/j.molp.2016.07.003

Wang Y, Wang Q, Gao L et al (2017a) Parsing the regulatory network between small RNAs and target genes in ethylene pathway in tomato. Front Plant Sci 8:1–13. https://doi.org/10.3389/fpls.2017.00527

Wang Y, Zhang J, Cui W et al (2017b) Improvement in fruit quality by overexpressing miR399a in woodland strawberry. J Agric Food Chem 65:7361–7370. https://doi.org/10.1021/acs.jafc.7b01687

Wu G, Poethig RS (2006) Temporal regulation of shoot development in Arabidopsis thaliana by miR156 and its target SPL3. Development 133:3539–3547. https://doi.org/10.1242/dev.02521

Wu G, Park MY, Conway SR et al (2009) The sequential action of miR156 and miR172 regulates developmental timing in Arabidopsis. Cell 138:750. https://doi.org/10.1016/J.CELL.2009.06.031

Wu J, Wang D, Liu Y et al (2014) Identification of miRNAs involved in pear fruit development and quality. BMC Genomics 1–19

Xia R, Zhu H, An YQ et al (2012) Apple miRNAs and tasiRNAs with novel regulatory networks. Genome Biol 13:R47. https://doi.org/10.1186/gb-2012-13-6-r47

Xia R, Ye S, Liu Z et al (2015) Novel and recently evolved miRNA clusters regulate expansive F-box gene networks through phasiRNAs in wild diploid strawberry. Plant Physiol. https://doi.org/10.1104/pp.15.00253

Xiang Y, Huang CH, Hu Y et al (2017) Evolution of rosaceae fruit types based on nuclear phylogeny in the context of geological times and genome duplication. Mol Biol Evol 34:262–281. https://doi.org/10.1093/molbev/msw242

Xing S, Salinas M, Garcia-Molina A et al (2013) SPL8 and miR156-targeted SPL genes redundantly regulate Arabidopsis gynoecium differential patterning. Plant J 75:566–577. https://doi.org/10.1111/tpj.12221

Xu J, Li J, Cui L et al (2017) New insights into the roles of cucumber TIR1 homologs and miR393 in regulating fruit/seed set development and leaf morphogenesis. BMC Plant Biol 17:1–14. https://doi.org/10.1186/s12870-017-1075-6

Xu L, Hu Y, Cao Y et al (2018) An expression atlas of miRNAs in *Arabidopsis thaliana*. Sci China Life Sci 61:178–189. https://doi.org/10.1007/s11427-017-9199-1

Yant L, Mathieu J, Dinh TT et al (2010) Orchestration of the floral transition and floral development in Arabidopsis by the bifunctional transcription factor APETALA2. Plant Cell 22:2156–2170. https://doi.org/10.1105/tpc.110.075606

Yao J-L, Tomes S, Xu J, Gleave AP (2016) How microRNA172 affects fruit growth in different species is dependent on fruit type. Plant Sig Behav 11:e1156833. https://doi.org/10.1080/15592324.2016.1156833

Zeng S, Liu Y, Pan L et al (2015) Identification and characterization of miRNAs in ripening fruit of *Lycium barbarum* L. using high-throughput sequencing. Front Plant Sci 6:1–15. https://doi.org/10.3389/fpls.2015.00778

Zhang JZ, Ai XY, Guo WW et al (2012) Identification of miRNAs and their target genes using deep sequencing and degradome analysis in trifoliate orange [*Poncirus trifoliate* (L.) Raf]. Mol Biotechnol 51:44–57. https://doi.org/10.1007/s12033-011-9439-x

Zhang YC, Yu Y, Wang CY et al (2013) Overexpression of microRNA OsmiR397 improves rice yield by increasing grain size and promoting panicle branching. Nat Biotechnol 31:848–852. https://doi.org/10.1038/nbt.2646

Zhang H, Yin L, Wang H et al (2017) Genome-wide identification of Hami melon miRNAs with putative roles during fruit development. PLoS ONE 12:20–24. https://doi.org/10.1371/journal.pone.0180600

Zuo J, Zhu B, Fu D et al (2012) Sculpting the maturation, softening and ethylene pathway: the influences of microRNAs on tomato fruits. BMC Genom 13:7. https://doi.org/10.1186/1471-2164-13-7

Chapter 6
microRNAs in Plant Embryogenesis

Ana Alves, Andreia S. Rodrigues and Célia Miguel

Abstract In higher plants, the development of a mature embryo from the zygote follows a synchronized sequence of cell division, growth and differentiation events ultimately regulated by a highly coordinated gene expression. Several genome-wide expression studies during embryogenesis in Arabidopsis have been reported, including high-resolution single-cell directed, but current knowledge is mainly based on the coding transcriptome. Despite the available state-of-the-art technologies for transcriptome sequencing, there is still a gap in our understanding of the complex regulatory networks involving small non-coding RNAs. While a few microRNAs of specific conserved families have been functionally characterized, the role played by a major part of the microRNA population during the plant life cycle, including embryo development, in both model and non-model plants are yet to be discovered. In this chapter, we review the current knowledge of the gene expression regulation of plant embryogenesis by microRNAs, and discuss future perspectives for advancing our knowledge on plant embryo development in the light of the latest discoveries in this area.

Keywords miRNA · Target transcript · Embryo development · Gene expression · Post-transcriptional regulation · Seed

A. Alves · C. Miguel (✉)
Faculdade de Ciências, BioISI—Biosystems & Integrative Sciences Institute, Universidade de Lisboa, Lisbon, Portugal
e-mail: cmmiguel@fc.ul.pt

A. S. Rodrigues · C. Miguel
Instituto de Biologia Experimental e Tecnológica (IBET), Oeiras, Portugal

A. S. Rodrigues
Instituto de Tecnologia Química e Biológica, Universidade Nova de Lisboa, Oeiras, Portugal

C. Miguel et al. (eds.), *Plant microRNAs*, Concepts and Strategies in Plant Sciences, https://doi.org/10.1007/978-3-030-35772-6_6

6.1 Introduction

6.1.1 Major Developmental Events During Plant Embryogenesis

In seed plants, embryogenesis is initiated after fertilization of the egg cell by one of the sperm cells carried by the pollen tube, giving rise to a diploid sporophyte, the zygote. Zygotic embryogenesis (or embryogeny) comprehends the developmental period from fertilization up to the desiccated and dormant embryo. This is a crucial period in the seed plant life cycle when the shoot–root body plan (apical-basal axis) is established and the embryo stores the food reserves needed later on during germination (Goldberg et al. 1994).

In *Arabidopsis thaliana* (Arabidopsis), the model angiosperm used in most plant embryogenesis studies, the events that lead to the formation of a mature embryo from the single-celled zygote follow a highly predictable pattern of cell division, growth and differentiation events (reviewed in ten Hove et al. 2015). The embryogenesis phase is usually divided into stages that reflect the morphology of the forming embryo, namely the zygotic, globular, heart, torpedo and mature embryo stages. The first zygotic division occurs after initial cell elongation, leading to the formation of two daughter cells with different developmental fates in subsequent development, a small apical cell and a longer basal cell. The whole mature embryo will be formed from the apical cell, with the exception of the distal part of the root apex including the quiescent center that results from the basal cell. Following the first asymmetric division, the apical cell undergoes two longitudinal divisions followed by a transverse division yielding an octant stage proembryo. The tangential division of all the eight cells gives rise to a 16-celled proembryo (dermatogen stage), with the superficial cell layer forming the protoderm. The epidermis will originate from the protoderm while all the other tissues (ground and vascular) originate from the inner cells. Meanwhile, the basal longer cell suffers a series of transverse divisions giving rise to a file of cells that will form the suspensor. Only the upper cell in this file, the hypophysis, becomes later integrated into the embryo. While the embryo continues to increase in size with further cell divisions during the early globular stage, the alteration in the membrane polar rearrangement of PIN1 and PIN7 auxin transporter proteins leads to a shift in auxin flow towards the hypophysis, a crucial event for the establishment of the distal part of the root apical meristem (RAM) containing the quiescent center. At the transition to the heart stage, another polar rearrangement of the PIN1 proteins in the epidermis ultimately results in the accumulation of auxin at the two sites of the future cotyledon primordia, in the center of which the shoot apical meristem (SAM) will form. The inner cells of the cotyledon primordia will contribute to auxin flow directed at the root pole. At the end of embryogenesis a basic body plan along the basal-apical and radial axes is established, bearing the SAM and RAM stem cell niches from which all the organs of the future plant will emerge.

Variation in this highly regular and predictable pattern of cell division exists in angiosperms other than Arabidopsis. Moreover, Arabidopsis has a limited utility as a model for gymnosperm embryogenesis. Gymnosperm (including conifers) and angiosperm lineages are estimated to have driven apart over 300 million years ago (Smith et al. 2010), and during the phases that comprise gymnosperm embryogenesis, namely proembryogeny, early embryogeny and late embryogeny (Singh 1978), important differences in embryonic patterning processes can be pointed out when comparing to angiosperms (reviewed in von Arnold et al. 2019). Commonly, several initial nuclear duplication events in the zygote, which are not immediately followed by cytokinesis, result in a free nuclear stage during the proembryogeny phase. The occurrence of cleavage polyembryogeny is also common in gymnosperms during early embryogeny, as well as the characteristic differentiation of multiple cotyledons around the SAM during late embryogeny conferring radial symmetry to the embryo. Importantly, a single fertilization event occurs in gymnosperms implying that a haploid megagametophyte surrounds the developing embryo(s), in contrast with the presence of a triploid endosperm that results from the double fertilization for seed formation in angiosperms. Within the gymnosperms, the development of pine embryos has also been categorized into nine stages according to a staging system based on clearly visible morphological features that allow easier identification and sampling by stage (Pullman and Webb 1994; Gonçalves et al. 2005).

6.1.2 *Somatic Embryogenesis*

Somatic embryogenesis (SE) is the process by which haploid or diploid somatic cells develop, without fusion of gametes, into differentiated plants through characteristic embryological stages in response to external stimuli. The process was first described for carrot (*Daucus carota*) in 1954 (Wiggans 1954) and later on for numerous angiosperm species. In gymnosperms, SE was first reported in the conifer Norway spruce (*Picea abies*) (Hakman and von Arnold 1985) and since then several SE protocols have been published for other conifer species, including multiple pine species. Somatic embryos characteristically develop through developmental stages that resemble their zygotic counterparts.

SE has become an important tool for mass propagation (Klimaszewska et al. 2016), mainly in conifer species with high economic impact, especially the *Picea* and *Pinus* genera (Smertenko and Bozhkov 2014; Zimmerman 1993), and enabling technology for genetic transformation and efficient cryopreservation in these species. Importantly, SE also provides a good experimental model system to unravel the complexities of gene expression during embryo development in conifers. Several important regulators, including microRNAs (miRNAs), of the transition from somatic to embryo identities and of subsequent embryo development have been identified in such experimental model systems, as discussed further ahead in the chapter.

6.1.3 Gene Expression During Embryo Development

Coordination of gene expression patterns is responsible for the establishment of cell identities throughout embryogenesis, either in angiosperms or gymnosperms. The first studies with embryo mutants and, more recently, several studies using genomic approaches, have greatly contributed to the elucidation of the molecular regulators involved in embryo development in angiosperms and gymnosperms. Even though Arabidopsis and conifer species belong to evolutionary distant plant taxonomic groups, many genes identified in conifer embryo development have orthologs in Arabidopsis and most identified biological processes seem to be conserved between them (Cairney and Pullman 2007; de Vega-Bartol et al. 2013).

The study of small RNAs (sRNAs) in the control of gene expression has been a rapidly evolving field. Since sRNAs were discovered, these molecules have been found across many eukaryotic organisms, from plants to mammals. In plants, sRNAs are produced as 21–24 nucleotides and can be categorized into different classes, such as miRNAs and short interfering RNAs (siRNAs), based on their origin, biogenesis and function (Axtell 2013). The primary miRNA transcript (pri-miRNA) is trimmed into a precursor miRNA (pre-miRNA) which folds into a hairpin-like structure that is subsequently recognized and processed to form a miRNA-miRNA star duplex by the RNaseIII enzyme DICER-LIKE1 (DCL1). The C2H2 zinc-finger protein SERRATE (SE) and the RNA-binding domain protein HYPONASTIC LEAVES 1 (HYL1) are also important for improving the accuracy of the miRNA biogenesis process, in cooperation with DCL1. miRNAs are then assembled into ARGONAUTE1 (AGO1) proteins and guide the silencing complex RISC to target the messenger RNA (mRNA) through base pairing to mediate post-transcriptional gene silencing by target cleavage or translation inhibition (D'Ario et al. 2017). The nearly perfect base-pairing of the miRNA with its target transcript ensures high regulatory specificity and allows to predict high-confidence target transcripts.

In the past 10 years, and despite the difficulties posed to experimental analyses by the deep localization of zygotic embryos within maternal tissues, several studies indicate that multiple miRNAs are required for embryo morphogenesis and pattern formation, as well as for the control of the embryo maturation timing. Over 400 miRNA targets seem to be expressed in the developing embryo (Willmann et al. 2011) underlining the relevant role of miRNAs in embryogenesis. A more recent study based on the high throughput RNA sequencing from different embryo developmental stages, from pre-globular to mature embryo, reported a set of 354 miRNAs expressed during embryogenesis and identified 59 embryonic miRNA target transcripts considered of high confidence, although only 22 yielded detectable miRNA-directed cleavage products in at least one embryo stage (Plotnikova et al. 2019). Profiling of the transcripts at a genome-wide level in *dcl1-5* globular embryos compared to wild-type embryos, revealed close to 3,000 differentially expressed genes at a two-fold threshold, suggesting that significant changes are evidenced at the transcriptomic level as a consequence of miRNA defective biogenesis (Plotnikova et al. 2019).

In this chapter, we will review the current knowledge on the functions of miRNAs throughout embryo development, starting with the early embryogenesis stages up to the embryo maturation. We will also highlight the main differences among different plant groups, including angiosperms *versus* gymnosperms and, within the angiosperms, dicots *versus* monocots. The reported roles of different miRNAs in somatic embryogenesis, from somatic embryo induction to embryo maturation will also be reviewed. Finally, perspectives for future advances in this field will be presented.

6.2 miRNAs Are Required in Embryo Patterning

The first hints pointing to the essential role of miRNAs in plant embryogenesis came from the analysis of mutants defective in the miRNA biogenesis pathway proteins. DCL1 can recognize the hairpin-like structure characteristically formed by the primary *MIRNA* (*MIR*) gene transcripts, and together with the accessory proteins SERRATE and HYL1 (Fang and Spector 2007; Dong et al. 2008), is responsible for the processing of most miRNAs. Several studies have shown that *dcl1* Arabidopsis mutants exhibit a number of defects and arrest early in embryo development. The identification of the first Arabidopsis *dcl1*-null mutants described the arrest of embryo development at the globular stage and abnormal cell divisions in the suspensor (Schwartz et al. 1994). Further support for the essential role of miRNAs in embryogenesis came from the identification of developmental embryo defects in mutants for additional genes involved in miRNA biogenesis, and the fact that miRNA target genes are needed for the normal development of the embryo as further described below. When miRNAs are completely absent, the adoption and maintenance of the identity of most embryo regions are not achieved (Seefried et al. 2014). Also recent studies reporting high throughput sequencing of miRNAs and their targets throughout Arabidopsis embryogenesis described the miRNA-directed dynamic cleavage and repression of a set of 59 target transcripts, including transcription factor transcripts from eight families with relevant developmental roles, namely containing ARF, GRAS, HD-ZIP, MADS-box, MYB, NAC, SBP and TCP domains (Plotnikova et al. 2019).

In 2010, the work by Nodine and Bartel provided important insights into the functions of miRNAs in embryogenesis demonstrating that DCL1 is needed for embryo pattern formation from as early as the octant embryo stage. However, more recent studies reporting the detailed phenotypic analysis of an allelic series of *dcl1*, *se-1* and *hyl1-2* mutants revealed that miRNAs are already required at the zygote stage (Armenta-Medina et al. 2017), although the penetrance of the observed phenotypes depend on the allele and on the ecotype. Relevant deviations from normal development observed at the very first stages included (1) the loss of zygote polarity in some *dcl1*, *serrate-1* and *hyl1* embryos, where vacuoles were observed both at apical or basal positions relative to the nucleus (2) the more symmetric pattern of the first division leading to altered length ratio between the resulting apical and basal cells, and (3) the subsequent horizontal instead of vertical division of the apical cell observed.

In octant wild-type embryos two developmental domains can be distinguished where the upper tier of cells will ultimately give origin to the aerial tissues and the lower tier to the hypocotyl and root tissues. The expression domains of the *WUSCHEL* related *HOMEOBOX2* (*WOX2*), *WOX8* and *WOX9* mark the apical domain, the suspensor and the basal domain, respectively (ten Hove et al. 2015). In the octant and dermatogen *dcl1* mutant embryos, the apical cell lineage already revealed defects related to the absence of the cell-specific expression of *WOX2* (Nodine and Bartel 2010). As a result of all the observed alterations at these early stages, the embryo proper in the studied mutants had a morphology that exhibited intermediate features between the suspensor and the embryo (Armenta-Medina et al. 2017).

Although to the best of our knowledge, miRNA profiling has not been performed at the zygote and octant stages of embryo development, spatiotemporal analysis of the expression patterns of miR165/166 in embryogenesis detected *MIR166G* and *MIR165B* expression in the uppermost suspensor cell of the two and eight-cell stage embryos, respectively (Miyashima et al. 2013). The miRNA165/166 targets the *HD-ZIP III* genes *PHABULOSA* (*PHB*), *PHAVOLUTA* (*PHV*), *REVOLUTA* (*REV*), *ARABIDOPSIS THALIANA HOMEOBOX 8* (*ATHB8*) and *ATHB15* (Mallory et al. 2004), which play major roles during embryogenesis and post-embryonic development and growth. In early embryo development, *HD-ZIP III* and *PLETHORA* (*PLT*) genes have antagonistic functions in the establishment of apical and basal polarity, being the *HD-ZIP III* genes involved in the specification of the apical fate. The *HD-ZIP III* are expressed during the early globular stage in the upper tier cells (Emery et al. 2003; ten Hove et al. 2015) being their expression domain restricted by the miR165/166 which are present in the basal–peripheral region of embryos. The interference with miR165/166 expression by the loss of *SERRATE* leads to non-specific expression of PHB and PHV in early embryos resulting in a lethal phenotype (Grigg et al. 2009). The ectopic expression of PHB, PHV and REV in the basal embryo domain results in the formation of a SAM instead of the RAM (Smith and Long 2010), and the phenotype exhibited by *phb-1d* embryos evidences a role in the establishment of embryo bilateral symmetry. The SAM formation seems to be also dependent on the correct expression of the miR394 (Knauer et al. 2013). This miRNA may have a non-cell-autonomous function because although being specifically expressed in the L1 layer during SAM formation where it targets the *LEAF CURLING RESPONSIVENESS* (*LCR*), evidence suggests that it also contributes to the specification of the underlying stem cell layers (Knauer et al. 2013).

Interestingly, the first tissue to be established in the embryo, the protoderm, was previously reported as unaffected in the studied *dcl1-5* and *dcl1-15* mutants (Nodine and Bartel 2010; Seefried et al. 2014). The specification of inner and outer cells at the transition from octant to dermatogen embryo is coincident with the change in the *MERISTEM LAYER1* (*ATML1*) expression which becomes restricted to the outer cells that form the protodemal layer and, as far as it is currently known, the regulation of this change in the gene expression domain is independent of miRNAs

(Nodine and Bartel 2010). However, both *ATML1* and another protoderm characteristic transcript, *PROTODERMAL FACTOR1* (*PDF1*), accumulate in the *dcl1-5* mutant embryo suspensor (Nodine and Bartel 2010), and this is in accordance with previous observations of the accumulation of starch grains and protein bodies in *dcl1* embryo suspensor (Schwartz et al. 1994). In a recent study by Takanashi et al. (2018), miR171a expression was detected in the egg cells and during *Arabidopsis* embryogenesis mainly at the basal half domain of the embryo proper in globular and later embryo stages. The authors further report that miR171a is expressed in the protoderm and epidermis of embryo and SAM, where it is suggested to target *HAIRY MERISTEM* (*HAM*) genes with a role in the establishment and maintenance of meristematic activities, acting as WUS and WOX5 interacting cofactors (Zhou et al. 2015).

Deviations from the wild-type embryo development are also observed in hypophysis division and differentiation of *dlc1* and *hyl1-2* and *serrate-1* early globular embryos (Nodine and Bartel 2010; Armenta-Medina et al. 2017), including misoriented cell divisions in the suspensor, hypophysis and embryo proper, and the precocious storage of lipid and carbohydrate reserves. Moreover, the establishment of the radial patterning axis is compromised in late globular *dcl1* embryos due to the absence of sub-protodermal divisions that result in a failure to form the vascular and ground tissue primordia.

In the follow-up of the RNA sequencing of globular stage *serrate-1* embryos in the work of Armenta-Medina et al. (2017), additional miRNAs with potentially important functions in early embryogenesis were highlighted, such as miR159, miR161, miR164, miR168, and others with still unknown functions during this important stage of the plant life cycle.

Another important developmental event occurring later in embryogenesis is the formation of the cotyledons that are initiated in the transition from the globular to the heart stage. The correct spatial patterning of the cotyledons depends on the correct expression of the CUC1 (CUP-SHAPED COTYLEDON1) and CUC2 NAC-domain transcription factors that are also involved in the SAM formation (Takada et al. 2001). The miR154 targets both *CUC1* and *CUC2* and blocking the regulation of *CUC1* by miR154 results in defects in cotyledon spatial arrangement (Mallory et al. 2004). On the other hand, the role of miR154 is consistent with the phenotype obtained by its overexpression, which is similar to the one exhibited by the double mutant *cuc1cuc2*. A member of the TEOSINTE BRANCHED1, CYCLOIDEA, and PCF (TCP) transcription factor family, TCP4, targeted by the miR319, has also been reported by Nodine et al. (2010) as one of the misregulated miRNA target transcripts in *dcl1* embryos, and indeed cotyledon growth defects were detected in miR319-resistant *TCP4* target transcript (Plotnikova et al. 2019).

6.3 miRNAs and Auxin Responses in Embryogenesis

Auxin biosynthesis, transport and response are key to the proper development of plant embryos from an early developmental stage. A number of mutants have been characterized whose phenotypes result from mutations in genes involved in these pathways (Möller and Weijers 2009). While the distribution and accumulation of auxin in specific cells or domains of the embryo result mainly from the function of polarly localized auxin efflux transporters, the PIN proteins, the response to varying levels of auxin depends on regulatory mechanisms involving the proteins AUXIN/INDOLE-3-ACETIC ACID INDUCIBLE (AUX/IAA), which are repressors of the AUXIN RESPONSE FACTOR (ARF) transcription factors. Upon ubiquitination and degradation of AUX/IAA proteins in the 26S proteosome promoted by high auxin levels, the ARF transcription factors are released from repression (Dharmasiri et al. 2005). The ARFs can then activate or repress the transcription of a large set of downstream auxin-responsive genes that translate the auxin level information into specific cellular functions. From the over 20 ARFs identified in Arabidopsis, 15 are expressed in the developing embryo (Seefried et al. 2014; Rademacher et al. 2011), some of which are described as miRNA target transcripts.

Increased levels of auxin response gene transcripts have been observed, including *ARF* genes such as *ARF6* and *ARF8* in *dcl1-5* globular embryos, *ARF16* in *dcl1-15* torpedo embryos and *ARF10//17* in both *dcl1-5* globular and *dcl1-15* torpedo embryos (Seefried et al. 2014). In later stages of embryogenesis even more dramatic changes in the expression of auxin response genes have been detected. Gain-of-function analyses have previously revealed that the expression of *ARF 10/16/17* is negatively regulated by miR160 (Mallory et al. 2005; Wang et al. 2005; Liu et al. 2007). Also, miR167 targets *ARF6* and *ARF8* (Jones-Rhoades and Bartel 2004), and miR390 direct the production of the trans-acting siRNA (ta-siRNA) *TAS3*, that in turn targets *ARF3* and *ARF4* (Hunter et al. 2006). Interestingly, although no miRNAs are known to target the well-characterized *ARF5/MONOPTEROS* (*MP*), which has crucial roles in embryogenesis being required for the initiation and maintenance of the ground tissue and of the embryonic root (Weijers et al. 2006; Möller et al. 2017), this ARF transcription factor has been recently shown to regulate miR390 by directly binding to its promoter (Dastidar et al. 2019). Another characterized ARF with direct functions in embryogenesis is *ARF7/NON-PHOTOTROPIC HYPOCOTYL4* (Hardtke et al. 2004), which up to now has been pointed out as a predicted miRNA target in *Lilium lancifolium* (He et al. 2018). Although a specific role for *ARF17* in embryo development has not been identified, it has been shown that the expression of miR160 and miR167-resistant versions of *ARF17* and *ARF8*, respectively, led to deviations in the expression of early auxin-responsive genes and embryo development (Mallory et al. 2005; Plotnikova et al. 2019). Also the auxin receptor genes *TIR1*, *AFB2* and *AFB3* are validated targets of miR393 (Navarro et al. 2006) and their levels are up-regulated in globular stage *dcl1-5* embryos (Armenta-Medina et al. 2017).

Overall, the mislocalization of auxin transport and distribution observed in *dcl1* embryos (Seefried et al. 2014) uncovers a relevant role for miRNAs in auxin-regulated embryo development, and important clues for miRNA roles in this context have been provided in studies using somatic embryogenesis, as discussed further below.

6.4 miRNAs in the Control of Embryo/Seed Maturation

A major role of miRNAs during embryogenesis is the repression of the embryo maturation gene expression program in early-stage embryos, in order to allow proper embryo patterning by preventing differentiation events characteristic of later embryogenesis stages (Nodine and Bartel 2010). The miR156, targeting *SQUAMOSA PROMOTOR BINDING PROTEIN LIKE* (*SPL*) transcription factor genes, was implicated as a major player in such a role as early as the eight-cell embryo given that *SPLs* were highly derepressed in mutant embryos. miRNA156 is one of the most conserved plant miRNAs families (Rhoades et al. 2002), being involved in the regulation of several important biological processes like fertility, juvenile to adult transition phase, development of roots, leaves and fruit, stress responses and secondary metabolism (Yu et al. 2015). The repression of *SPL10* and *SPL11* transcripts by miR156 was reported to inhibit the expression of seed maturation genes before the embryo reaches a proper developmental stage for that to occur, and it is also required for the correct division of RAM precursor cells (Nodine and Bartel 2010). Interestingly, although the levels of miR156 seem not to decrease in the transition to maturation, *SPL10* and *SPL11* increase their activity by a still not clear mechanism that may be related to the requirement of a certain transcript threshold.

More recently, Armenta-Medina et al. (2017) have reported, by using miRNA-resistant *SPL10* and *SPL11* constructs, abnormal cell divisions in hypophysis and suspensor of the recovered embryos. Moreover, besides *SPL11* also *SPL2* and *SPL3* transcripts increase in both *serrate-1* and *dcl1-5* mutants, implying that these *SPL* genes may play a still uncovered role in embryogenesis.

Other genes important during the embryo maturation period, some of which are considered major regulators of the synthesis and accumulation of protein and lipid storage reserves, include *FUSCA3* (*FUS3*), *LEAFY COTYLEDON2* (*LEC2*), *LEC1-LIKE* (*L1L*, *NF-YB6*) and some *MYB* and *bZIP* genes, which have been reported as up-regulated in the *dcl1-15* mutant (Willmann et al. 2011). These authors have made a series of observations that highlight this overall role of miRNAs. The observations include precocious chloroplast maturation and accumulation of reserves in the form of starch grains, lipid bodies and storage proteins in *dcl1-15* mutants. Moreover, the gene expression analysis of *dcl1-15* torpedo stage embryos by microarray hybridization revealed a transcriptomic signature characteristic of older embryos. In the same work, the authors suggest that the *ARABIDOPSIS 6B-INTERACTING PROTEIN1-LIKE1* (*ASIL1*) and *ASIL2* and the histone deacetylase *HDA6/SIL1*, which are down-regulated in *dcl1-15* torpedo embryo stage, might be downstream and be negatively regulated by the miRNA156-targeted *SPL* transcripts.

6.5 miRNAs Highlighted During Embryogenesis of Non-model Species

Although most studies of zygotic embryogenesis have been conducted in Arabidopsis, additional evidence of the role of several miRNAs in embryo development have been documented for other species, including other angiosperms of the monocots group, and gymnosperms mostly focusing on conifer species. Given the relevance of protein, lipid and carbohydrate storage processes during embryo/seed maturation in economically important food and oil crops worldwide, many of these studies have focused on the seed maturation phase highlighting the role of miRNAs in determining seed traits which are the improvement targets of several breeding programs. An overview of these studies, with the specific miRNAs identified and/or characterized is shown in Table 6.1.

In the particular case of conifers, miRNA functions have been pointed out in a *Pinus pinaster* genome-wide transcriptomics study, as important regulators of gene expression in particular from mid to late embryogenesis (de Vega-Bartol et al. 2013), and the time-course analysis of functional gene categories highlighted functions associated to *miRNA metabolic process*, *siRNA and miRNA binding*, and *gene silencing by miRNA*. In the same study, several transcripts related to sRNA biogenesis pathways were identified among the differentially expressed genes, such as a putative *DAWDLE* (*DDL*), also up-regulated from mid to late embryogenesis, that may act through DCL, by facilitating access or recognition of pri-miRNAs (Yu et al. 2008). Also, a putative homolog of *HYL1* was found up-regulated in the same developmental period. A very recent report of the sRNA transcriptome of *P. pinaster* embryos at the same developmental stages has provided an important resource for the future characterization of the functional roles of sRNAs in conifer embryogenesis (Rodrigues et al. 2019). The authors identified a total of 215 conserved miRNAs corresponding to 40 conserved miRNA families, including conifer-specific miRNAs (miR946, miR947, miR950, miR951, miR1311, miR1312, miR1313, miR1314, miR1315, miR1316, miR3699, miR3701 and miR3711), and 212 high-confidence novel miRNAs. A shortlist of highly expressed miRNAs was also identified which included 20 conserved miRNAs such as several isoforms of miR159 and miR166, as well as miR162 and miR319, and 21 novel miRNAs. The miR159, identified as possibly targeting *KASII* (also known as *FAB1*) and involved in the plastid fatty acid biosynthesis pathway during seed maturation in *Brassica napus* (Wang et al. 2016), was one of the most represented families, showing increased expression across embryogenesis. Interestingly, also in *P. pinaster* embryos, a homolog of *KASII*, encoding '3-ketoacyl-ACP synthase II', was

Table 6.1 List representing an overview of the miRNAs that have been identified/characterized as relevant in developing plant embryos, either zygotic or somatic

miRNA	Role in embryogenesis	Target gene	Plant group	Species	References
miR156	Globular embryo development	*SPL*	Dicot	*Gossypium hirsutum*	Yang et al. (2013)
	Embryogenic callus differentiation		Monocot	*Oryza sativa*	Luo et al. (2006)
				Zea mays	Chen et al. (2014)
	Enhances SE capability		Dicot	*Citrus*	Long et al. (2018)
	Regulates cell identity in the hypophysis and embryo maturation		Dicot	*Arabidopsis thaliana*	Nodine and Bartel (2010), Willmann et al. (2011), Seefried et al. (2014)
	Cotyledonary embryo development		Conifer	*Larix leptolepis*	Zhang et al. (2012)
	Tissues development and maturation		Conifer	*Picea abies*	Yakovlev and Fossdal (2017)
miR159	Seed germination and vegetative transition	*MYB* family	Dicot	*Arabidopsis thaliana*	Szyrajew et al. (2017
			Dicot	*Citrus*	Wu et al. (2011)
	Embryogenic callus differentiation		Conifer	*Larix leptolepis*	Li et al. (2012)
			Conifer	*Picea abies*	Zhang et al. (2010)
	Embryo development regulation	*MYB33*	Conifer	*Pinus taeda*	Oh et al. (2008)
miR160	Promotes auxin responses and early pattern formation. Connected with miR167	*ARF10, ARF16, ARF17*	Conifer	*Picea abies*	Liu et al. (2007)
			Dicot	*Dimocarpus logan*	Lin et al. (2015)
			Dicot	*Arabidopsis thaliana*	Liu et al. (2010)
miR162	Regulatory function during SE	*DCL1*	Conifer	*Larix leptolepis*	Li et al. (2013)
				Picea abies	Zhang et al. (2012)
			Monocot	*Lilium pumilum*	Siddiqui et al. (2018)
			Dicot	*Arabidopsis thaliana*	Siddiqui et al. (2018)
miR164	Polar auxin transport	*NAC* domain containning family	Dicot	*Citrus sinensis*	Wu et al. (2015)
	Regulates cotyledon formation			*Arabidopsis thaliana*	Mallory et al. (2004)

(continued)

Table 6.1 (continued)

miRNA	Role in embryogenesis	Target gene	Plant group	Species	References
miR165/166	Promotes shoot apical meristem (SAM) and cotyledon fate and embryo maturation. Plays a crucial role in LEC2 mediated embryogenic response	*HD-ZIP III*	Dicot	*Arabidopsis thaliana*	Liu et al. (2009), Tang et al. (2012), Miyashima et al. (2013), Wójcik et al. (2017)
	Polar auxin transport and maintenance of shoot apical meristem and vascular patterning		Dicot	*Citrus*	Wu et al. (2011)
			Conifer	*Picea abies*	Yakovlev and Fossdal (2017)
	Inhibits cotyledon formation		Conifer	*Larix leptolepis*	Li et al. (2016)
miR166	Embryo development regulation	*HB15*	Conifer	*Pinus taeda*	Oh et al. (2008)
miR167	Cotyledonary embryo development	*ARF6, ARF8*	Dicot	*Dimocarpus logan*	Lin et al. (2015)
			Dicot	*Gossypium hirsutum*	Yang et al. (2013)
			Dicot	*Citrus sinensis*	Wu et al. (2011)
	Controls stress/temperature and auxin-responsive signaling during development, SE	*ARF6, ARF8, IAR3*	Dicot	*Arabidopsis thaliana*	Wu et al. (2006), Su et al. (2016), Yao et al. (2019)
	Embryo development regulation	*ARF8*	Conifer	*Pinus taeda*	Oh et al. (2008)
miR168	AGO1-mediated post-transcriptional silencing	*AGO1*	Conifer	*Picea abies*	Jagtap and Shivaprasad (2014)
	Regulatory function during SE		Conifer	*Larix leptolepis*	Li et al. (2016)
miR171	Exerts function on proembryogenic masses (PEMs)	unknown	Conifer	*Larix leptolepis*	Li et al. (2016)
	Acts on the induction process				
		SCL/GRAS family	Dicot	*Arabidopsis thaliana*	Armenta-Medina et al. (2017)

(continued)

Table 6.1 (continued)

miRNA	Role in embryogenesis	Target gene	Plant group	Species	References
	Embryo development regulation	*SCR*/GRAS family	Conifer	*Pinus taeda*	Oh et al. (2008)
	Responses to exogenous auxin and very early cellular differentiation and development	*GRAS* family	Conifer	*Picea abies*	Yakovlev and Fossdal (2017)
miR172	Embryo development regulation	*AP2* domain	Conifer	*Pinus taeda*	Oh et al. (2008)
	Seed development and maturation		Conifer	*Picea abies*	Zhang et al. (2010)
miR319	Proper cotyledon formation	*TCP4*	Dicot	*Arabidopsis thaliana*	Palatnik et al. (2003)
	Versatile functions in multiple aspects of plant growth and development	*Myb* and *TCP* family	Conifer	*Picea abies*	Reichel and Miller (2015)
miR393	Induces embryogenic transition of somatic tissue by modification of explant sensitivity to auxin	*TIR1, AFB1, AFB2, AFB3*	Dicot	*Arabidopsis thaliana*	Wójcik and Gaj (2016)
miR394	Confers stem cell competence to the SAM	*LEAF CURLING RESPONSIVENESS (LCR)*	Dicot	*Arabidopsis thaliana*	Knauer et al. (2013)
	Regulation of the cell cycle	*HAT* family, LRR domains	Conifer	*Picea abies*	Yakovlev and Fossdal (2017)
miR396	Cell proliferation	F-box family, UDP-glucosyl transferases	Conifers	*Picea abies*	Yakovlev and Fossdal (2017)
miR397	Cell wall thickening	Laccases	Dicot	*Citrus*	Wu et al. (2011)
			Conifer	*Larix leptolepis*	Zhang et al. (2012)
miR398	SE induction	*CSD1* (Cu/Zn superoxided is mutase 1)	Dicot	*Arabidopsis thaliana*	Szyrajew et al. (2017)
	Modulation of PEM propagation and transition to single embryo	Plastocyanin	Conifer	*Larix leptolepis*	Zhang et al. (2012)

predicted as a target of miR159 (Rodrigues et al. 2019). Furthermore, the majority of the differentially expressed miRNAs exhibited an expression peak late in embryogenesis, at the cotyledonary and mature embryo stages (Rodrigues et al. 2019), in accordance with a previous report referring that although sRNA pathways are differentially regulated across embryogenesis, miRNA functions predominate in mid to late embryogenesis (de Vega-Bartol et al. 2013). Still, each stage of embryogenesis appeared associated with regulation by a characteristic set of miRNAs. For instance, the analysis of predicted targets highlighted *transporter activity* in pre-cotyledonary embryos, and *sulfur metabolism*, for which a link to abscisic acid metabolism has been described in Arabidopsis (Cao et al. 2014), also appeared associated with very early or mid-late embryo developmental stages.

An interesting aspect related to the epigenetic memory formation in seeds has been investigated in Norway spruce embryos and a link to sRNAs has been reported (Yakovlev and Fossdal 2017). The authors refer to the identification of over 1000 highly expressed miRNAs in developing somatic embryos, including both conserved and novel sequences, and the expression of 654 of those miRNAs was dependent on epitype inducing (EpI) temperature. Among the putative targets of 124 differentially expressed miRNAs over 200 were found to be epigenetic regulators that may be involved in the formation of epigenetic marks, and thus the authors propose that miRNA regulation provides control over the developmental program and the epigenetic memory formation. For more details on this subject please refer to Chap. 4 of this book.

6.6 miRNAs in the Induction and Development of Somatic Embryos

As mentioned in the introduction, SE is considered a model system for studying zygotic embryogenesis because it can provide easier access to large numbers of embryos at several developmental stages, in contrast with the, sometimes difficult, manipulation and isolation of zygotic embryos from the seed tissues. However, since somatic embryogenic cultures in most species are often obtained after treatment of explants with phytohormones, especially auxin, assumptions based on gene expression data should be made with care because, as described above, auxin promotes transcriptional changes at a genome-wide level mediated by the AUX/IAA and ARF transcription factors. Still, SE is extremely useful and the identification of miRNAs involved in SE can significantly extend our still scarce knowledge about the molecular mechanisms controlling embryogenic transition in somatic plant cells. Indeed, the molecular mechanisms that control the transition of specialized somatic cells into pluripotent cells with the ability to develop into an embryo, are still one of the most relevant and least understood topics in plant development (Vogel 2005).

The roles of miRNAs in SE have been studied in several species, both during the induction phase and the somatic embryo development (Kou et al. 2012; Li et al. 2013;

Luo et al. 2006; Szittya et al. 2008; Vernoux and Benfey 2005; Wu et al. 2011; Yang et al. 2013), with low representation of some plant groups such as gymnosperms (Lin and Lai 2013; Oh et al. 2008).

During the induction of somatic embryogenesis, miR156 has been identified as relevant in several species. By comparing the miRNA expression profiles in embryogenic and non-embryogenic callus of rice, Luo et al. (2006) reported a few miRNAs potentially important during the transition to an embryogenic status, such as the miR156, miR319, miR397 and miR398. The miR156 has also been associated to somatic embryogenesis induction in sweet orange, and miR397 to the globular embryo formation and the lack of embryogenic competence of non-embryogenic callus (Wu et al. 2011). The citrus csi-miR156a, was the one with the highest expression within the miR156 family in the embryogenic cells and during SE (Wu et al. 2015), and its overexpression and corresponding downregulation of *SPLs* was confirmed to play an important role in citrus SE, involving stress responses and hormone signaling pathways (Long et al. 2018). In the case of miR397, it is possible that the mediated downregulation of laccase genes (Constabel et al. 2000) is important for keeping the characteristic thin cell walls of meristematic cells as pointed out by Luo et al. (2006). In *Larix kaempferi* SE the level of miR397 levels slowly decreased upon transfer to embryo maturation medium while the laccase mRNA increased, putatively enabling the cell wall thickening (Zhang et al. 2012).

The *LEC2* gene mentioned above as required for the maturation phase seems to be essential also for SE induction. Its auxin-dependent regulation was found related to SE induction in Arabidopsis whereas its expression was much lower in non-embryogenic callus and hormonal mutants with low SE capacity (Ledwoń and Gaj 2009). Despite the regulatory mechanism of *LEC2* transcription during SE has remained largely unknown, the HD-ZIP III transcription factor genes *PHB*/*PHV* targeted by miR165/166 are likely involved candidates as they act as positive regulators of *LEC2* transcription in Arabidopsis vegetative development (Yang et al. 2019). Recently, Wójcik et al. (2017) showed that both miR160 and miR165/166 may also be involved in induction of SE in Arabidopsis through an auxin-dependent mechanism. The authors found that if miR165/166 and miR160 are misexpressed, an accumulation of auxin occurs in cultured somatic tissues leading to spontaneous SE, and postulated that both miRNA families impact transition into an embryogenic status through *LEC2* expression and alterations in auxin signaling in in vitro cultured somatic tissues involving the up-regulation of *YUC* genes and activation of auxin biosynthesis. In the case of miR160, different isoforms may contribute differently to SE regulation. It is also suggested that an interaction exists between the miR160 and the miR165/166 pathways during SE induction since the *ARF10* and *ARF16* transcript levels were up-regulated in *phb, phv, phb1-d* and *STTM165/166* lines. Additionally, *ARF10* and *ARF16* are possible negative regulators of *PHB* since *PHB* transcript levels were increased in *arf10arf16*, and down-regulated in *mARF16, miR160b* and *miR160c* (Wójcik et al. 2017). However, the exact nature of miR160 and miR165/166 pathways interaction needs to be elucidated in future studies.

The down-regulation of miR160 and consequent impact on *ARF10/16/17* regulation by endogenous target mimics have been reported to play a role also in the SE in *Dimocarpus logan* (Lin et al. 2015). In conifers such as *Larix kaempferi*, expression of different *ARF* genes, possibly regulated by miR160 or miR167, was shown to increase up to the early SE cotyledonary stage and then significantly decreased (Zhang et al. 2012). In this group of plants, several miRNAs were proposed as potential biomarkers of somatic embryogenesis such as miR159, miR166, miR168, miR171, miR156, miR167, miR397, miR398, which might contribute to normal and synchronized development of somatic embryos (Miguel et al. 2016, Zhang et al. 2012, Zhang and Wang 2015). The overexpression of miR166 in somatic embryos of *Larix leptolepis* yielded some alterations in the SAM that became wider and flatter (Li et al. 2016). Furthermore, it also promoted WOX up-regulation, which may mean that it plays a role in SAM development through *WOX* genes.

Also in conifers, different isoforms of miR171 may play differential roles in SE. Zhang et al. (2012) reported that miR171a/b may affect the function of conifers proembryogenic masses (PEMs) and miR171c may, on the other hand, be involved in SE induction. While investigating the underlying mechanisms of embryogenic competence loss during prolonged subculturing of embryogenic cultures of *L. kaempferi*, Li et al. (2012) reported that miRNA159 targeting of *LaMYB33* might be involved in the maintenance of such potential and in somatic embryo development. Both miR159 and miR171 have been associated to embryogenecity (Miguel et al. 2016). Apart from the *HAM* genes mentioned previously, other targets of miR171 are the *SCARECROW-LIKE* (*SCL*) transcription factor genes, which are interesting candidates to further explore through functional studies (Zhu et al. 2015). In *P. pinaster*, MAP kinases (MPK) were also identified as potential miR171 targets (Rodrigues et al. 2019), which is consistent with the reported diversification of *MIR171* genes (Zhu et al. 2015).

The fact that somatic embryos are not developing within an endosperm or megagametophyte raises other interesting questions regarding the mechanisms that may be differentially regulated in the somatic embryo to compensate for the lack of such embryo surrounding tissues. In a recent study in *P. pinaster*, the megagametophyte tissues corresponding to mid and late stages of embryo development were also studied by RNA sequencing, in addition to zygotic and somatic embryos (Rodrigues et al. 2019). While some miRNA families were found mostly in embryos, either zygotic or somatic, such as miR159, miR171 and miR394, other families were more represented in somatic embryos and megagametophytes but not in zygotic embryos, such as miR397 and miR408. Most miRNAs expressed in megagametophytes appeared to peak at the late embryo developmental stage, and a gene enrichment analysis of the putative miRNA targets in these tissues highlighted biological processes that included *response to stimulus*, *cell communication*, *signaling*, and *glucose metabolism*, among others. miR408 targets transcripts encoding copper-containing proteins which seem to be regulated by sucrose, with the involvement of miR408 and miR398. In the presence of high sucrose levels these proteins are positively regulated by SPL7 (Ren and Tang 2012, Yamasaki et al. 2009). Interestingly, in the endosperm of maize sucrose treatment significantly affected miR408 expression (Huang et al. 2016). It can be speculated that the increased expression of both miR408 and miR397 in somatic embryos might be related to the high levels of sucrose used in the SE culture media

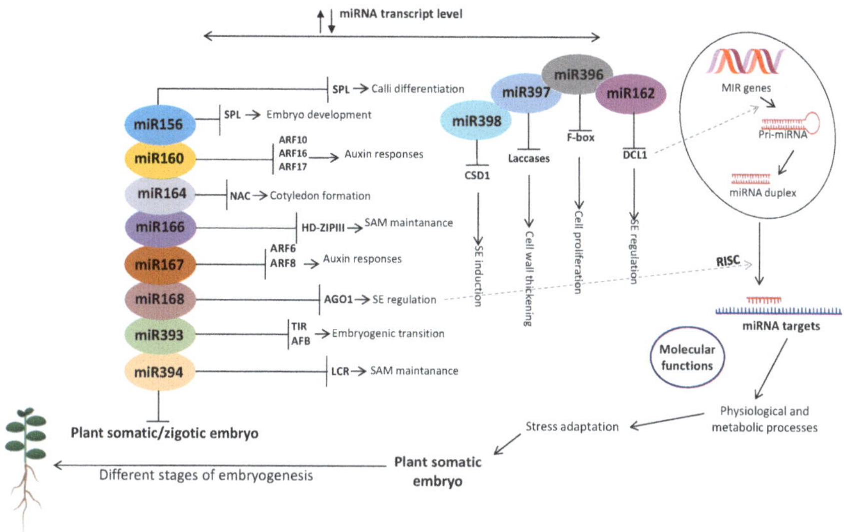

Fig. 6.1 Schematic representation showing the current understanding of a set of miRNAs important for the regulation of plant embryogenesis

(Rodrigues et al. 2019). For a summarized overview of the current understanding of a set of miRNAs important for the regulation of plant embryogenesis refer to Fig. 6.1.

6.7 Conclusions

The gene expression regulation layer provided by miRNAs during embryogenesis is of capital importance. While a large body of knowledge has been generated over the years on the molecular mechanisms underlying major events during embryo development, especially in the model plant Arabidopsis, many questions remain. The current pace at which new genomes and transcriptomic datasets are being released provides opportunities for the discovery of still unsuspected regulators of plant embryogenesis, among which are probably many miRNAs whose targets and associated functions need to be experimentally characterized. Given the relevance of the identification of the specific domains of expression of miRNAs that exclude the activity of their target genes, often confined to a specific cell or a group of cells, it will be important to adopt methodologies allowing single-cell analyses, including RNA sequencing. Extending these state-of-the-art technologies to a broad range of species and plant taxonomic groups will be required if this knowledge is to be applied to crops and non-crop species like conifer trees. Also, the tools for identification and annotation of miRNAs, as well as for integrated analysis of miRNAs and their targets, have to evolve in order to allow more powerful biological insights from large-scale datasets. Significant progress in this research area is expected in the near future, although the limiting step will be the extremely laborious and time-consuming experimental studies for miRNA functional validation and discovery, without which their application in agriculture and forestry will be difficult to achieve.

Acknowledgements We acknowledge the funding support provided by Fundaçãopara a Ciência e a Tecnologia (FCT), through grants UID/Multi/04046/2013 to BioISI (Biosystems and Integrative Sciences Institute)] and GREEN-it (UID/Multi/04551/2013), and the doctoral fellowships SFRH/BD/79779/2011 (to ASR) and SFRH/BD/128827/2017 (to AA).

References

Armenta-Medina A, Lepe-soltero D, Xiang D, Datla R (2017) *Arabidopsis thaliana* miRNAs promote embryo pattern formation beginning in the zygote. Dev Biol 431:145–151

Axtell MJ (2013) Classification and comparison of small RNAs from plants. Annu Rev Plant Biol 64(1):137–159

Cairney J, Pullman GS (2007) The cellular and molecular biology of conifer embryogenesis. New Phytol 176:511–536

Cao M-J, Wang Z, Zhao Q, Mao JL, Speiser A, Wirtz M, Hell R, Zhu JK, Xiang CB (2014) Sulfate availability affects ABA levels and germination response to ABA and salt stress in *Arabidopsis thaliana*. Plant J 77:604–615

Chen J, Zeng B, Zhang M, Xie S, Wang G, Hauck A, Lai J (2014) Dynamic transcriptome landscape of maize embryo and endosperm development. Plant Physiol 166:252–264

Constabel CP, Yip L, Patton JJ, Christopher ME (2000) Polyphenol oxidase from hybrid poplar. Cloning and expression in response to wounding and herbivory. Plant Physiol 124:285–295

D'Ario M, Griffiths-Jones S, Kim M (2017) Small RNAs: big impact on plant development. Trends Plant Sci 22(12):1056–1068

Dastidar MG, Scarpa A, Magele I, Ruiz-Duarte P, Born P, Bald L, Jouannet V, Maizel A (2019) ARF5/MONOPTEROS directly regulates miR390 expression in *Arabidopsis thaliana* primary root meristem. Plant Direct https://doi.org/10.1101/463943

de Vega-Bartol J, Simões M, Lorenz WW, Rodrigues AS, Alba R, Dean JFD, Miguel C (2013) Transcriptomic analysis highlights epigenetic and transcriptional regulation during zygotic embryo development of Pinuspinaster. BMC Plant Biol 13:123

Dharmasiri N, Dharmasiri S, Weijers D, Lechner E, Yamada M, Hobbie L, Ehrismann JS, Jürgens G, Estelle M (2005) Plant development is regulated by a family of auxin receptor F box proteins. Dev Cell 9:109–119

Dong Z, Han MH, Fedoroff N (2008) The RNA-binding proteins HYL1 and SE promote accurate in vitro processing of pri-miRNA by DCL1. Proc Natl Acad Sci 105(29):9970–9975

Emery JF, Floyd SK, Alvarez J, Eshed Y, Hawker NP, Izhaki A, Baum SF, Bowman JL (2003) Radial patterning of Arabidopsis shoots by class III HD-ZIP and KANADI genes. Curr Biol 13:1768–1774

Fang Y, Spector DL (2007) Identification of nuclear dicing bodies containing proteins for microRNA biogenesis in living Arabidopsis plants. Curr Biol 17(9):818–823

Goldberg RB, de Paiva G, Yadegari R (1994) Plant embryogenesis: zygote to seed. Science 266(5185):605–614

Gonçalves S, Cairney J, Maroco J, Oliveira MM, Miguel C (2005) Evaluation of control transcripts in real-time RT-PCR expression analysis during maritime pine embryogenesis. Planta 222:556–563

Grigg SP, Galinha C, Kornet N, Canales C, Scheres B, Tsiantis M (2009) Report repression of apical homeobox genes is required for embryonic root development in Arabidopsis. Curr Biol 19:1485–1490

Hakman I, von Arnold S (1985) Plantlet regeneration through somatic embryogenesis in *Picea abies* (Norway Spruce). J Plant Physiol 121(2):149–158

Hardtke CS, Ckurshumova W, Vidaurre DP, Singh S, Stamatiou G, Tiwari SB, Hagen G, Guilfoyle TJ, Berleth T (2004) Overlapping and non-redundant functions of the Arabidopsis auxin response factors MONOPTEROS and NONPHOTOTROPIC HYPOCOTYL 4. Development 131(5):1089–1100

He X, Shenkute AG, Wang W, Xu S (2018) Characterization of conserved and novel miRNAs in *Lilium lancifolium* Thunb. by high-throughput sequencing. Sci Rep 8:2880

Huang H, Long J, Zheng L, Li Y, Hu Y, Yu G et al (2016) Identification and characterization of microRNAs in maize endosperm response to exogenous sucrose using small RNA sequencing. Genomics 108:216–222

Hunter C, Willmann MR, Wu G, Yoshikawa M, Gutiérrez-Nava M, Poethig RS (2006) Trans-acting siRNA-mediated repression of ETTIN and ARF4 regulates heteroblasty in Arabidopsis. Development 133:2973–2981

Jagtap S, Shivaprasad PV (2014) Diversity, expression and mRNA targeting abilities of Argonaute-targeting miRNAs among selected vascular plants. BMC Genom 15:1049

Jones-Rhoades MW, Bartel DP (2004) Computational identification of plant microRNAs and their targets, including a stress-induced miRNA. Mol Cell 14:787–799

Klimaszewska K, Hargreaves C, Trontin J (2016) Advances in conifer somatic embryogenesis since year 2000. In: Germana MA, Lambardi M (eds) In vitro embryogenesis in higher plants. Methods in molecular biology, vol 1359. Humana Press, New York

Knauer S, Holt AL, Rubio-Somoza I, Tucker EJ, Hinze A, Pisch M, Javelle M, Timmermans MC, Tucker MR, Laux T (2013) A protodermal miR394 signal defines a region of stem cell competence in the Arabidopsis shoot meristem. Dev Cell 24:125–132

Kou SJ, Wu XM, Liu Z, Liu YL, Xu Q, Guo WW (2012) Selection and validation of suitable reference genes for miRNA expression normalization by quantitative RT-PCR in citrus somatic embryogenic and adult tissues. Plant Cell Rep 31:2151–2163

Ledwoń A, Gaj MD (2009) LEAFY COTYLEDON2 gene expression and auxin treatment in relation to embryogenic capacity of Arabidopsis somatic cells. Plant Cell Rep 28:1677–1688

Li WF, Zhang SG, Han SY, Wu T, Zhang JH, Qi LW (2012) Regulation of LaMYB33 by miR159 during maintenance of embryogenic potential and somatic embryo maturation in Larixkaempferi (Lamb.) Carr Plant Cell Tissue Organ Cult 113(1):131–136

Li D, Wang L, Liu X, Cui D, Chen T, Zhang H, Jiang C, Xu C, Li P, Li S, Zhao L, Chen H (2013) Deep sequencing of maize small RNAs reveals a diverse set of microRNA in dry and imbibed seeds. PLoS ONE 8:1–14

Li ZX, Li SG, Zhang L, Han S, Li WF, Xu H, Yang W, Liu Y, Fan Y, Qi LW (2016) Over-expression of miR166a inhibits cotyledon formation in somatic embryos and promotes lateral root development in seedlings of *Larix leptolepis*. Plant Cell Tissue Organ Cult 127:461–473

Lin Y, Lai Z (2013) Comparative Analysis Reveals Dynamic Changes in miRNAs and Their Targets and Expression during Somatic Embryogenesis in Longan (Dimocarpus longan Lour.). PLoS ONE 8 (4):e60337

Lin Y, Lai Z, Tian Q, Lin L, Lai R, Yang M, Zhang D, Chen Y, Zhang Z (2015) Endogenous target mimics down-regulate miR160 mediation of ARF10, -16, and -17 cleavage during somatic embryogenesis in *Dimocarpus longan* Lour. Front Plant Sci 6:956

Liu PP, Montgomery TA, Fahlgren N, Kasschau KD, Nonogaki H, Carrington JC (2007) Repression of AUXIN RESPONSE FACTOR10 by microRNA160 is critical for seed germination and post-germination stages. Plant J 52:133–146

Liu X, Huang J, Wang Y, Khanna K, Xie Z, Owen H, Zhao D (2010) The role of floral organs in carpels, an Arabidopsis loss-of-function mutation in MicroRNA160a, in organogenesis and the mechanism regulating its expression. Plant J 62:416–428

Liu Q, Yao X, Pi L, Wang H, Cui X, Huang H (2009) The ARGONAUTE10 gene modulates shoot apical meristem maintenance and establishment of leaf polarity by repressing miR165/166 in Arabidopsis. Plant J 58 (1):27-40

Long JM, Liu CY, Feng MQ, Liu Y, Wu XM, Guo WW (2018) MiR156-SPL modules regulate induction of somatic embryogenesis in citrus callus. J Exp Bot 69:2979–2993

Luo YC, Zhou H, Li Y, Chen JY, Yang JH, Chen YQ, Qu LH (2006) Rice embryogeniccalli express a unique set of microRNAs, suggesting regulatory roles of microRNAs in plant post-embryogenic development. FEBS Lett 580:5111–5116

Mallory AC, Reinhart BJ, Jones-Rhoades MW, Tang G, Zamore PD, Barton MK, Bartel DP (2004) MicroRNA control of PHABULOSA in leaf development: importance of pairing to the microRNA 5′ region. EMBO J 23:3356–3364

Mallory AC, Bartel DP, Bartel B (2005) MicroRNA-directed regulation of Arabidopsis AUXIN RESPONSE FACTOR17 is essential for proper development and modulates expression of early auxin response genes. Plant Cell 17:1360–1375

Miguel CM, Rupps A, Raschke J, Rodrigues AS and Trontin JF (2016) Impact of molecular studies on somatic embryogenesis development for implementation in conifer multi-varietal forestry. In: Park YS, Bonga JM, Moon HK (eds) Vegetative propagation of forest trees. National Institute of Forest Science, pp 373–421

Miyashima S, Honda M, Hashimoto K, Tatematsu K, Hashimoto T, Sato-Nara K, Okada K, Nakajima K (2013) A comprehensive expression analysis of the Arabidopsis MICRORNA165/6 gene family during embryogenesis reveals a conserved role in meristem specification and a non-cell-autonomous function. Plant Cell Physiol 54(3):375–384

Möller BK, Hove CA, Xiang D, Williams N, López LG, Yoshida S, Smit M, Datla R, Weijers D (2017) Auxin response cell-autonomously controls ground tissue initiation in the early Arabidopsis embryo. PNAS 114(12):E2533–2539

Möller B, Weijers D (2009) Auxin control of embryo patterning. Cold Spring Harb Perspect Biol 1(5):a001545

Navarro L, Dunoyer P, Jay F, Arnold B, Dharmasiri N, Estelle M, Voinnet O, Jones JD (2006) A plant miRNA contributes to antibacterial resistance by repressing auxin signaling. Science 312:436–439

Nodine MD, Bartel DP (2010) MicroRNAs prevent precocious gene expression and enable pattern formation during plant embryogenesis. Genes Dev 24(23):2678–2692

Oh TJ, Wartell RM, Cairney J, Pullman GS (2008) Evidence for stage-specific modulation of specific microRNAs (miRNAs) and miRNA processing components in zygotic embryo and female gametophyte of loblolly pine (*Pinus taeda*). New Phytol 179:67–80

Palatnik JF, Allen E, Wu X, Schommer C, Schwab R, Carrington JC, Weigel D (2003) Control of leaf morphogenesis by microRNAs. Nature 425:257–263

Plotnikova A, Kellner MJ, Mosiolek M, Schon MA, Nodine MD (2019) MicroRNA dynamics and functions during Arabidopsis embryogenesis. Plant Cell https://doi.org/10.1105/tpc.19.00395

Pullman GS, Webb DT (1994) An embryo staging system for comparison of zygotic and somatic embryo development. In: Proceedings of the TAPPI R &D division biological sciences symposium, Minneapolis, MN, 3–6 Oct 1994. Technical Association of the Pulp and Paper Industry Press, Atlanta, pp 31–34

Rademacher EH, Möller B, Lokerse AS, Llavata-Peris CI, van den Berg W, Weijers D (2011) A cellular expression map of the Arabidopsis AUXIN RESPONSE FACTOR gene family. Plant J 68:597–606

Reichel A, Miller AA (2015) Specificity of plant miRNA target MIMIcs: Cross-targeting of miR1559 and miR319. J Plant Physiol 180:45–48

Ren L, Tang G (2012) Identification of sucrose-responsive microRNAs reveals sucroseregulated copper accumulations in an SPL7-dependent and independent manner in *Arabidopsis thaliana*. Plant Sci 187:59–68

Rhoades MW, Reinhart BJ, Lim LP, Burge CB, Bartel DP (2002) Prediction of plant microRNA targets. Cell 110:513–520

Rodrigues AR, Chaves I, Costa BV, Lin Y-C, Lopes S, Milhinhos A, Van de Peer Y, Miguel CM (2019) Small RNA profiling in *Pinus pinaster* reveals the transcriptome of developing seeds and highlights differences between zygotic and somatic embryos. Sci Rep 9:11327 https://doi.org/10.1038/s41598-019-47789-y

Schwartz BW, Yeung EC, Meinke DW (1994) Disruption of morphogenesis and transformation of the suspensor in abnormal suspensor mutants of Arabidopsis. Development 120:3235–3245

Seefried WF, Willmann MR, Clausen RL, Jenik PD (2014) Global regulation of embryonic patterning in Arabidopsis by MicroRNAs. Plant Physiol 165:670–687

Siddiqui ZH, Abbas ZK, Ansari MW, Khan MN (2018) The role of miRNA in somatic embryogenesis. Genomics 111(5):1026–1033

Singh H (1978) Embryology of gymnosperms. Encyclopedia of plant anatomy XII, Embryol gymnosperms Encycl plant Anat XII

Smertenko A, Bozhkov PV (2014) Somatic embryogenesis: life and death processes during apical-basal patterning. J Exp Bot 65(5):1343–1360

Smith ZR, Long JA (2010) Control of Arabidopsis apical–basal embryo polarity by antagonistic transcription factors. Nature 464:423–426

Smith SA, Beaulieu JM, Donoghue MJ (2010) An uncorrelated relaxed-clock analysis suggests an earlier origin for flowering plants. Proc Natl Acad Sci U S A 107:5897–5902

Su YH, Liu YB, Zhou C, Li XM, Zhang XS (2016) The miRNA167 controls somatic embryogenesis in Arabidopsis through regulating its target genes ARF6 and ARF8. Plant Cell Tiss Organ Cult 124:405–417

Szittya G, Moxon S, Santos MD, Jing R, Fevereiro MP, Moulton V, Dalmay T (2008) High-throughput sequencing of Medicago truncatula short RNAs identifies eight new miRNA families. BMC Genomics 9(1):593

Szyrajew K, Bielewicz D, Dolata J, Wójcik AM (2017) MicroRNAs are intensively regulated during induction of somatic embryogenesis in Arabidopsis. Front Plant Sci 8:1–16

Takada S, Hibara K, Ishida T, Tasaka M (2001) The CUP-SHAPED COTYLEDON1 gene of Arabidopsis regulates shoot apical meristem formation. Development 128:1127–1135

Takanashi H, Sumiyoshi H, Mogi M, Hayashi Y, Ohnishi T, Tsutsumi N (2018) miRNAs control HAM1 functions at the single-cell-layer level and are essential for normal embryogenesis in Arabidopsis. Plant Mol Biol 96:627–640

Tang X, Bian S, Tang M, Lu Q, Li S, Liu X, Tian G, Nguyen V, Tsang EWT, Wang A, Rothstein SJ, Chen X, Cui Y (2012) MicroRNA-mediated repression of the seed maturation program during vegetative development in Arabidopsis. PLoS Genet 8:20–22

ten Hove CA, Lu K-J, Weijers D (2015) Building a plant: cell fate specification in the early Arabidopsis embryo. Development 142:420–430

Vernoux T, Benfey PN (2005) Signals that regulate stem cell activity during plant development. Curr Opin Genet Dev 15:388–394

Vogel G (2005) How does a single somatic cell become a whole plant? Science 309:86

von Arnold S, Clapham D, Abrahamsson M (2019) Embryology in conifers. In: Cánovas FM (ed) Molecular physiology and biotechnology of trees. Advances in botanical research, vol 89. Elsevier, Amsterdam, pp 157–184

Wang JW, Wang LJ, Mao YB, Cai WJ, Xue HW, Chen XY (2005) Control of root cap formation by MicroRNA-targeted auxin response factors in Arabidopsis. Plant Cell 17:2204–2216

Wang J, Jian H, Wang T, Wei L, Li J, Li C, Liu L (2016) Identification of microRNAs actively involved in fatty acid biosynthesis in developing *Brassica napus* seeds using high-throughput sequencing. Front Plant Sci 7:1570

Weijers D, Schlereth A, Ehrismann JS, Schwank G, Kientz M, Jurgens G (2006) Auxin triggers transient local signaling for cell specification in Arabidopsis embryogenesis. Dev Cell 10(2):265–270

Wiggans SC (1954) Growth and organ formation in callus tissues derived from *Daucus carota*. Am J Botany 41(4):321

Willmann MR, Mehalick AJ, Packer RL, Jenik PD (2011) MicroRNAs regulate the timing of embryo maturation in Arabidopsis. Plant Physiol 155:1871–1884

Wójcik AM, Gaj MD (2016) miR393 contributes to the embryogenic transition induced in vitro in Arabidopsis via the modification of the tissue sensitivity to auxin treatment. Planta 244:231–243

Wójcik AM, Nodine MD, Gaj MD (2017) miR160 and miR166/165 contribute to the LEC2-mediated auxin response involved in the somatic embryogenesis induction in Arabidopsis. Front Plant Sci 8:1–17

Wu MF, Tian Q, Reed JW (2006) Arabidopsis microRNA167 controls patterns of ARF6 and ARF8 expression, and regulates both female and male reproduction. Development 133:4211–4218

Wu XM, Liu MY, Ge XX, Xu Q, Guo WW (2011) Stage and tissue-specific modulation of ten conserved miRNAs and their targets during somatic embryogenesis of Valencia sweet orange. Planta 233:495–505

Wu XM, Kou SJ, Liu YL, Fang YN, Xu Q, Guo WW (2015) Genomewide analysis of small RNAs in nonembryogenic and embryogenic tissues of citrus: microRNA-and siRNA-mediated transcript cleavage involved in somatic embryogenesis. Plant Biotechnol J 13:383–394

Yakovlev IA, Fossdal CG (2017) In silico analysis of small RNAs suggest roles for novel and conserved miRNAs in the formation of epigenetic memory in somatic embryos of Norway spruce. Front Physiol 8:674

Yamasaki H, Hayashi M, Fukazawa M, Kobayashi Y, Shikanai T (2009) SQUAMOSA promoter binding protein-like7 is a central regulator for copper homeostasis in Arabidopsis. Plant Cell 21:347–361

Yang X, Wang L, Yuan D, Lindsey K, Zhang X (2013) Small RNA and degradome sequencing reveal complex miRNA regulation during cotton somatic embryogenesis. J Exp Bot 64:1521–1536

Yang T, Wang Y, Teotia S, Wang Z, Shi C, Sun H, Gu Y, Zhang Z, Tnag G (2019) The interaction between miR160 and miR165/ 166 in the control of leaf development and drought tolerance in Arabidopsis. Sci Rep 9:2832

Yao X, Chen J, Zhou J, Yu H, Ge C, Zhang M, Gao X, Dai X, Yang ZN, Zhao Y (2019) An essential role for miRNA167 in maternal control of embryonic and seed development. Plant Physiol 180:453–464

Yu B, Bi L, Zheng B, Ji L, Chevalier D, Agarwal M, Ramachandran V, Li W, Lagrange T, Walker JC, Chen X (2008) The FHA domain proteins DAWDLE in Arabidopsis and SNIP1 in humans act in small RNA biogenesis. ProcNatlAcadSci USA 105:10073–10078

Yu N, Niu QW, Ng KH, Chua NH (2015) The role of miR156/SPLs modules in Arabidopsis lateral root development. Plant J 83:673–685

Zhang S, Zhou J, Han S, Yang W, Li W, Wei H, Li X, Qi L (2010) Four abiotic stress-induced miRNA families differentially regulated in the embryogenic and non-embryogenic callus tissues of *Larix leptolepis*. Biochem Biophys Res Commun 398:355–360

Zhang J, Zhang S, Han S, Wu T, Li X, Li W, Qi L (2012) Genome-wide identification of microRNAs in larch and stage-specific modulation of 11 conserved microRNAs and their targets during somatic embryogenesis. Planta 236:647–657

Zhou Y, Liu X, Engstrom EM, Nimchuk ZL, Pruneda-Paz JL, Tarr PT, Yan A, Kay SA, Meyerowitz EM (2015) Control of plant stem cell function by conserved interacting transcriptional regulators. Nature 517:377–380

Zhu X, Lenf X, Sun X, Mu Q, Wang B, Li X, Wang C, Fang J (2015) Discovery of conservation and diversification of genes by phylogenetic analysis based on global genomes. Plant Genome 8(2)

Zimmerman JL (1993) Somatic embryogenesis: a model for early development in higher plants. Plant Cell 5(10):1411–1423

Zhang B, Wang Q (2015) MicroRNA-based Biotechnology for Plant Improvement. J Cell Physiol 230(1):1-15

Chapter 7
Emerging of microRNAs as Key Regulators in Plant Secondary Metabolism

Abdul Fatah A. Samad, Muhammad Sajad and Ismanizan Ismail

Abstract microRNAs (miRNAs) are well-known gene regulators of various biological processes in plants. Despite the growing knowledge of the miRNAs as main players in genetic systems, their involvement in regulating plant secondary metabolite (SM) biosynthesis is still poorly understood. In plants, these metabolites provide protection against various types of stresses that arise from unfavorable conditions. These plant-derived compounds are also significantly used in pharmaceuticals, food additives, and fragrances industries. Hence, there is a pressing need to explore how their biosynthetic pathways are being regulated, especially at post-transcriptional level with the involvement of miRNAs. Here, we discuss the recent progress on the post-transcriptional regulation by miRNAs of SMs biosynthetic pathways, namely terpenoids, flavonoids and alkaloids. We also summarize the key points to utilize miRNAs for genetic engineering to manipulate SMs production in plants.

Keywords microRNA · Post-transcription · Plant secondary metabolite · Terpenoid · Flavonoid · Alkaloid

---

A. F. A. Samad · M. Sajad · I. Ismail (✉)
Institute of Systems Biology, Universiti Kebangsaan Malaysia, 43600 Bangi, Selangor, Malaysia
e-mail: maniz@ukm.edu.my

A. F. A. Samad
Department of Biosciences, Faculty of Science, Universiti Teknologi Malaysia, 81310 Skudai, Johor, Malaysia

M. Sajad
Department of Plant Breeding and Genetics, University College of Agriculture and Environmental Sciences, The Islamiah University of Bahawalpur, Bahawalpur, Punjab, Pakistan

I. Ismail
Department of Biological Sciences and Biotechnology, Faculty of Science and Technology, Universiti Kebangsaan Malaysia, 43600 Bangi, Selangor, Malaysia

C. Miguel et al. (eds.), *Plant microRNAs*, Concepts and Strategies in Plant Sciences, https://doi.org/10.1007/978-3-030-35772-6_7

7.1 Introduction

Discovery of regulatory genetic networks in higher organisms like plants, animals, and humans, involving several regulation layers has contributed to a new paradigm shift in molecular biology. Gene expression regulation can take place at different levels including the epigenetic level that involves histone modification and DNA methylation, the pre-mRNA processing involving splicing, methylation and polyadenylation, and among others. miRNAs belong to a group of small RNA molecules that have well-established roles at post-transcriptional level in regulating a number of developmental and defense mechanisms in plants (Samad et al. 2017). Recent findings have also shown the ability of miRNAs to regulate certain biosynthetic pathways. This ability has attracted the attention of plant researchers to the potential of miRNAs as powerful genetic tools in manipulating whole biosynthetic pathways. Initially, the investigation of the regulatory roles of these tiny molecules began in some model plants like *Arabidopsis thaliana* and *Glycine max* (soybean) (Gou et al. 2011; Tuteja et al. 2009). However, the number of studies in this research topic kept increasing and extended to non-model plants as well, mostly plants with high economic value. High value products from plants are usually related to plant SMs like terpenoids, flavonoids, and alkaloids. At present, around 200,000 SM compounds have been characterized in plants. Although this is a huge number, it actually represents only a small fraction if compared to the total number of identified plant species, which is approximately 391,000 (Kessler and Kalske 2018).

The regulation of SMs biosynthesis is under the precise control of different gene regulators that turn on and/or turn off various genes at particular time and situations throughout plant growth and development and under different environmental stresses. This regulation leads to changes in abundance, either increase or decrease, of SMs which are usually present in low amounts in the plant organs, and occurs mostly according to the specific need of plants at a particular time. SMs molecules may be of significant importance to mankind due to their pharmaceutical, agricultural, and food industrial uses. Thus, plant biotechnologists have to look for different ways to produce the particular compounds in higher quantity. Among the suggested approaches are the use of miRNAs to induce RNA interference (RNAi) and gene silencing, which appear as one of the most reliable methodologies for genetic manipulation purposes (Bulgakov and Avramenko 2015; Hannon 2002). However, the exact target identification for each miRNA is a prerequisite for effectiveness of this technology. False target identification may result in undesirable traits in plants. Here, we have summarized the current status on how miRNAs regulate SMs synthesis by targeting certain genes involved in a particular pathway. Generally, the miRNA-mediated biosynthetic regulation pathways of SMs were identified using both wet and dry laboratory approaches. Through this chapter, we have gathered useful information which will help to improve our current understanding of the regulation of SMs biosynthesis by miRNAs.

7.2 miRNA Roles in the Biosynthesis of Terpenoids

Terpenoids are the most diverse group of plant secondary metabolites. These compounds are made up of five carbon isoprene units (Dubey et al. 2003). Basically, these compounds are synthesized from a condensation process between isopentenyl diphosphate (IPP) and dimethylallyl diphosphate (DMAPP). Both of these intermediates are derived from two different biosynthetic pathways, mevalonate (MVA) and methylerythritol phosphate (MEP) which are localized in the cytosol and plastid, respectively (Nagegowda 2010; Tholl 2015). Although these compounds have various types of structures, the fundamental backbone is based on the isoprene rule as suggested by Dubey et al. (2003), and the chemical structure is summarized as $(C_5H_8)_n$. This rule leads to the nomenclature system according to the number of isoprenes present in the compound, namely monoterpenoid, sesquiterpenoid, diterpenoid, sesterpenoid, triterpenoid, tetraterpenoid, and polyterpenoid. In plants, these compounds have many important roles, mainly in defense against a variety of biotic stresses, pollinator's attraction, and seed disseminators (Byers et al. 2014; Huang et al. 2012; Li et al. 2015b).

In *A. thaliana* and *Pogostemon cablin*, transcription factor SPL9 and SPL10 are targeted by miR156 (Yu et al. 2014). Both of these transcription factors are activators for terpene synthase 21 gene (*TPS21*). Overexpression of miR156 resulted in the downregulation of sesquiterpene level in both plant species (Yu et al. 2014). In *Xanthium strumarium*, some miRNAs were predicted to be involved in sesquiterpene biosynthesis pathway via computational approaches (Fan et al. 2015). The identified genes which encode the upstream enzymes, 1-deoxy-D-xylulose-5-phosphate synthase (DXS) and IPP isomerase (IDI), were targeted by miR7539 and miR5021, respectively. In addition, two more miRNAs, miR1134 and miR5021, were found to regulate the 3-hydroxy-3-methylglutaryl-coenzyme A reductase (*HMGR*) gene. Besides, various miRNAs including miR5255, miR5491, miR6435, miR6449, and miR7540 targeted genes encoding for downstream enzymes such as squalene epoxidase, beta-amyrin synthase, gemacrene A oxidase, ent-kaurene synthase, and R-linalool synthase, respectively (Fan et al. 2015). In *Mentha* spp., miR156 and miR414 were involved in terpenoid biosynthesis pathway by targeting *DXS* and *TPS21* genes, respectively. In addition, miR5021 in *Mentha* spp. targets two genes involved in terpenoid biosynthetic pathway, encoding geranyl geranyl diphosphate synthase (GGDS) and IDI (Singh et al. 2016b). Interestingly, miR5021 in *Catharanthus roseus* also targeted two genes, *GGDS* and *GCPE*, which are involved in the terpenoid biosynthesis pathway (Pani and Mahapatra 2013). In onion, miR838 targets a gene encoding cytochrome 71 (CYP71) (Singh et al. 2016a). In *Podopyhllum hexandrum*, two miRNAs, miR5021 and miR5538, were involved in regulating the terpenoid pathway by targeting the mevalonate diphosphate decarboxylase (MVD) and protein-*S*-isoprenylcysteine *O*-methyltransferase encoding genes (Biswas et al. 2016). In *Salvia miltiorrhiza*, miR5072 was shown to be involved in regulating tanshinones (a type of terpenoid compound) by targeting acetyl-CoA C-acetyl transferase (AACT) (Xu et al. 2014). In *Taxus mairei*, two biosynthetic genes of paclitaxel (diterpenoid), taxane 13α hydroxylase and taxane 2α-*O*-benzoyltransferase, were targeted by miR164 and miR171, respectively (Hao et al. 2012). In cotton variety KV-1, miR7508 was involved in diterpenoid biosynthesis by targeting 2-oxoglutarate-dependent dioxygenase (He et al. 2014b). In *Picrorhiza kurroa*, miR4995 was reported to be involved in picrosides biosynthesis by targeting 3-deoxy-7-phosphoheptulonate synthase (Vashisht et al. 2015). In the pharmaceutically important plant, *Chlorophytum borivilianum*, a number of miRNAs were found to be involved upstream and downstream of terpenoid backbone biosynthesis. Among those investigated miRNAs, the most important included miR172d-3p.2 targeting DXS, miR172c targeting farnesyl diphosphate synthase (FDS) and miR172c-5p targeting 4-hydroxy-3-methylbut-2-enyl diphosphate reductase (Kajal and Singh 2017). Recently, it has been reported that miR5023 is involved in terpenoid biosynthesis by targeting CYP71 in both *Papaver somniferum* L. and *Papaver bracteatum* L. (Davoodi Mastakani et al. 2018). A recent study in the terpenoid-rich plant, *Persicaria minor*, showed six miRNAs that target both terpenoid pathways, MVA and MEP (Samad et al. 2019). Interestingly, two miRNAs, pmi-miR6300 and pmi-miR396a, are targeting the rate-limiting enzyme from both pathways, HMGR and DXS, respectively. Other terpenoid related miRNAs are pmi-miR398f/g targeting DXR, pmi-miR530 targeting MVD, pmi-miR6173 targeting FDS1 and sesquiterpene synthase and a novel miRNA, pmi-nov_13 targeting mevalonate kinase (MVK) (Samad et al. 2019). Both, Table 7.1 and Fig. 7.1a, b show the current status on the knowledge of miRNAs regulating terpenoid biosynthesis pathways in plants.

7.3 miRNA Roles in the Biosynthesis of Flavonoids

Flavonoids are plant secondary metabolites that belong to a group of natural compounds with diverse phenolic structures. The basic structure consists of at least one aromatic ring attached to hydroxyl groups. Currently, around 10,000 flavonoids have been identified which contribute to the colorful pigments in fruits, vegetables, medicinal plants, and herbs (Mathesius 2018). Based on their molecular structures, flavonoids can be divided into different classes: flavones, flavonols, flavanones, flavanonols, flavanols or catechins, anthocyanins, and chalcones (Panche et al. 2016). In plants, flavonoids play crucial roles in plant defense response during biotic and abiotic stresses (Lin et al. 2016). Interestingly, vicenin-2, a well-known flavonoid, acts as a unique sunlight barrier, especially against UV-B radiation (Silva et al. 2014). Other functions of flavonoids include signalling, antioxidant, and antimicrobial properties (Panche et al. 2016).

Flavonoids are synthesized through the phenylpropanoid pathway by the activity of a multienzyme complex (Petrussa et al. 2013). In plants, this pathway is one of the most extensively studied pathways concerning transcriptional regulation of flavonoid biosynthesis (Falcone Ferreyra et al. 2012; Zhang and Schrader 2017). For

Table 7.1 Roles of miRNAs in terpenoid biosynthesis

Plant species	miRNA	Target	Function of target	Approaches	References
A. thaliana	miR156	SPL9	Activates *TPS21* gene in terpenoid biosynthesis	Transgenic	Yu et al. (2014)
P. cablin	miR156	SPL10			
X. strumarium	miR7539	DXS	Catalyzes conversion of 1-deoxy-D-xylulose 5-phosphate into pyruvate and D-glyceraldehyde 3-phosphate	Small RNA sequencing and computational	Fan et al. (2015)
	miR5021	IDI	Catalyzes conversion of isopentenyl pyrophosphate to dimethylallyl pyrophosphate		
	miR1134	HMGR	Catalyzes the conversion of HMG CoA into mevalonate		
	miR5021				
	miR5255	Squalene epoxidase	Catalyzes the oxidation of squalene to 2,3-oxidosqualene		
	miR5491	Beta-amyrin synthase	Catalyzes the conversion of (3S)-2,3-epoxy-2,3-dihydrosqualene into beta-amyrin		
	miR6435	Gemacrene A oxidase	Catalyzes the oxidation of gemacrene A to gemacrene A acid		
	miR6499	Ent-kaurene synthase	Catalyzes bidirectional conversion of ent-copalyl diphosphate to ent-kaurene		
	miR7540	R-linalool synthase	Catalyzes bidirectional conversion of geranyl diphosphate into (3R)-linalool		

(continued)

Table 7.1 (continued)

Plant species	miRNA	Target	Function of target	Approaches	References
Mentha spp.	miR156	DXS	Catalyzes the conversion of 1-deoxy-D-xylulose 5-phosphate into pyruvate and D-glyceraldehyde 3-phosphate	Computational	Singh et al. (2016b)
	miR414	TPS21	Activates TPS21 gene in terpenoid biosynthesis		
	miR5021	GGDS	Catalyzes the synthesis of GGPP from farnesyl diphosphate and isopentenyl diphosphate		
	miR5021	IDI	Catalyzes the conversion of isopentenyl pyrophosphate to dimethylallyl pyrophosphate		
C. roseus	miR5021	GGDS	Catalyzes the synthesis of GGPP from farnesyl diphosphate and isopentenyl diphosphate	Computational	Pani and Mahapatra (2013)
	miR5021	GCPE	Catalyzes the conversion of CDP-ME 2-phosphate and 2-C-methyl-D-erythritol 2,4-cyclodiphosphate (Me-cPP) to hydroxymethylbutenyl 4-diphosphate		
Z. officinale	miR838	CYP71	Involved in menthofuran synthase activity	Computational	Singh et al. (2016a)
P. hexandrum	miR5021	MVD	Catalyzes the conversion of mevalonate diphosphate to isopentenyl diphosphate	Small RNA sequencing and computational	Biswas et al. (2016)
	miR5538	Protein-*S*-isoprenylcysteine *O*-methyltransferase	Catalyzes the post-translational methylation of isoprenylated C-terminal cysteine residue		

(continued)

Table 7.1 (continued)

Plant species	miRNA	Target	Function of target	Approaches	References
S. miltiorrhiza	miR5072	AACT	Catalyzes the conversion of acetyl-CoA to acetoacetyl-CoA	Small RNA and degradome sequencing	Xu et al. (2014)
T. mairei	miR164	Taxane 13α hydroxylase	Involved in taxane biosynthesis	Small RNA and degradome sequencing	Hao et al. (2012)
	miR171	Taxane 2α-*O*-benzoyltransferase	Involved in taxane biosynthesis		
Cotton varieties KV-1	miR7580	2-oxoglutarate-dependent dioxygenase	Involved in oxygenation/hydroxylation reaction in diterpene biosynthesis	Small RNA sequencing and computational	He et al. (2014b)
P. kurroa	miR4995	3-deoxy-7-phosphoheptulonate synthase	Catalyzes the synthesis of 3-deoxy-7-phosphoheptulonate from phosphoenolpyruvate and erythrose 4-phosphate	Computational	Vashisht et al. (2015)
C. borivilianu	miR172d-3p.2	DXS	Catalyzes the conversion of 1-deoxy-D-xylulose 5-phosphate into pyruvate and D-glyceraldehyde 3-phosphate	Small RNA sequencing and computational	Kajal and Singh (2017)
	miR172c	Farnesyl diphosphate synthase	Catalyzes the synthesis of farnesyl diphosphate from isopentenyl diphosphate and dimethylallyl diphosphate		
	miR172c-5p	4-hydroxy-3-methylbut-2-enyl diphosphate reductase	Catalyzes the conversion of 1-hydroxy-2-methyl-2-(E)-butenyl 4-diphosphate (HMBPP) into a mixture of isopentenyl diphosphate (IPP) and dimethylallyl diphosphate (DMAPP)		

(continued)

Table 7.1 (continued)

Plant species	miRNA	Target	Function of target	Approaches	References
P. soniferum L. and *P. bracteatum L.*	miR5023	CYP71	Involved in menthofuran synthase activity	Computational	Davoodi Mastakani et al. (2018)
Taxus sp.	miR8154	WRKY33	Activates taxol biosynthetic genes	Transgenic	Zhang et al. (2015)
	miR5298b	WRKY33	Activates taxol biosynthetic genes		
P. minor	miR6300	HMGR	Catalyzes the conversion of HMG CoA into mevalonate	Small RNA sequencing and computational	Samad et al. (2019)
	miR396a	DXS	Catalyzes the conversion of 1-deoxy-D-xylulose 5-phosphate into pyruvate and D-glyceraldehyde 3-phosphate		
	miR398f/g	DXR	Catalyzes the conversion of DXP to MEP		
	miR530	MVD	Catalyzes the conversion of mevalonate diphosphate to isopentenyl diphosphate		
	miR6173	FDS1	Catalyzes the synthesis of farnesyl diphosphate from isopentenyl diphosphate and dimethylallyl diphosphate		
		Sesquiterpene synthase	Catalyzes the synthesis of sesquiterpene from farnesyl diphosphate		
	Pmi-nov_13	MVK	Catalyzes the phosphorylation of mevalonate to mevalonate 5-phosphate		

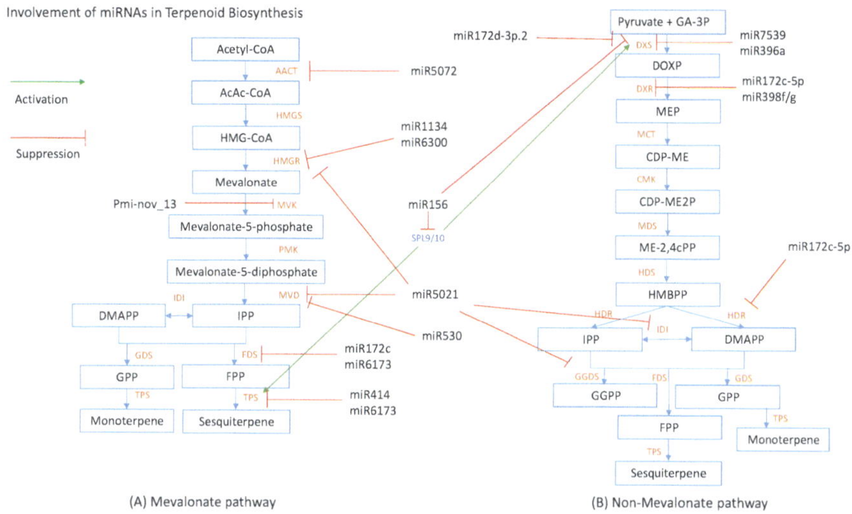

Fig. 7.1 Schematic overview of terpenoid biosynthetic pathway and its possible interaction with miRNAs: Acetoacetyl-CoA thiolase (AACT); Acetoacetyl-CoA (AcAc-CoA); 4-(cytidine 5′-diphospho)-2-*C*-methyl-D-erythritol (CDP-ME); 4-(cytidine 5′-diphospho)-2-*C*-methyl-D-erythritol phosphate (CDP-ME2P); CDP-ME kinase (CMK); Dimethylallyl diphosphate (DMAPP); 1-deoxy-D-xylulose 5-phosphate (DOXP); DOXP reductoisomerase (DXR); DOXP synthase (DXS); Farnesyl diphosphate synthase (FDS); Farnesyl diphosphate (FPP); Glyceraldehyde-3-phosphate (GA-3P); Geranyl diphosphate synthase (GDS); Geranyl geranyl diphosphate synthase (GGDS); Geranyl geranyl diphosphate (GGPP); Geranyl diphosphate (GPP); (*E*)-4-hydroxy-3-methylbut-2-enyl diphosphate reductase (HDR); (*E*)-4-hydroxy-3-methylbut-2-enyl diphosphate synthase (HDS); (*E*)-4-hydroxy-3-methylbut-2-enyl diphosphate (HMBPP); 3-hydroxy-3-methylglutaryl-CoA (HMG-CoA); HMG-CoA reductase (HMGR); HMG-CoA synthase (HMGS); Isopentenyl diphosphate isomerase (IDI); Isopentenyl diphosphate (IPP); 2-*C*-methyl-D-erythritol 4-phosphate cytidylyltransferase (MCT); 2-*C*-methyl-D-erythritol 2,4-cyclodiphosphate synthase (MDS); 2-*C*-methyl-D-erythritol 2,4-cyclodiphosphate (ME-2,4cPP); 2-*C*-methyl-D-erythritol 4-phosphate (MEP); Mevalonate diphosphate decarboxylase (MVD); Mevalonate kinase (MVK); Phosphomevalonate kinase (PMK); Terpene synthase (TPS)

post-transcriptional regulation, some miRNAs were identified to be involved in regulating the phenylpropanoid pathway (Fig. 7.2). In *A. thaliana*, miR156 regulates the flavonoid biosynthesis pathway by targeting SPL9 resulting in a high accumulation of anthocyanin at the junction between stem and rosette leaf (Gou et al. 2011). This particular part showed the highest expression of miR156 which leads to the down regulation of SPL activity. The interaction between miR156-SPL activates anthocyanin biosynthetic genes like the ones encoding anthocyanidin synthase (ANS), flavanone 3-hydroxylase (F3H), dihydroflavonol 4-reductase (DFR) and UDP-glucosyl transferase 75C1 (UGT75C1). Interestingly, at the inflorescence part, where SPL level is high, these anthocyanin biosynthetic genes were repressed, thus preventing the accumulation of anthocyanin. Further, Gou et al. (2011) suggested that the transcription factor SPL9 prevents the expression of anthocyanin biosynthetic genes through destabilization of MYB-bHLH-WD40 transcriptional activation complex. In another

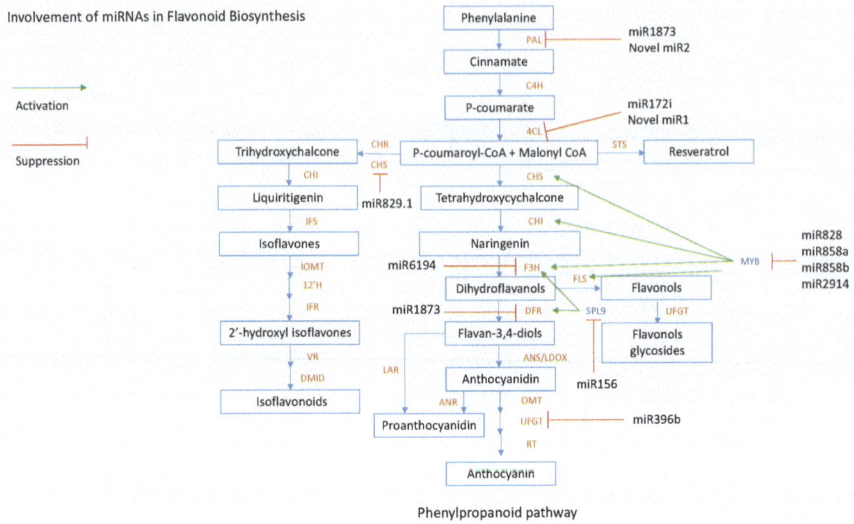

Fig. 7.2 Schematic overview of flavonoid biosynthetic pathway and its possible interaction with miRNAs: Phe ammonia-lyase (PAL); Cinnamate-4-hydroxylase (C4H); 4-coumaroyl:CoA ligase (4CL); chalcone reductase (CHR); Chalcone synthase (CHS); Stilbene synthase (STS); Chalcone isomerase (CHI); Flavanone 3-hydroxylase (F3H); Isoflavone synthase (IFS); Dihydroflavonol 4-reductase (DFR); Isoflavone O-methyltransferase (IOMT); Isoflavone 2′-hydroxylase (I2′H); Isoflavone reductase (IFR); Vestitone reductase (VR); 2′-dihydroxy, 49-methoxyisoflavanol dehydratase (DMID); Leucoanthocyanidin dioxygenase (LDOX); O-methyltransferase (OMT); UDPG-flavonoid glucosyl transferase (UFGT); Rhamnosyl transferase (RT); Flavonol synthase (FLS); Leucoanthocyanidin reductase (LAR); Anthocyanidin reductase (ANR); Anthocyanidin synthase (ANS)

study, carried out in the endangered medicinal plant species *Podophyllum hexandrum*, numerous miRNAs were predicted to be involved in the phenylpropanoid pathway via computational approaches. Those predicted miRNAs included miR172i, miR829.1, miR1438, miR1873 and miR5532 which seem to target mRNAs coding for biosynthetic enzymes such as 4-coumarate–CoA ligase, chalcone synthase, caffeoyl-CoA *O*-methyl transferase, dihydroflavonol 4-reductase C and 2-hydroxyisoflavanone dehydratase, respectively (Biswas et al. 2016). In long-term *Taxus* subculture, overexpression of miR8154 and miR5298b enhances the biosynthesis of taxol, phenylpropanoids and flavonoids, suggesting a potential role as key factors in the regulation of the entire metabolic pathway (Zhang et al. 2015). However, both miR8154 and miR5298b were not found to target any genes related to SMs biosynthetic pathway (Zhang et al. 2015). In oriental persimmon (*Diospyros kaki* Thunb.; CPCNA), miR395p-3p and miR858b were discovered to regulate proanthocyanidin biosynthesis by targeting bHLH and MYB transcription factors, respectively (Luo et al. 2015).

In *Rauvolfia serpentine*, both miR396b and miR828a were computationally identified to target kaempferol 3-*O*-beta-D-galactosyltransferase and anthocyanin regulatory C1 protein encoding genes, respectively (Prakash et al. 2016). In a study with *A. thaliana* transgenic plants over-expressing miR858a (miR858OX) and artificial miR858 target mimic (MIM858), miR858a was shown to target flavonol-specific MYBs which were involved in regulating flavonoid biosynthetic pathway at upstream level (Sharma et al. 2016). It was further deduced that the interaction between miR858a-*MYB* was an additional level of regulation that might coordinate biosynthesis of flavonoids and lignin in plants (Sharma et al. 2016). Similarly, miR2914–MYB was reported to regulate the phenylpropanoid pathway in *Papaver somniferum* (opium) and *Papaver bracteatum* (Persian poppies); whereas miR828-MYB had a similar role in tomato (Cardoso et al. 2018; Davoodi Mastakani et al. 2018). In *Halostachys caspica*, miR6194 was involved in regulating the flavonoid biosynthesis by targeting F3H (Yang et al. 2015). In pear, miR1061 targets naringenin 3-dioxygenase encoding gene which is involved in the flavonoid pathway (Wu et al. 2014). In a recent study, an interaction between rice and brown planthopper (BPH) has revealed that overexpression of one BPH-responsive target gene of OsmiR396, *growth-regulating factor 8* (*OsGRF8*), displayed resistance to BPH. In addition, the abundance of flavonoid increased in both the OsmiR396-sequestered and the OsGRF8 over-expressing plants (Dai et al. 2019). In *Cucumis sativus* (cucumber) infected with *Corynespora cassiicola*, two novel miRNAs, Novel-miR1 and Novel-miR2, were discovered to regulate the phenylpropanoid pathway by targeting 4-coumarate-CoA ligase and phenylalanine ammonia-lyase encoding genes, respectively (Wang et al. 2018). The regulation by miRNAs of the flavonoid biosynthesis has been summarized in Table 7.2.

7.4 miRNA Roles in the Biosyhthesis of Alkaloids and Nitrogen-Containing Secondary Compounds

In general, alkaloids can be defined as nitrogen-containing compounds with low molecular weight synthetized from amino acids and exist in a biologically active state (Alves de Almeida et al. 2017). The compounds of this group are synthesized from diverse metabolic pathways which operate in unicellular and multicellular organisms ranging from bacteria, fungi, plants and animals (Cushnie et al. 2014). The latest update showed more than 20,000 alkaloid compounds have been identified from various plant species (Alves de Almeida et al. 2017). The basic structure of alkaloids comprises nitrogen as functional group, which does not include nitrogen in amide or peptide bond. There are three types of alkaloids, which have been recognized as true alkaloids, protoalkaloids and pseudoalkaloids. Molecular structures are the main factors that differentiate these compounds. True alkaloids consist of heterocyclic ring with nitrogen; whereas, in protoalkaloid, the nitrogen atom derived from amino acids is not part of the heterocyclic ring. The pseudoalkaloids are a group not originated

Table 7.2 Roles of miRNAs in flavonoid biosynthesis

Plant species	miRNA	Target	Function of target	Approaches	References
A. thaliana	miR156	SPL9	Destabilizes MYB-bHLH-WD40 complex	Transgenic	Guo et al. (2011)
P. hexandrum	miR172i	4-coumarate CoA ligase	Catalyzes the activation of 4-coumarate and 4-hydroxycinnamates to their thiol esters	Small RNA sequencing and computational	Biswas et al. (2016)
	miR829.1	Chalcone synthase	Catalyzes the synthesis of naringenin chalcone from 4-coumaroyl-CoA and malonyl-CoA		
	miR1438	Caffeoyl-CoA *O*-methyl transferase	Catalyzes the methylation of caffeoyl-CoA		
	miR1873	Dihydroflavonol 4-reductase C	Involved in flavanone 4-reductase activity		
	miR5532	2-hydroxyisoflavanone dehydratase	Catalyzes the conversion of 2,7,4′-trihydroxyisoflavanone into diadzein		
Z. officinale	miR1873	Phenylalanine ammonia-lyase	Synthesizes trans-cinnamate from L-phenylalanine	Computational	Singh et al. (2016a)
Taxus sp.	miR8154	WRKY33	Activates flavonoid biosynthetic gene	Transgenic	Zhang et al. (2015)
	miR5298b	WRKY33	Activates flavonoid biosynthetic gene		

(continued)

Table 7.2 (continued)

Plant species	miRNA	Target	Function of target	Approaches	References
D. kaki	miR395p-3p	bHLH	Regulates proanthocyanidin biosynthetic gene	Small RNA sequencing and computational	Luo et al. (2015)
	miR858b	MYB	Regulates proanthocyanidin biosynthetic gene		
R. serpentine	miR396b	Kaempferol 3-*O*-beta-D-galactosyltransferase	Involved in transferase activity	Computational	Prakash et al. (2016)
	828a	Anthocyanin regulatory C1 protein	Binding of DNA/chromatin		
A. thaliana	miR858a	MYB	Regulates genes in flavonoid biosynthetic pathway	Transgenic	Sharma et al. (2016)
S. lycopersicum	miR828	MYB	Regulates genes in flavonoid biosynthetic pathway	Small RNA sequencing and computational	Cardoso et al. (2018)
S. pennellii	miR828	MYB	Regulates genes in flavonoid biosynthetic pathway	Small RNA sequencing and computational	Cardoso et al. (2018)
P. somniferum L. and *P. bracteatum L.*	miR2914	MYB	Regulates genes in flavonoid biosynthetic pathway	Small RNA sequencing and computational	Davoodi Mastakani et al. (2018)
H. caspica	miR6194	F3H	Catalyzes the conversion of flavanone into dihydroflavonol	Small RNA sequencing and computational	Yang et al. (2015)

(continued)

Table 7.2 (continued)

Plant species	miRNA	Target	Function of target	Approaches	References
Pyrus spp.	miR1061-3p	Naringenin 3-dioxygenase	Catalyzes the 3-beta-hydoxylation of 2S-flavanones to 2R, 3R-dihydroflavonols	Small RNA sequencing and computational	Wu et al. (2014)
O. sativa	miR396	GRF8	Involved in flavonoid biosynthesis	Transgenic	Dai et al. (2019)
C. sativus	Novel-miR1	4-coumarate CoA ligase	Synthesizes 3,4′,5-trihydroxystilbene from trans-4-coumarate	Small RNA sequencing and computational	Wang et al. (2018)
	Novel-miR2	Phenylalanine ammonia-lyase	Synthesizes trans-cinnamate from L-phenylalanine		

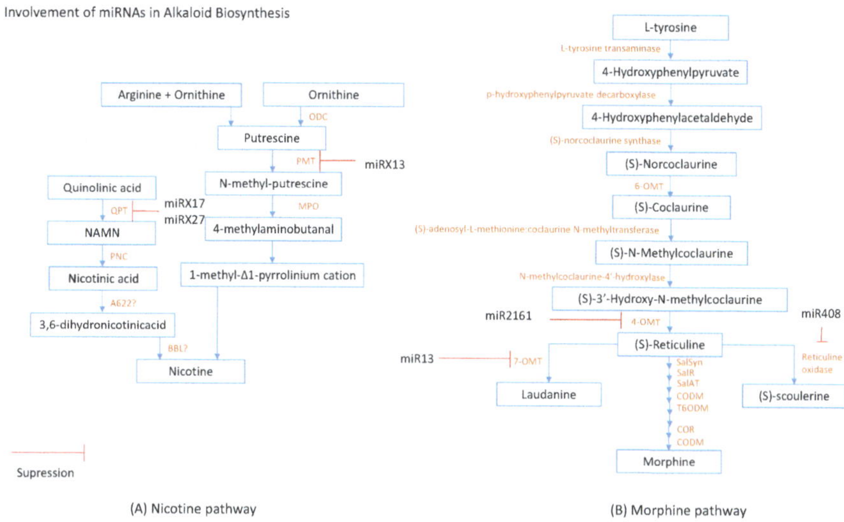

Fig. 7.3 Schematic overview of alkaloid biosynthetic pathway and its possible interaction with miRNAs: **a** nicotine pathway: ornithine decarboxylase (ODC); Putrescine N-methyltransferase (PMT); Quinolinic acid phosphoribosyltransferase (QPT); Pyridine nucleotide cycle (PNC); Nicotinic acid mononucleotide (NAMN); Isoflavone reductase-like protein (A622); Berberine bridge enzyme-like (BBL); N-methylputrescine oxidase (MPO). **b** Morphine pathway: (*S*)-adenosyl-L-methionine:norcoclaurine 6-*O*-methyltranferase (6-OMT); 30-hydroxy-*N*-methylcoclaurine 4-*O*-methyltransferase (4-OMT); Norreticuline-7-*O*-methyltransferase (7-OMT); Salutaridine synthase (SalSyn); Salutaridine reductase (SalR); Salutaridine 7-*O*-acetyltransferase (SalAT); Thebaine 6-*O*-demethylase (T6ODM); Codeinone reductase (COR); Codeine *O*-demethylase (CODM)

from amino acids, such as terpene-like, purine-like and steroid-like alkaloids (Alves de Almeida et al. 2017). These compounds have been widely used as stimulants, narcotics, pharmaceuticals, and poisons. Recently, post-transcriptional studies were carried out, and miRNAs involved in the alkaloid biosynthetic pathway have been identified (Fig. 7.3 and Table 7.3).

In silico and experimental characterization of miRNAs involved in alkaloid biosynthetic pathways have been reported in several plants. In *C. roseus,* miR5021 was predicted to target UDP-glucose iridoid glucosyltransferase which was involved in indole alkaloid pathway (Pani and Mahapatra 2013). In tobacco, nta-miRX17, nta-miRX27, nta-miRX20 and nta-miRX13 target quinolinic acid phosphoribosyltransferase 1 (*QPT1*), quinolinic acid phosphoribosyltransferase 2 (*QPT2*), *CYP82E4*, and putrescine N-methyltransferase 2 (*PMT2*) genes, respectively, which are involved in nicotine biosynthesis (Li et al. 2015a). In addition, Li and colleagues have demonstrated that inhibition of nta-eTMX27expression by nta-miRX27 led to an increase in the abundance of nicotine biosynthesis. In opium, three miRNAs (pso-miR13, pso-miR408, and pso-miR2161) were recognized as potential miRNAs involved in the alkaloid biosynthetic pathway by targeting mRNA of 7-*O*-methyltransferase (7-OMT), reticuline oxidase-like and *S*-adenosyl-L-methionine:

Table 7.3 Roles of miRNAs in alkaloid and nitrogen-containing secondary compound biosynthesis

Plant species	miRNA	Target	Function of target	Approaches	References
C. roseus	miR5021	UDP-glucose iridoid glucosyltransferase	Involved in transferase activity	Computational	Pani and Mahapatra (2013)
N. tabacum	miRX17	QPT1	Catalyzes the conversion of quinolinic acid into NAMN	Small RNA sequencing, computational, and transgenic	Li et al. (2015a, b)
	miRX27	QPT2	Catalyzes the conversion of quinolinic acid into NAMN		
	miRX20	CYP82E4	Catalyzes the conversion of nicotine into nornicotine		
	miRX13	PMT2	Catalyzes the conversion of putrescine into *N*-methylputrescine		
P. sonifereum L.	miR13	7-OMT	Catalyzes the conversion of *S*-reticuline into morphinan alkaloids	Small RNA sequencing and computational	Boke et al. (2015)
	miR408b	Reticuline oxidase-like protein	Catalyzes the conversion of *S*-reticuline into (*S*)-scoulerine		
	miR2161	4-OMT	Catalyzes the conversion of *S*-norcoclaurine into *S*-reticuline		
A. thaliana	miR156	SPL9	Involved in glucosinolate regulation	Transgenic	Mao et al. (2017)
A. thaliana	miR826	AOP2	Involved inside chain modification of Met-derived glucosinolate	Transgenic	He et al. (2014)
	miR5090	AOP2	Involved inside chain modification of Met-derived glucosinolate		

3'-hydroxy-*N*-methylcoclaurine 4'-*O*-methyltransferase 2 (4-OMT) (Boke et al. 2015). Similarly to the terpenoid pathway, the interaction between miRNAs and transcription factor transcripts could also affect the change in abundance of nitrogen-containing compounds. In *A. thaliana*, miR156 seems to regulate glucosinolate biosynthesis by targeting SPL9 (Mao et al. 2017). Glucosinolate is a nitrogen-containing compound involved in plant defense against pathogens and herbivores. The transcription factor SPL9 binds with JAZ3, a JA ZIM-domain (JAZ) protein, and regulates the jasmonate biosynthesis. High level of jasmonate has resulted in enhanced production of glucosinolate (Mao et al. 2017). Nitrogen starvation in *A. thaliana* was shown to induce the expression of miR826 and miR5090 which resulted in the downregulation of *ALKENYL HYDROXYALKYL PRODUCING2* (*AOP2*) gene expression and subsequently led to the decrease in glucosinolate accumulation (He et al. 2014).

7.5 Potential of miRNAs as Genetic Tool to Manipulate Secondary Metabolite (SM) Biosynthesis

Plants synthesize primary metabolites which include carbohydrates, proteins, lipids and nucleic acids which are essential for plant's normal growth and development. On the other hand, secondary metabolites are numerous biomolecules and once considered as waste by products with no significant role in plant growth and development (Hartmann 2007). However, in the past few decades, extensive research has been carried out in investigating the role of SMs which made it clear that these compounds are not waste by-products, rather they have important roles in plant growth and development and defense against various stresses (Hartmann 2007). In addition, plant SMs also have paramount importance in pharmacological and food industries. These compounds are used for treatment of several animal diseases from skin disorders like itchy to chronic disease like cancer. Due to these reasons, plant biotechnologists have devised several ways to improve the production of the plant SMs. These approaches include optimization of plant culture conditions, generation of hairy roots, precursor feeding, elicitation by biotic and abiotic factors, and introducing new biosynthetic genes (Hussain et al. 2012).

Recently, miRNAs have emerged as a new tool in engineering plant species to produce desirable products. This approach can be utilized to generate superior plants with high yields and increased tolerance against various stresses (Bulgakov and Avramenko 2015). Here, we reviewed the current knowledge on the regulation of plant SM biosynthesis by miRNAs showing that miRNAs can be used as a genetic tool to enhance the SM production in plants through two main approaches. In the first approach, miRNAs can be used to directly target a specific gene of a certain metabolite biosynthetic pathway. In a second approach, the miRNA can be manipulated to indirectly regulate the SM pathways by targeting other regulatory molecules such as transcription factor genes. The first approach should be more efficient since the effect is immediate and the validation procedure is comparatively simpler when compared with the second indirect approach which is more challenging unless the role of the transcription factor is well known in regulating the SM pathway. For instance, in the previous section (miRNA Roles in the Biosynthesis of Terpenoids), we have discussed that miR5072 can directly target AACT in terpenoid biosynthesis in *S. miltiorrhiza* (Xu et al. 2014), while miR156 regulates a transcription factor, SPL9, in flavonoid biosynthesis in *A. thaliana* (miRNA Roles in the Biosynthesis of Flavonoids) (Gou et al. 2011).

In addition, the behavior of the miRNA should also be carefully considered before any gene manipulating approaches are deployed. Since numerous miRNAs are negative regulators of the SM pathways, this means that higher expression of miRNAs may lead to decreased metabolite content. In this situation, miRNA knock-down approaches may be used to increase the desirable product such as target mimicry and CRISPR-cas9 technologies (Chang et al. 2016; Franco-Zorrilla et al. 2007). On the other hand, overexpression of miRNA is required if miRNA acts as a positive regulator, where the high expression of the miRNA leads to the higher production of plant secondary metabolites. If specific silencing of the target gene is desired, artificial miRNA (amiRNA) might be another option to be considered (Schwab et al. 2006). A previous study in *A. thaliana* showed the ath-miR159a had successfully reduced 15% of its target expression, chalcone synthase, with a corresponding decrease in the anthocyanin content (Niemeier et al. 2010).

7.6 Conclusion

The current knowledge on the roles of miRNAs in regulating the SMs biosynthesis pathways is at an early stage as compared with the known functions of miRNAs during specific stresses and plant developmental processes. One of the reasons for this is probably due to the neglected importance of these small regulator molecules in the regulation of such biosynthetic pathways at the post-transcriptional level. The miRNAs may open a new era in understanding the regulation of SMs biosynthesis in plants. However, miRNA target identification is still a major challenge regardless of the development of numerous target prediction tools. In accordance with Tables 7.1, 7.2, and 7.3, most of the miRNAs studies were based on in silico approaches. Thus, further validations are required prior to any genetic manipulation attempt which is carried out. miRNA target validation involves the detection of miRNA cleavage sites to ensure the mRNA is the true target for a particular miRNA. This type of validation can be achieved through RLM-RACE, poly (A) polymerase-mediated (PPM)-RACE, qRT-PCR, and degradome sequencing (Wang and Fang 2015).

Identification of miRNAs and their targets at all possible steps in a particular SM pathway may be the key to generate a biofactory, where plants produce a desirable product. However, the detailed characterization of the specific miRNA regulatory nodes needs to be performed, so the researchers can design adequate strategies to generate the biofactory. The regulatory roles of miRNAs in the regulation of different classes of SMs have been investigated in different plants. Among the most studied SMs, terpenoids have a well-documented role in defense against a variety of biotic and abiotic stresses. Various miRNAs involved in the terpenoids biosynthesis have been studied, and the involvement was found at different levels of regulation, i.e., successfully targeted transcripts encoding transcription factors, terpenoids synthases, and some rate-limiting enzymes in the terpenoids biosynthetic pathways. Similarly, roles of miRNAs in flavonoids, alkaloids, and nitrogen-containing secondary compounds biosynthesis have been discovered, but the understanding of miRNA involvement in the regulation of these SMs is still at the preliminary stages. The main hindrance in the study pace of miRNAs is the prerequisite of a comprehensive understanding of the miRNA and target gene interaction. Further discoveries in this area may pave a way to create novel biomolecules by manipulating SM biosynthetic pathways.

References

Alves de Almeida AC, de-Faria FM, Dunder RJ, Manzo LPB, Souza-Brito ARM, Luiz-Ferreira A (2017) Recent trends in pharmacological activity of alkaloids in animal colitis: potential use for inflammatory bowel disease evidence-based complementary and alternative medicine. eCAM 2017:8528210. https://doi.org/10.1155/2017/8528210

Biswas S, Hazra S, Chattopadhyay S (2016) Identification of conserved miRNAs and their putative target genes in Podophyllum hexandrum (Himalayan Mayapple) Plant Gene 6:82–89. http://dx.doi.org/10.1016/j.plgene.2016.04.002

Boke H, Ozhuner E, Turktas M, Parmaksiz I, Ozcan S, Unver T (2015) Regulation of the alkaloid biosynthesis by miRNA in opium poppy. Plant Biotechnol J 13:409–420. https://doi.org/10.1111/pbi.12346

Bulgakov VP, Avramenko TV (2015) New opportunities for the regulation of secondary metabolism in plants: focus on microRNAs. Biotech Lett 37:1719–1727. https://doi.org/10.1007/s10529-015-1863-8

Byers KJ, Bradshaw Jr. HD, Riffell JA (2014) Three floral volatiles contribute to differential pollinator attraction in monkeyflowers (Mimulus). J Experim Biol 217:614–623. https://doi.org/10.1242/jeb.092213

Cardoso TCS et al (2018) New insights into tomato microRNAs. Sci Reports 8:16069. https://doi.org/10.1038/s41598-018-34202-3

Chang H, Yi B, Ma R, Zhang X, Zhao H, Xi Y (2016) CRISPR/cas9, a novel genomic tool to knock down microRNA in vitro and in vivo Sci Rep 6:22312–22312. https://doi.org/10.1038/srep22312

Cushnie TP, Cushnie B, Lamb AJ (2014) Alkaloids: an overview of their antibacterial, antibiotic-enhancing and antivirulence activities. Int J Antimicrob Agents 44:377–386. https://doi.org/10.1016/j.ijantimicag.2014.06.001

Dai Z et al. (2019) The OsmiR396-OsGRF8-OsF3H-flavonoid pathway mediates resistance to the brown planthopper in rice (*Oryza sativa*). Plant Biotech J. https://doi.org/10.1111/pbi.13091

Davoodi Mastakani F, Pagheh G, Rashidi Monfared S (2018) Identification and expression analysis of a microRNA cluster derived from pre-ribosomal RNA in *Papaver somniferum* L. and *Papaver bracteatum* L. Plos One 13:e0199673. https://doi.org/10.1371/journal.pone.0199673

Dubey VS, Bhalla R, Luthra R (2003) An overview of the non-mevalonate pathway for terpenoid biosynthesis in plants. J Biosci 28:637–646

Falcone Ferreyra ML, Rius SP, Casati P (2012) Flavonoids: biosynthesis, biological functions, and biotechnological applications Front Plant Sci 3:222. https://doi.org/10.3389/fpls.2012.00222

Fan R, Li Y, Li C, Zhang Y (2015) Differential microRNA analysis of glandular trichomes and young leaves in *Xanthium strumarium* L. reveals their putative roles in regulating terpenoid biosynthesis. PLoS One 10:e0139002. https://doi.org/10.1371/journal.pone.0139002

Franco-Zorrilla JM et al (2007) Target mimicry provides a new mechanism for regulation of microRNA activity. Nat Genet 39:1033–1037. https://doi.org/10.1038/ng2079

Gou JY, Felippes FF, Liu CJ, Weigel D, Wang JW (2011) Negative regulation of anthocyanin biosynthesis in Arabidopsis by a miR156-targeted SPL transcription factor. Plant Cell 23:1512–1522 https://doi.org/10.1105/tpc.111.084525

Hannon GJ (2002) RNA interference Nature 418:244–251. https://doi.org/10.1038/418244a

Hao DC, Yang L, Xiao PG, Liu M (2012) Identification of Taxus microRNAs and their targets with high-throughput sequencing and degradome analysis. Physiol Plant 146:388–403. https://doi.org/10.1111/j.1399-3054.2012.01668.x

Hartmann T (2007) From waste products to ecochemicals: fifty years research of plant secondary metabolism. Phytochemistry 68:2831–2846. https://doi.org/10.1016/j.phytochem.2007.09.017

He H, Liang G, Li Y, Wang F, Yu D (2014) Two young MicroRNAs originating from target duplication mediate nitrogen starvation adaptation via regulation of glucosinolate synthesis in *Arabidopsis thaliana*. Plant Physiol 164:853–865. https://doi.org/10.1104/pp.113.228635

He X et al (2014b) Identification of novel microRNAs in the Verticillium wilt resistant upland cotton variety KV-1 by high-throughput sequencing. Springer Plus 3:564–564. https://doi.org/10.1186/2193-1801-3-564

Huang M, Sanchez-Moreiras A, Abel C, Sohrabi R, Lee S, Gershenzon J, Tholl D (2012) The major volatile organic compound emitted from *Arabidopsis thaliana* flowers, the sesquiterpene (E)-beta-caryophyllene, is a defense against a bacterial pathogen. New Phytol 193:997–1008

Hussain MS, Fareed S, Ansari S, Rahman MA, Ahmad IZ, Saeed M (2012) Current approaches toward production of secondary plant metabolites. J Pharm Bioallied Sci 4:10–20. https://doi.org/10.4103/0975-7406.92725

Kajal M, Singh K (2017) Small RNA profiling for identification of miRNAs involved in regulation of saponins biosynthesis in *Chlorophytum borivilianum*. BMC Plant Biol 17:265. https://doi.org/10.1186/s12870-017-1214-0

Kessler A, Kalske A (2018) Plant secondary metabolite diversity and species interactions. Ann Rev Ecol Evol System 49:115–138. https://doi.org/10.1146/annurev-ecolsys-110617-062406

Li F, Wang W, Zhao N, Xiao B (2015a) Regulation of nicotine biosynthesis by an endogenous target mimicry of microRNA in tobacco. Plant Physiol 169:1062–1071. https://doi.org/10.1104/pp.15.00649

Li R, Tee C-S, Jiang Y-L, Jiang X-Y, Venkatesh PN, Sarojam R, Ye J (2015b) A terpenoid phytoalexin plays a role in basal defense of *Nicotiana benthamiana* against Potato virus X. Sci Rep 5:9682. https://doi.org/10.1038/srep09682. https://www.nature.com/articles/srep09682#supplementary-information

Lin D et al (2016) An overview of plant phenolic compounds and their importance in human nutrition and management of Type 2 diabetes. Molecules (Basel, Switzerland) 21. https://doi.org/10.3390/molecules21101374

Luo Y, Zhang X, Luo Z, Zhang Q, Liu J (2015) Identification and characterization of microRNAs from Chinese pollination constant non-astringent persimmon using high-throughput sequencing. BMC Plant Biol 15:11. https://doi.org/10.1186/s12870-014-0400-6

Mao YB et al (2017) Jasmonate response decay and defense metabolite accumulation contributes to age-regulated dynamics of plant insect resistance. Nat Commun 8:13925. https://doi.org/10.1038/ncomms13925

Mathesius U (2018) Flavonoid functions in plants and their interactions with other organisms. Plants (Basel, Switzerland) 7:30. https://doi.org/10.3390/plants7020030

Nagegowda DA (2010) Plant volatile terpenoid metabolism: biosynthetic genes, transcriptional regulation and subcellular compartmentation. FEBS Lett. 584:2965–2973. http://dx.doi.org/10.1016/j.febslet.2010.05.045

Niemeier S, Alves Junior L, Merkle T (2010) Improvement of the design and generation of highly specific plant knockdown lines using primary synthetic microRNAs (pri-smiRNAs). BMC Res Notes 3:59. https://doi.org/10.1186/1756-0500-3-59

Panche AN, Diwan AD, Chandra SR (2016) Flavonoids: an overview. J Nutr Sci 5:e47. https://doi.org/10.1017/jns.2016.41

Pani A, Mahapatra RK (2013) Computational identification of microRNAs and their targets in *Catharanthus roseus* expressed sequence tags. Genom Data 1:2–6. https://doi.org/10.1016/j.gdata.2013.06.001

Petrussa E, Braidot E, Zancani M, Peresson C, Bertolini A, Patui S, Vianello A (2013) Plant flavonoids–biosynthesis, transport and involvement in stress responses. Int J Mol Sci 14:14950–14973. https://doi.org/10.3390/ijms140714950

Prakash P, Rajakani R, Gupta V (2016) Transcriptome-wide identification of *Rauvolfia serpentina* microRNAs and prediction of their potential targets. Comput Biol Chem 61:62–74. https://doi.org/10.1016/j.compbiolchem.2015.12.002

Samad AFA, Rahnamaie-Tajadod R, Sajad M, Jani J, Murad AMA, Noor NM, Ismail I (2019) Regulation of terpenoid biosynthesis by miRNA in Persicaria minor induced by Fusarium oxysporum. BMC Genomics 20:586. https://doi.org/10.1186/s12864-019-5954-0

Samad AFA, Sajad M, Nazaruddin N, Fauzi IA, Murad AMA, Zainal Z, Ismail I (2017) MicroRNA and transcription factor: key players in plant regulatory network. Front Plant Sci 8. https://doi.org/10.3389/fpls.2017.00565

Schwab R, Ossowski S, Riester M, Warthmann N, Weigel D (2006) Highly specific gene silencing by artificial microRNAs in Arabidopsis. Plant Cell 18:1121–1133. https://doi.org/10.1105/tpc.105.039834

Sharma D, Tiwari M, Pandey A, Bhatia C, Sharma A, Trivedi PK (2016) MicroRNA858 is a potential regulator of phenylpropanoid pathway and plant development in Arabidopsis Plant. Physiology. https://doi.org/10.1104/pp.15.01831

Silva DB, Turatti ICC, Gouveia DR, Ernst M, Teixeira SP, Lopes NP (2014) Mass spectrometry of flavonoid vicenin-2, based sunlight barriers in *Lychnophora* species. Sci Rep 4:4309–4309. https://doi.org/10.1038/srep04309

Singh N, Srivastava S, Sharma A (2016a) Identification and analysis of miRNAs and their targets in ginger using bioinformatics approach. Gene 575:570–576. https://doi.org/10.1016/j.gene.2015.09.036

Singh N, Srivastava S, Shasany AK, Sharma A (2016b) Identification of miRNAs and their targets involved in the secondary metabolic pathways of *Mentha* spp. Comput Biol Chem 64:154–162. https://doi.org/10.1016/j.compbiolchem.2016.06.004

Tholl D (2015) Biosynthesis and biological functions of terpenoids in plants. Adv Biochem Eng Biotechnol 148:63–106. https://doi.org/10.1007/10_2014_295

Tuteja JH, Zabala G, Varala K, Hudson M, Vodkin LO (2009) Endogenous, tissue-specific short interfering RNAs silence the chalcone synthase gene family in glycine max seed coats. Plant Cell 21:3063–3077. https://doi.org/10.1105/tpc.109.069856

Vashisht I, Mishra P, Pal T, Chanumolu S, Singh TR, Chauhan RS (2015) Mining NGS transcriptomes for miRNAs and dissecting their role in regulating growth, development, and secondary metabolites production in different organs of a medicinal herb, *Picrorhiza kurroa*. Planta 241:1255–1268. https://doi.org/10.1007/s00425-015-2255-y

Wang C, Fang J (2015) RLM-RACE, PPM-RACE, and qRT-PCR: an integrated strategy to accurately validate miRNA target genes. Methods Mol Biol 1296:175–186. https://doi.org/10.1007/978-1-4939-2547-6_16

Wang X, Zhang D, Cui N, Yu Y, Yu G, Fan H (2018) Transcriptome and miRNA analyses of the response to *Corynespora cassiicola* in cucumber. Sci Rep 8:7798. https://doi.org/10.1038/s41598-018-26080-6

Wu J, Wang D, Liu Y, Wang L, Qiao X, Zhang S (2014) Identification of miRNAs involved in pear fruit development and quality. BMC Genom 15:953. https://doi.org/10.1186/1471-2164-15-953

Xu X et al (2014) Deep sequencing identifies tissue-specific micrornas and their target genes involving in the biosynthesis of tanshinones in *Salvia miltiorrhiza*. PLoS ONE 9:e111679. https://doi.org/10.1371/journal.pone.0111679

Yang R, Zeng Y, Yi X, Zhao L, Zhang Y (2015) Small RNA deep sequencing reveals the important role of microRNAs in the halophyte *Halostachys caspica*. Plant Biotechnol J 13:395–408. https://doi.org/10.1111/pbi.12337

Yu ZX, Wang LJ, Zhao B, Shan CM, Zhang YH, Chen DF, Chen XY (2014) Progressive regulation of sesquiterpene biosynthesis in Arabidopsis and Patchouli (*Pogostemon cablin*) by the miR156-targeted spl transcription factors. Mol Plant. https://doi.org/10.1093/mp/ssu127

Zhang B, Schrader A (2017) TRANSPARENT TESTA GLABRA 1-Dependent regulation of flavonoid biosynthesis. Plants 6. https://doi.org/10.3390/plants6040065

Zhang M, Dong Y, Nie L, Lu M, Fu C, Yu L (2015) High-throughput sequencing reveals miRNA effects on the primary and secondary production properties in long-term subcultured Taxus cells. Front Plant Sci 6:604. https://doi.org/10.3389/fpls.2015.00604

Chapter 8
microRNA-Mediated Regulation of Plant Vascular Development and Secondary Growth

Ana Milhinhos, Susana Lopes and Célia Miguel

Abstract Secondary or lateral growth, is the biological process that confers girth to stems and roots in plants, which is essential to structurally sustain organs, for water and nutrients transport and continuous plant growth. This process occurs at the expense of the cell division activity in two main lateral meristems—the vascular cambium and the cork cambium. From the activity of cambia, one of the main sources of biomass on Earth, wood, is produced. Given its economic, environmental and societal relevance a great deal of attention has been given to finding regulators of vascular cambium activity and wood formation. Some of the regulatory networks found are under post-transcriptional regulation of gene expression by microRNAs during vasculature formation, cambium activity, and wood formation and in several aspects of vascular development. In this chapter, we will briefly review the current knowledge on microRNAs roles during plant vascular development with a focus on recent work on miRNAs activity in secondary growth in plants.

Keywords miRNA · Vascular development · Secondary growth · Stem cells · Cambium · Wood · Cork · Xylem · Trees · Arabidopsis · Populus · miR165/166

8.1 Plant Vascular Development: From Primary to Secondary Growth

Plant organs differentiate from niches of stem cells—the meristems, i.e. populations of undifferentiated cells with dual properties: on the one hand, these cells retain the ability to maintain their pluripotent status to renew themselves 'eternally' while

A. Milhinhos (✉) · S. Lopes
Instituto de Tecnologia Química e Biológica, Universidade Nova de Lisboa, Oeiras, Portugal
e-mail: milhinho@itqb.unl.pt

A. Milhinhos · S. Lopes · C. Miguel
BioISI—Biosystems & Integrative Sciences Institute, Faculdade de Ciências, Universidade de Lisboa, Lisbon, Portugal

S. Lopes · C. Miguel
iBET, Instituto de Biologia Experimental e Tecnológica, Oeiras, Portugal

C. Miguel et al. (eds.), *Plant microRNAs*, Concepts and Strategies in Plant Sciences, https://doi.org/10.1007/978-3-030-35772-6_8

on the other, they also maintain the differentiation potential to become specialised cell types (Weigel and Jurgens 2002). These organizing centres in the plant body are present since plant embryogenesis, within the shoot apical meristem (SAM) and the root apical meristem (RAM) but organise also during the post-embryonic life of higher plants, for instance, in the form of vascular and cork cambia. In Arabidopsis (*Arabidopsis thaliana*), the first vascular precursor cells are formed at the early globular stage during embryogenesis (Yoshida et al. 2014) and, through several rounds of periclinal cell divisions, a diarch patterned vascular bundle with a central xylem axis and two phloem poles at the RAM is formed (Scheres et al. 1994). The shoot vascular tissues derive from the SAM. The procambium, which contains the vascular stem cells in the SAM, develops into the vascular bundles in the shoot. In Arabidopsis, the vascular bundles of the shoot have a collateral pattern, where xylem organizes towards inside and phloem towards the outside of the bundle. Later on, in many herbaceous and especially in woody stems, the differentiation of interfascicular cambium from parenchymatic ground tissue in between the vascular bundles unites to procambial tissue of the vascular bundles and forms a ring-like meristem, the vascular cambium. Through periclinal cell divisions of the vascular cambium, conductive secondary xylem and secondary phloem tissues arise and enlarge the stems in a process known as secondary growth, producing wood. In addition, secondary growth derives from another concentric cylindrical meristem, the phellogen, or cork cambium, which undergoes periclinal cell divisions to produce an outer protective tissue, phellem (or cork), covering the surface of the stems, and another inwardly, the phelloderm. Secondary growth also occurs in root tissues. In the Arabidopsis root, the vascular and cork cambia derive from the procambium and xylem pole pericycle cells (Smetana et al. 2019). Only recently have cell-lineage-tracing experiments demonstrated that the cells adjacent to the xylem are the only cells able to generate cambial stem cells (Smetana et al. 2019).

Several challenges pose to the study of vascular and cork meristems in woody stems, mainly due to the seclusion of these tissues in inner layers of the plant body, making them inaccessible for phenotyping unless the plant is damaged. Also, implementation of genetic studies in perennials is much more challenging as many years are often required just for flowering to occur; there are few characterized mutant lines; mutagenesis screens are also difficult to attain and as a result, most of the molecular mechanisms and regulatory networks so far described during plant vascular development are based on studies of the Arabidopsis simple and well mapped vascular architecture in young roots, leaves and shoots. At later stages of Arabidopsis life cycle, the large expansion of secondary xylem and secondary phloem tissues observed in the hypocotyl is turning this embryonic organ into an important model system to study secondary or radial growth (Ragni et al. 2011; Liebsch et al. 2014; Felipo-Benavent et al. 2018; Wunderling et al. 2018; Smetana et al. 2019; Milhinhos et al. 2019; Thamm et al. 2019). Novel approaches are emerging that include somatic transformation of stem cells with GUS tagged genes to follow cell lineages (Bossinger and Spokevicius 2018) or fluorescent genetic cell lineage markers (Shi et al. 2019; Smetana et al. 2019), which we anticipate will bring exciting views on cambium activity and wood formation in the next following years.

8.2 miRNAs as Key Regulators of Plant Vascular Development

MicroRNAs (miRNAs) are endogenous ~21–24 nt RNAs that operate by inhibiting the translation or cleaving the mRNA complementary to its mature sequence, thus affecting mRNA translation or stability (Rhoades et al. 2002). Regulation of protein-coding genes expression by miRNAs is crucial to control development over time, in different organs and at certain (developmental) phase changes that occur in plants. In plants, miRNAs have a set of particular properties: they often target transcription factors; are commonly found in multigenic families; only act negatively on their targets; may be sequestered by long non-coding RNAs or ARGONAUTE proteins and their primary transcripts may encode small regulatory peptides (Couzigou and Combier 2016). miRNA-mediated regulation of stem cell maintenance and cell fate in the vascular tissues of the stem and root has been subject of intensive study in the last couple of decades. A summary of miRNAs involved in vascular development and secondary growth can be found in Table 8.1. One of the best-studied miRNA-mediated regulatory pathways operating above and below ground tissues of Arabidopsis involves CLASS III HOMEODOMAIN-LEUCINE ZIPPER (HD-ZIP III) transcription factor proteins relationship with the miRNA165/166 (miR165/166) module. The HD-ZIP III family of transcription factors comprises the genes: *INTERFASCICULAR FIBERLESS/REVOLUTA (IFL/REV)*, *PHABULOSA (PHB)/ATHB14*, *PHAVOLUTA (PHV)/ATHB9*, *INCURVATA4/CORONA/ATHB15* and *ATHB8* (McConnell and Barton 1998; McConnell et al. 2001; Baima et al. 2001; Mallory et al. 2004; for a recent review on HD Zip III functions see Ramachandran et al. 2017). All the *HD-ZIP III* mRNA sequences contain target sites for miR165/166, and the suppression of *HD-ZIP III* expression by the miR165/166 action was found critical for proper meristem function and vascular development of plants. In the miR165/166 gene family, the mature sequences of species 165 and 166 diverge only at one nucleotide, and there are two miR165 members (miR165a and miR165b) and seven miR166 members (miR166a to miR166g) that operate in different pathways depending on the spatial and developmental context (Mallory et al. 2004). Expression of the *HD-ZIP III* genes together with local maxima of auxin (indole-acetic acid, IAA) were shown to promote the quiescence of the organizer cells and the proliferation of xylem cells (Baima et al. 2001; Ohashi-Ito and Fukuda 2003; Emery et al. 2003; Mallory et al. 2004; Carlsbecker et al. 2010; Smetana et al. 2019). Auxin accumulation, mi165/166-restricted expression of *HD-ZIP IIIs*, movement of miR165/166 themselves and the antagonistic interplay of HD-ZIP III and other elements, such as the GARP KANADI (KAN) or GRAS SHORT ROOT (SHR) families of transcription factors, have proven essential during patterning of xylem and the establishment of polarity in vascular tissues in SAM and RAM of Arabidopsis (Emery et al. 2003; Prigge et al. 2005; Izhaki and Bowman 2007; Kerstetter et al. 2001; Ilegems et al. 2010; Carlsbecker et al. 2010). Regulation of cell fate in meristem maintenance and development by miR165/166 has recently been added a layer of complexity because miRNAs are recruited into AGO-centered RNA induced silencing complexes (RISCs) to repress expression of

Table 8.1 miRNAs in plant vascular development and secondary growth

miRNA	Potential or validated target genes	Associated function and tissues	Species	References
miR156	*SPL* gene family	Vegetative phase change (juvenile to adult transition)	*Arabidopsis thaliana*	Wu and Poethig (2006), Wu et al. (2009)
			Populus x canadensis	Wang et al. (2011)
	Unknown	Cambium dormancy release (dormancy-active growth transition); reactivating cambium	*Cunninghamia lanceolata*	Quan et al. (2016)
			Populus alba × *P. glandulosa*	Ding et al. (2014)
	SPL genes, *Nitrate transporter*	Tension wood formation	*Populus trichocarpa*	Lu et al. (2005)
	Pto-SPL15, *Pto-SPL20*, *Pto-SPL25*	Tree growth and wood properties	*Populus tomentosa*	Quan et al. (2016)
miR159	*MYB33/65; miR156* (via MYB33)	Vegetative phase change (juvenile to adult transition)	*Arabidopsis thaliana*	Guo et al. (2017)
	MYB33/65	Cambium dormancy release (dormancy-active growth transition)	*Populus alba* × *P. glandulosa*	Ding et al. (2014)
	MYB, *Asparagine synthase*, (1-4)-*β-mannan endohydrolase*	Tension wood formation	*Populus trichocarpa*	Lu et al. (2005), Chen et al. (2015a), Chen et al. (2016a)
	Unknown	Leaves and phellem tissues	*Quercus suber*	Chaves et al. (2014)
miR160	*ARFs (ARF10, ARF16, ARF17)*	Tree growth and wood formation	*Populus tomentosa*	Tian et al. (2016)
		Cambium dormancy release	*Populus alba* × *P. glandulosa*	Ding et al. (2014)

(continued)

Table 8.1 (continued)

miRNA	Potential or validated target genes	Associated function and tissues	Species	References
		Auxin signaling	*Arabidopsis thaliana*	Rhoades et al. (2002), Mallory et al. (2005)
miR164		Cambium dormancy release (dormancy-active growth transition); reactivating cambium	*Populus alba* × *P. glandulosa*	Ding et al. (2014)
	NAC-domain protein	Tension wood formation	*Populus trichocarpa*	Lu et al. (2005), Chen et al. (2015a), Chen et al. (2016a)
	Unknown	Cambium dormancy release (dormancy-active growth transition); reactivating cambium	*Cunninghamia lanceolata*	Quan et al. (2016)
miR165/166	*HD-Zip III* genes	Vascular development, xylem differentiation, radial patterning, organ polarity, cell to cell communication during xylem differentiation	*Arabidopsis thaliana*	Ratcliffe et al. (2000), Otsuga et al. (2001), McConnell and Barton (1998), McConnell et al. (2001), Reinhart et al. (2002), Rhoades et al. (2002), Emery et al. (2003), Mallory et al. (2004), Prigge et al. (2005), Izhaki and Bowman (2007), Kerstetter et al. (2001), Ilegems et al. (2010), Carlsbecker et al. (2010), Miyashima et al. (2011), Zhong and Ye (1999, 2004), Kim et al. (2005)
	POPREVOLUTA	Primary and secondary growth (polarity)	*Populus tremula x P. alba*	Robischon et al. (2011)

(continued)

Table 8.1 (continued)

miRNA	Potential or validated target genes	Associated function and tissues	Species	References
	PtaHB1	Secondary xylem, seasonal growth	*Populus tremula x P. alba*	Ko et al. (2006)
	POPCORONA	Lignification of xylem and phloem fibers	*Populus tremula x P. alba*	Du et al. (2011)
	PtrHB8/PttHB8	Control of thermospermine levels, secondary growth	*Populus trichocarpa/Populus tremula x tremuloides*	Zhu et al. (2013), Milhinhos et al. (2013)
	PtrHB4	Interfascicular cambium differentiation	*Populus trichocarpa*	Zhu et al. (2018)
		Cambium dormancy release (dormancy-active growth transition)	*Populus alba × P. glandulosa*	Ding et al. (2014)
	Unknown	Leaves and phellem tissues	*Quercus suber*	Chaves et al. (2014)
miR167	*ARFs* (*ARF6*, *ARF8*)	Auxin signaling	*Arabidopsis thaliana*	Rhoades et al. (2002), Mallory et al. (2005)
	Pto-ARF8, *lncRNA ARFRL*	Auxin signaling during active growth, tree growth and wood formation	*Populus tomentosa*	Quan et al. (2018)
		Cambium dormancy release (dormancy-active growth transition)	*Populus alba × P. glandulosa*	Ding et al. (2014)
miR168	*ARGONAUTE1* (*AGO1*)	Cambium dormancy release (dormancy-active growth transition)	*Populus alba × P. glandulosa*	Ding et al. (2014)

(continued)

Table 8.1 (continued)

miRNA	Potential or validated target genes	Associated function and tissues	Species	References
miR171	*Scarecrow-Like Proteins (SLP)*	Cambium dormancy release	*Populus alba* × *P. glandulosa*	Ding et al. (2014)
miR172	*APETALA2 (AP2)*	Cambium dormancy release	*Populus alba* × *P. glandulosa*	Ding et al. (2014), Cui et al. (2019)
			Cunninghamia lanceolata	Qiu et al. (2015)
		Juvenile-to-adult phase transition in woody perennials	*Arabidopsis thaliana*	Wu and Poethig (2006), Wu et al. (2009)
			Populus x canadensis	Wang et al. (2011)
miR257	*PtoDUF248-1*, *PtoDUF248-2*, *PtoAGR1*, *PtoGH9C1*, *PtoCKX6*, *PtoPSKR2-1*, *PtoPSKR2-2*, *PtoRBK1-like*, *PtoFBT8*, *PtoSepR*, *PtoDUF620*, *PtoAcot13-like*	Growth and wood properties (contents in holocellulose, hemicellulose and α-cellulose)	*Populus tomentosa*	Chen et al. (2016b)
miR319	*TCP* genes	Cambium dormancy release	*Populus alba* × *P. glandulosa*	Ding et al. (2014)
miR390	*TAS3/ARF*	Cambium dormancy release (auxin pathway during active growth)	*Populus alba* × *P. glandulosa*	Ding et al. (2014), He et al. (2018)

(continued)

Table 8.1 (continued)

miRNA	Potential or validated target genes	Associated function and tissues	Species	References
	Unknown	Leaves and phellem tissues	*Quercus suber*	Chaves et al. (2014)
miR396	*GRFs*	Cell division in tacs in the RAM	*Arabidopsis thaliana*	Rodriguez et al. (2015)
		Tension wood formation	*Populus trichocarpa*	Lu et al. (2005)
			Populus tomentosa	Chen et al. (2016b)
		Cambium dormancy release	*Populus alba* × *P. glandulosa*	Ding et al. (2014), Cui et al. (2019)
		Abundant in cambium cells	*Cunninghamia lanceolata*	Qiu et al. (2015)
	Unknown	Leaves and phellem tissues	*Quercus suber*	Chaves et al. (2014)
miR397	*LACCASE* genes (*LAC4*, *LAC17*)	Modulation of lignin biosynthesis	*Arabidopsis thaliana*	Wang et al. (2014)
	LACCASE genes		*Populus trichocarpa*	Lu et al. (2005, 2013)
	HSP40, *LEA*, *SPRY*	Associates with tree growth and wood-property traits	*Populus tomentosa*	Chen et al. (2015b)
miR475	*PPR1*, *PPR2*, *PPR3*, *PPR4*	Associates with tree growth and wood-property traits	*Populus tomentosa*	Xiao et al. (2017)
miR530	*BREVIPEDICELLUS* (*BP/KNAT1*)	Fiber differentiation, xylem expansion	*Populus tomentosa*	Yang et al. (2015)
miR857	*LACCASE-7*	Modulation of lignin biosynthesis (SPL7 activates miR857 and repress LACCASE7)	*Populus trichocarpa*	Lu et al. (2013), Zhao et al. (2015)

target genes. So, while miRNA165/166 maintains the *HD-ZIP III* levels under tight control following the developmental and spatial cues to control xylem differentiation, the AGO1 and AGO10 maintain tight control over miRNA165/166. It was found that while AGO1 recruits miR165/166 to silence *HD-ZIP III* genes (Zhang and Zhang 2012), AGO10 competes with AGO1 to sequester miR165/166 and antagonize the silencing promoted by AGO1-miRNA165/166 complex. By sequestering miR165b, miR166a, miR166b, and miR166g, AGO10 inhibits miR165/166 and enriches the SAM with HD-ZIP III transcripts to promote SAM maintenance (Ji et al. 2011; Zhu et al. 2011; Zhou et al. 2015).

The first described mutant alleles for *HD-ZIP III* harboured synonymous (silent) mutations at the lipid or steroid-binding START (Steroidogenic acute regulatory protein-related lipid transfer) domain (McConnell and Barton 1998; McConnell et al. 2001). These mutant alleles resulted in similar gain of function phenotypes for HD-ZIP IIIs PHB (*phb-1d*), PHV (*phv-1d*) and REV (*rev-10d*). Vascular bundles in Arabidopsis inflorescence stems typically have xylem to the centre and phloem towards the outside, but *phb-1d*, *phv-1d* and *rev-10d* displayed abnormal amphivasal vascular bundles, with xylem surrounding phloem (McConnell et al. 2001; Reinhart et al. 2002; Rhoades et al. 2002; Emery et al. 2003; Mallory et al. 2004) whereas the loss of HD-ZIP III function resulted in phloem surrounding xylem in Arabidopsis (Emery et al. 2003). These gain-of-function mutations in HD-ZIP IIIs were found located at the complementary site for miR165/166 sequences. The fact that similar phenotypic effects could be observed in gain of *HD-ZIPIII* function mutants and loss of *mir165/166* function mutants suggested miR165/166-mediated cleavage of *HD-ZIP III* and their interconnected roles in vascular cells fate (Emery et al. 2003; Mallory et al. 2004). The correct patterning of adaxial-abaxial domains in the organs of plants is essential for the initiation of a polar vascular system early during plant development (McConnell et al. 2001; Emery et al. 2003). Thus, miR165/166 function to maintain such abaxial (lower/ventral, peripheral) fate of the vasculature and the lateral organs by restricting *HD-ZIPIII* expression to the adaxial (upper/dorsal, central) domains (Emery et al. 2003; Zhou et al. 2007).

Besides the polarity of vascular tissues, miR165/166-REV was shown to control meristem function, vascular architecture and development (Ratcliffe et al. 2000; Otsuga et al. 2001; Emery et al. 2003) and to regulate interfascicular fiber differentiation (Zhong and Ye 1999, 2004). The miRNA165/166-ATHB8 module functions as a positive regulator of procambium and cambium development in vascular tissues (Baima et al. 1995, 2001). Overexpression of *ATHB8* promotes xylem differentiation, however the *athb8* knock out mutant lacks phenotypes in secondary cell wall development. In addition to the control over *REV* and *ATHB8* expression levels, the miR166 regulates the vasculature in inflorescence stems of Arabidopsis by targeting also *ATHB15/CNA* for cleavage (Kim et al. 2005). Activation tagged alleles of mi165/166 have increased our understanding of ATHB15 role in vascular development in Arabidopsis. An activation tagged allele of miR166a, named *meristem enlargement 1* (*men1*), in which *MIR166A* was overexpressed, down-regulated its target *ATHB15*, while it increased differentiation of xylem tissues in the stem (Kim et al. 2005). More recently, Du et al. (2015) showed that activation tagging of miR165b results in mutant plants (*stp-2d*) with secondary wall thickening in the cells of the pith (STP). It was observed that the *miR165b* expression was increased by the tagging in *stp-2d* dominant mutant, which, in turn, lowered the expression levels of targets *PHB*, *PHV* and *AtHB15* (but not of *AtHB8*, nor *REV*) and resulted in secondary cell wall biosynthesis in pith cells, which are located in the centre of the stem in Arabidopsis and are typically parenchymatic without a secondary cell wall (Du et al. 2015). The overexpression of a miR165b-resistant *ATHB15* reverses the *stp-2d* phenotype to wild-type, which confirms miR165b-AtHB15 role in repressing secondary wall development in the pith cells (Du et al. 2015). Thus, the miR165/166 seems to impact on well-known regulatory pathways and switches of secondary cell wall formation through ATHB15. For instance, the *athb15* mutants show up-regulation of NAM, ATAF1/2, and CUC2 (NAC) master switches of secondary wall biosynthesis SND1/NST3 (secondary wall-associated NAC domain protein 1) and NST2 (NAC secondary wall thickening promoting factor 2) (Wang et al. 2010; Du et al. 2015). Mutations of stem-expressed *WRKY* transcription factor genes that are known to consequently up-regulate genes encoding the NAC and C3H zinc finger transcription factors, also activate secondary wall synthesis (Wang et al. 2010). Similarities between phenotypes in the *stp-2d* and *athb15* to *wrky12* mutants hint that their relationships and theirs to miRNA165 indirectly function in negatively regulating secondary cell wall development, but further studies are needed to fully understand miRNAs action in this regulatory pathway (Du et al. 2015). Hence, it seems clear that miRNA165/166 functions upstream of the NAC master switches to regulate secondary cell wall biosynthesis.

Due to the simple Arabidopsis primary root architecture which enables tracking of cell lineages during differentiation, many researchers since the 90s' have used this model system to study plant development (Benfey et al. 1993; Dolan et al. 1993). Therefore, it comes as no surprise that it was in Arabidopsis root cells that miR165/166 were found to be mobile signals that set boundaries and cell fate domains during root vascular development. The primary root tissues are organised in a radial pattern of concentric cell layers (outwards: pericycle, endodermis, cortex and epidermis) that start from the central vascular stele. In the primary root, the miR166 is produced in the endodermis layer of cells and moves to establish a gradient in the direction of the stele. The miR166 consequently creates a gradient of *HD-ZIPIII* genes expression that is inverse to miR166 own gradient across the root cell layers (Carlsbecker et al. 2010). This gradient of miR166 activity is created via a bidirectional movement of the miR166 and the SHR transcription factor. While SHR is produced in the stele, it travels to the endodermis, where it stimulates the expression of *MIR165a* and *MIR166b*. The miR165/166 then diffuses and moves out of the endodermis to form a gradient across the root that will, in turn, down-regulate the *HD-ZIPIIIs* (*PHB*) and restrict and direct cell fate decisions in the xylem, in a non-cell autonomous, dose-dependent manner (Carlsbecker et al. 2010; Miyashima et al. 2011). Also, in situ hybridization with locked nucleic acid (LNA) probes showed that *HD-ZIP III* mRNA levels in the RAM were increased in the centre of the xylem axis and decreased towards the stele periphery (Dastidar et al. 2016). While *shr* and miR-resistant *phd* mutants often show protoxylem replacement by metaxylem cells, the simultaneous *phv*-13 *can-2 rev-6 athb8-1* mutations result in over-differentiation of protoxylem instead of metaxylem cells. The quintuple *hd-zip iii* mutant shows no xylem formation at all. Overall, miR165/166-mediated suppression of HD-ZIP III in the stele periphery seems necessary for proper pattern formation in the xylem (Carlsbecker et al. 2010; Miyashima et al. 2011; Sakaguchi and Watanabe 2012).

In the RAM of Arabidopsis, it was recently found that the transition from stem cells into differentiated cells is regulated very early in the differentiation process by miR396 interaction with *GROWTH-REGULATING FACTORs* (*GRFs*) (Rodriguez et al. 2015). Before differentiation, the derivatives of cambial cells go through a phase where cell divisions are more actively occurring, as means to ensure proper organ growth. The dividing cells, called transit-amplifying cells (TACs) need highly expressed *GRFs* to actively divide whereas in stem cells the opposite happens, and GRFs are inhibited by miR396. Furthermore, it was described that while GRFs negatively regulate the *PLETHORA* (*PLT*) genes domain of expression, *PLT* positively feedback to activate miR396 in the stem cells to repress GRFs (Rodriguez et al. 2015). It is likely but unknown so far if the regulation and balance of meristem size *vs* periclinal formative cell divisions during the transition of stem cells at the stem cell niches to transit-amplifying-like cells could also take place in tissues other than the RAM, such as in the vascular cambium.

8.3 miRNAs and the Regulation of Secondary Growth

Radial or secondary growth, consequence of the bidirectional activity of the cambia, which confers thickness to tree stems and roots also occurs in Arabidopsis. Very recently, radial growth was granted much attention to work that uncovered some mechanisms of plant-thickening. It was first demonstrated that xylem and phloem are differentiated from the same lineage of stem cells. By using fluorescent proteins to label different cell types in the cambium of Arabidopsis hypocotyl Thomas Greb group (Heidelberg, Germany) identified a file of bifacial stem cells that is the single source of both xylem and phloem (Shi et al. 2019). Bossinger and Spokevicius (2018) using somatic transformation with a GUS reporter line had earlier shown that a single cell layer of stem cells in wounded stems of Populus (*P. alba* '*pyramidalis*' L.) was regenerating the vascular cambium. Smetana et al. (2019) also revealed the existence of a single file of bifacial stem cells in the cambium and in addition identified the organising centre of radial cambium patterning in adjacent developing xylem cells. These works elucidated the bidirectional mode of xylem and phloem production from the vascular cambium. At the same time, Miyashima et al. (2019) characterized the role of the newly described set of mobile *PEAR* genes in maintenance of procambial cell divisions in the Arabidopsis root. The authors show that cambium proliferation relies on the feedback loop between PEAR and HD-ZIP IIIs. PEAR proteins are a group of six DNA-BINDING ONE ZINC FINGER (DOF) transcription factors that are antagonised by HD-ZIP IIIs (together with auxin and mobile miR165/166). It was found that the PEAR proteins act locally to promote the HD-ZIP IIIs which in turn

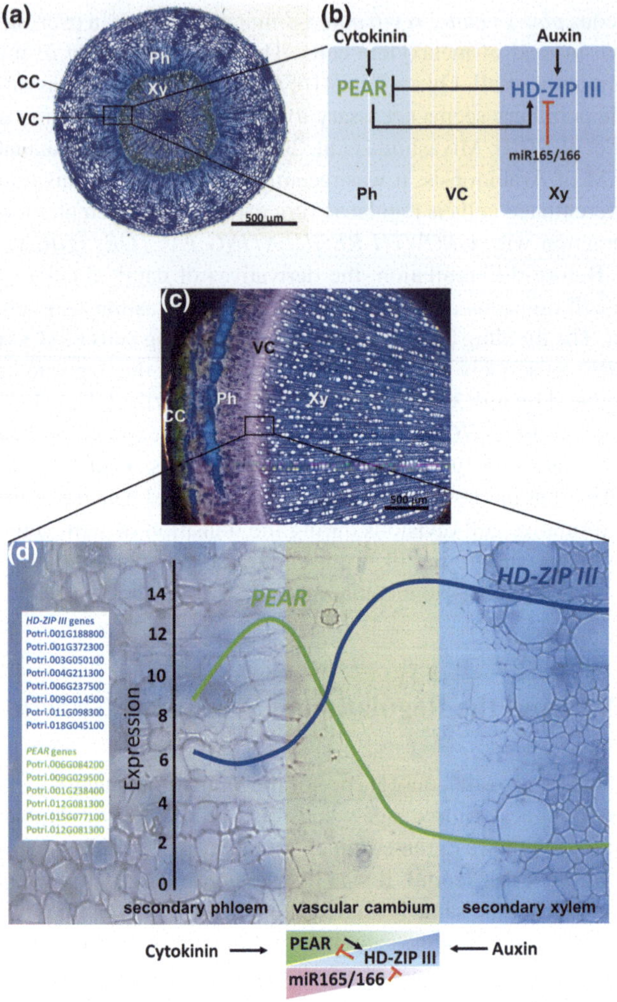

repress *PEAR* genes, in a negative-feedback loop that establishes the boundaries of concentration gradients of HD-ZIP IIIs, and sets the cell division zone that is needed to establish the cambial domain. The authors reason that the spatial information of auxin domain and of miR165/166 gradients during procambial development in the RAM, establish the cell dividing zones, that will be the foundation for further secondary growth (Fig. 8.1a, b; Miyashima et al. 2019). The aftermath of these findings will be interesting to follow in the future, especially if experiments to understand whether these components operate in a similar manner in initiation and maintenance of secondary growth in trees are ensued. It likely is the case, given the opposite PEAR and HD-ZIP III expression profiles across the Populus stems (Fig. 8.1c, d).

◂**Fig. 8.1 Secondary growth observed in cross-sections of the Arabidopsis hypocotyl and Populus stem.** **a** The Arabidopsis hypocotyl in 6-week-old adult plants shows extensive production of secondary xylem and secondary phloem tissues derived from the vascular cambium file of cells. **b** At the vascular cambium, PEAR proteins migrate from the phloem to adjacent cells and activate HD-ZIP III that in turn feedback negatively to PEAR proteins (Miyashima et al. 2019; Smetana et al. 2019). The miR165/166 regulates HD-ZIP III in the xylem to promote xylem differentiation. **c** The vasculature in the Populus stem shows remarkable similarities to the vascular tissue arrangement in the hypocotyl of Arabidopsis. Expression of putative homolog *PEAR* and *HD-ZIP III* genes in the Populus stem (as taken from Aspwood database, Sundell et al. 2017) follows similar gradients as to those observed during Arabidopsis cambial domain establishment and maintenance, where PEAR genes expression is found increased in phloem cells whereas it is gradient-reduced across the vascular cambium and secondary xylem cells. Antagonistically, the *HD-ZIP III* genes expression is gradient-increased from the phloem towards the xylem cells. Expression gradients for miR165/166 are suggested, though not currently available for the Populus stem, these are expected to follow the opposite expression profile to that of *HD-ZIP III* genes since the same miRNA regulatory pathway seems to operate in Populus and Arabidopsis during cambial activity and xylem differentiation. Xy—xylem, Ph—phloem, VC—vascular cambium, CC—cork cambium

MiRNA-mediated regulation of HD-ZIPIII genes hence also occurs in trees during the vascular tissue expansion that gives rise to secondary growth. In *Populus trichocarpa*, eight HD-Zip III genes have been found (Ko et al. 2006) and over 20 members of miR165/166 (Barakat et al. 2007). For miR166 species alone, there are 17 members described from xylem libraries (Ptc_miR166a-Ptc-miR166q; Puzey et al. 2012; Kozomara and Griffiths-Jones 2014) and Ptc-miR166 members are among the most abundant found also in SAM tissues in *Populus tomentosa* (Cui et al. 2019). In *Populus tremula* x *Populus alba* trees, Ko et al. (2006) first showed that miR166 accumulation was inversely correlated to the expression of *PtaHB1* gene (*REV* ortholog) and to the growth of secondary xylem. It was interesting to find miRNA166-REV seasonally regulated in trees. Cambial activity in Populus woody stems follows the seasons and it becomes dormant during the winter months. In the winter, a peak of *miR166* expression maintains the low levels of PtaHB1, suggesting miR166 has perhaps broader roles in wood development in trees, such as its involvement in the control of seasonal cessation of growth or dormancy release. Several pleiotropic effects of inhibiting the miR166 binding site in *popREVOLUTA* (*PtaREV, PRE*) gene in *P. tremula* x *P. alba* were also described (Robischon et al. 2011; Porth et al. 2014). Transgenic trees expressing a miRNA-resistant gain of function *popREVOLUTA* version had perturbed primary and secondary growth and reverse polarity in secondary vascular tissues derived from abnormal formation of cambia within cortical parenchyma (Robischon et al. 2011). Since these abnormalities probably are common among land plants, *REV* is viewed as a key regulatory gene in the control of plant morphogenesis, but novel tree-specific roles related to tree wood development (such as the described in seasonal growth) are likely to exist.

In Populus trees, POPCORONA, one ortholog of Arabidopsis CORONA/AtHB15 has also been implicated in secondary growth. By creating an artificial miR166 that specifically suppressed *POPCORONA*, Du et al. (2011) observed that the pith of the stems was atypically lignified, whereas overexpression of a miRNA-resistant version of *POPCORONA* delayed the lignification of xylem and phloem fibers in the woody stems. There are two AtHB8 orthologs in Populus, the PtrHB7 and PtrHB8. Like in Arabidopsis, PtrHB8 and PtrHB7 play important roles in promoting vascular cambium differentiation during secondary growth (Zhu et al. 2013; Milhinhos et al. 2013). IAA, PttHB8 and poplar thermospermine synthase coding ACL5 (POPACAULIS5) are part of a feedback loop where POPACAULIS5 represses *PttHB8* expression through its action on auxin levels, while in contrast overexpression of miR165/166-resistant *PttHB8* promotes *POPACAULIS5* (Milhinhos et al. 2013). Besides, PtrHB7 was shown to be required for a balanced differentiation between secondary xylem and phloem tissues during Populus secondary growth (Zhu et al. 2013). Recently, another regulatory interaction was found where Aux/IAA9-ARF5 module is required for auxin signaling to regulate wood formation via orchestrating the expression of *P. tomentosa PtoHB7/PtoHB8* (Xu et al. 2018). Another *HD-ZIP III* gene in Populus, the *PtrHB4* (with homology to *PHV/PHB* genes) was recently implicated in secondary growth. PtrHB4 repression caused defects in interfascicular cambium formation in *Populus* while the miR165/166-resistant *PtrHB4* overexpression induced cambium activity and xylem differentiation during secondary growth (Zhu et al. 2018).

More work will be needed to find how miR165/166 or other miRNAs operate in these regulatory pathways to regulate wood formation without the use of constitutive promoters that result in pleiotropic effects early in transgenic trees growth, but so far there seems to be a spatial-temporal distribution of the HD-ZIP IIIs functions in trees secondary growth. While PtaREV/POPREVOLUTA seems to operate early during primary growth being required for patterning and initiation of cambium (Robischon et al. 2011) PtrHB7 seems to operate by promoting cambium activity (Zhu et al. 2013), POPCORONA/PtrHB5 inhibits cambium activity and xylem differentiation (Du et al. 2011; Zhu et al. 2013) and PtrHB4 was found required for interfascicular cambium formation (Zhu et al. 2018).

MiRNAs, other than miR165/166, have been reported to operate during secondary growth. Zhao et al. (2015) described how miR857 regulates the transcription of LACCASE7 (LAC7) gene, reducing its activity and affecting the lignin content in the cell wall of the secondary xylem in Arabidopsis. Plant laccase enzymes belong to the blue copper oxidase family and their function is to oxidise the polymerization of monolignols into lignin. The authors further describe a regulator of miR857, the SQUAMOSA PROMOTER BINDING PROTEIN-LIKE7 (SPL7) that activates miR857 under low copper conditions to regulate secondary growth (Zhao et al. 2015). LACCASE4 (LAC4), that controls lignin biosynthesis and seed yield, was also found regulated by a miRNA in Arabidopsis (Wang et al. 2014). Overexpression of *miR397b* resulted in reduced lignin deposition, reducing the thickness of secondary cell walls in xylem vessels and fibers, which led to weak vascular tissues. On the other hand, the miRNA-resistant version of LAC4 results in quite the opposite phenotype, with additional inflorescence stems also being formed (Wang et al. 2014). Regulation of lignin biosynthesis by miR397-LAC module was in fact first described in *P. trichocarpa*, when it was demonstrated that overexpression of *Ptr-miR397a* in poplar downregulated 17 PtrLACs, reducing the transgenic trees lignin content without affecting their monolignols biosynthesis (Lu et al. 2013). This suggested that LACs and miR397a are components of one and same regulatory pathway of lignin biosynthesis which could be altered for biotechnological ends, such as to produce reduced-lignin feedstocks. Curiously, in flax (*Linum usitatissimum*), the localization of miR397 mRNA by in situ hybridization is found in the developing xylem cells where bulk lignification of cell walls is yet to occur (Le Roy et al. 2017) suggesting existence of an articulated spatial-temporal action of miR397 in suppressing lignin biosynthesis during xylem development. In addition to targeting LACs, other putative targets of miR397a include *HSP40* (*HEAT SHOCK PROTEIN 40*), *LEA* (*LATE EMBRYOGENESIS ABUNDANT*), and *SPRY* (*SPla and RYanodine RECEPTOR*) which are associated to wood properties in *P. tomentosa* (Chen et al. 2015b).

SQUAMOSA PROMOTER BINDING PROTEIN-LIKE (*SPL*) genes encode plant-specific transcription factors involved in different stages of plant development, namely during vegetative-reproductive phase transition. In Arabidopsis, miR156 regulation of *SPL* genes and downstream activity of miR172 is needed for the correct timing in juvenile to adult phase transition (Wu and Poethig 2006; Wu et al. 2009). Interestingly, the vegetative and reproductive phase transition coincides with the triggering of bulk lignified xylem expansion, in hypocotyl tissues, which has been previously associated to flowering (Ragni et al. 2011) but the exact role of such miRNAs or their involvement in cambium activity and xylem expansion in hypocotyls of Arabidopsis is not clear. Curiously, it was work reported in leaves of Populus and of other trees together with reports on reactivation of cambium activity in a conifer that have described the links between miR156, miR172 and cambium activity (Wang et al. 2011; Qiu et al. 2015). In Chinese fir (*Cunninghamia lanceolata*), an economically important conifer, smallRNAome analysis to cryo-sectioned dormant, reactivating, and active cambium was used to study the regulatory network of miRNAs during vascular cambium development (Qiu et al. 2015). Amongst the most abundant miRNAs in the cambium cells of Chinese fir were miR156, miR164, miR166, miR168, miR172, miR396, and miR535. While cln-miR156 were highly abundant in reactivating cambium and low in the reactivating-to-active growth transition, cln-miR172 showed the opposite expression pattern, suggesting inverse roles during cambium dormancy break in a similar manner as in juvenile-adult vegetative phase transition. Taken together, the facts that (i) miR156 regulates *SPL* genes to control the juvenile-to-adult transition timing, and (ii) the miR172 (which targets *APETALA2-like* (*AP2-like*) regulates cell division and elongation that results in production of organ cell lineages (Aukerman and Sakai 2003; Huijser and Schmid 2011), suggest that the transition from dormancy to active growth in vascular cambium is regulated by miR156 and miR172 (Qiu et al. 2015). Other miRNAs are likely important during the cambium dormancy and dormancy release. Ding et al. (2014) performed large scale small RNA sequencing and profiled miRNAs in the cambial meristem during dormancy and active growth transition in poplar. Among the well-known development-related miRNA families, the authors found eight families differentially expressed, enriched in the active growth phase: a list that included mentioned miR156 and miR172; and also miR164, miR396 and miR319, that function in cell proliferation; miR166, in vascular development; miR168, in miRNA biogenesis and miR171 (Ding et al. 2014). Highly expressed during active growth were also auxin-related miR167 and miR390, and miR160 was found activated by the dormancy release, but not operating during active cambial activity in Populus. Disparate MIR169 family members expression profiles suggest functional diversity of miR169 (that is implicated in ABA and stress responses; Potkar et al. 2013) during dormancy-active growth transition, while miR159 (implicated in gibberellin signaling) was increasingly expressed during the transition and especially during cambium activity (Ding et al. 2014). Studies in Arabidopsis have often laid the grounds to uncover miRNAs roles during secondary growth in trees, but more and more often, miRNAs are described involved in wood formation related processes in woody species. New regulatory nodes where these miRNAs operate during vascular development in trees are likely to be described in the future, but direct evidences are lacking. However, identification of the small RNAs in woody plants has only just begun.

8.4 miRNAs and Their Targets in Association to Wood Formation and Wood Properties

Genome-wide profiling of Populus miRNAs was reported soon after the sequencing of the *P. trichocarpa* genome (Klevebring et al. 2009). Ever since various articles have reported association of small RNAs to secondary growth traits in Populus. Since the basis of miRNA-mediated regulation is the interaction of miRNAs and their target sites, if mutations occur in the mature miRNAs or their binding sites, the disruption of sequence complementarity or compromised binding affinity can originate phenotypic variation (Todesco et al. 2012). Like in many species, phenotypic variations may result from gene expression alterations caused by these allelic mutations in Populus (Evans et al. 2014). Considerable amount of work from Deqiang Zhang lab (Beijing Forestry University, China) has been since dedicated to the natural variation found in miRNAs and their targets in the genomes of Populus populations and how it may affect secondary growth. Thus, transcript profiling and multi-SNP association studies have been used to explore the genetic regulatory relationship between miRNAs and their targets. One of the first such studies in a forest tree species looked into allelic variants of Pto-miR530a and its *HOMEOBOX PROTEIN KNOTTED-1-LIKE 1* target, *Pto-KNAT1*, in 460 unrelated individuals of *P. tomentosa* which associated with eight traits related to growth and wood properties (Yang et al. 2015). KNAT1 interacts with DELLAs, negative regulators of gibberellins signalling, to regulate the differentiation of xylem fibers in the Arabidopsis hypocotyl (Liebisch et al. 2014; Felipo-Benavent et al. 2018; Milhinhos et al. 2019). Although a miRNA-mediated mechanism for this pathway in trees is yet to be reported, its existence is plausible, given gibberellins known roles in differentiation of xylem (Wareing 1958; Digby and Wareing 1966; Ridoutt et al. 1996; Mauriat and Moritz 2009; Ragni et al. 2011; Felipo-Benavent et al. 2018).

Besides gibberellin, phytohormone auxin that forms a gradient across the cambial zone and peaks in cambial initials and xylem derivatives in the *Populus* stems (Uggla et al. 1996; Tuominen et al. 1997) has determinant roles in growth and wood formation in trees. Auxin signaling is reported as miRNA-mediated. Several *AUXIN RESPONSIVE FACTOR* (*ARF10*, *ARF16* and *ARF17*) genes are targeted by miR160 and act in regulating auxin signaling (Rhoades et al. 2002; Mallory et al. 2005) while miR167 regulates *ARF6* and *ARF8* (Rhoades et al. 2002; Bartel and Bartel 2003). MiR160-resistant versions of *ARF17* have increased accumulation of auxin-conjugating proteins (Mallory et al. 2005). The regulation of ARFs by miR160 suggests the important roles for miR160 and ARFs in plant embryogenesis and growth, and this same pathway seems to operate during wood formation as well (Tian et al. 2016). In *P. tomentosa*, expression of *Pto-miR160* was found to negatively correlate with that of its target *Pto-ARF16* in cambium, developing and mature xylem tissues; and SNPs found in *Pto-miR160a* and *Pto-ARF16* associated with the same tree growth and wood properties traits, suggesting they play a role in the same pathway (Tian et al. 2016). MirR160 and miR167 coordinately modulate expression of repressing and activating ARF proteins. Besides miR160, miR390 and miR165/166 have also been shown to interact and move across cells to set the developmental boundaries in leaves to establish leaf polarity (Marin-Gonzalez and Suarez-Lopez 2012). It is possible that we will find these sRNAs movement and actions implicated in meristem maintenance and secondary growth in trees.

Another level of regulation in miRNA mediated control of expression was found in what are called sponge RNAs, which contain miRNA binding sites and regulate the expression of corresponding genes by competing for interactions with miRNA (Ebert and Sharp 2010). In plants, the long non-coding RNAs (lncRNAs) are a type of sponge RNA, modulating the expression of miRNA targets. In *P. tomentosa*, interaction between Pto-miR167a and its targets, sponge lncRNA *auxin response factors-related lncRNA* (*ARFRL*), and *Pto-ARF8*, was understood by degradome sequencing and psRNA Target prediction (Quan et al. 2018). *Pto-ARF8* was identified as a target gene of Pto-miR167a and of lncRNA ARFRL. SNP-based association studies (additive, dominant, and epistatic) in a population of 435 unrelated individuals deciphered the genetic interactions of SNPs within Pto-miR167a and lncRNA ARFRL, and *Pto-ARF8* targets, indicating their potential regulatory roles in tree growth and wood formation (Quan et al. 2018). Another lncRNA, called NERDL (from Needed for RDR2-independent DNA methylation (NERD)-related lncRNA) regulates *Pto-NERD* expression. Gene expression analysis and association mapping in 435 *P. tomentosa* unrelated individuals found correlation between NERDL and PtoNERD and 30 SNPs associated with five tree growth and wood properties traits (Shi et al. 2017).

Similar to Arabidopsis, genetic interaction between miRNA156 and *SPL* genes was also found in Populus (Quan et al. 2016). In *Populus trichocarpa,* there are 28 *SPL* genes that have different expression profiles in different tissues and organs, which indicate these genes functional diversity in trees (Li and Lu 2014). The miR156 family in *P. trichocarpa* has 12 members (Ptr-miR156a-m) with a particularly high abundance of Ptr-miR156c member in xylem (Kozomara and Griffiths-Jones 2014).

In *P. tomentosa* the corresponding Pto-miR156c is found abundant in developing xylem, mature xylem, cambium and phloem vascular tissues (Quan et al. 2016). Through bioinformatic prediction and degradome sequencing, *Pto-SPL15*, *Pto-SPL20* and *Pto-SPL25* were identified as putative targets of Pto-miR156c. Using SNP-based association mapping to a natural population of *P. tomentosa*, tree growth and wood properties were significantly associated to SNPs within Pto-miR156c and *Pto-SPLs* (Quan et al. 2016).

Pto-miR257, a highly expressed miRNA in mature xylem tissues of *P. tomentosa*, has about 12 predicted targets and overlapping expression profiles suggest it may function in wood formation (Chen et al. 2016b). Nonsynonymous nucleotide substitutions within the target genes of Pto-miR257 (*PtoDUF248-1*, *PtoDUF248-2*, *PtoAGR1*, *PtoGH9C1*, *PtoCKX6*, *PtoPSKR2-1*, *PtoPSKR2-2*, *PtoRBK1-like*, *PtoFBT8*, *PtoSepR*, *PtoDUF620* and *PtoAcot13-like*) significantly associated with growth and wood properties, such as contents in holocellulose, hemicellulose and α-cellulose (Chen et al. 2016b). A large set of predicted target genes of lncRNAs involved in processes such as cellulose, lignin and gibberellin biosynthesis, indicates the lncRNAs may have diverse roles during wood formation (Chen et al. 2015a). Similar association mapping studies from D. Zhang lab associated Pto-miR475b and *Pto-PPR1*, *Pto-PPR2*, *Pto-PPR3*, and *Pto-PPR4* targets to nine traits, with potential roles in tree growth and wood formation (Xiao et al. 2017). Pto-miR397a has also been associated with *HSP40*, *LEA*, and *SPRY* SNPs and tree growth and wood-property traits (Chen et al. 2015b).

One good system to study the genetic regulation of components in cell wall biosynthesis and wood formation is the tension wood system. In angiosperms, tension wood is formed in the upper side of a bent stem or branch and produces a cellulose-rich G-layer inside the secondary wall, and low lignin contents. Lu et al. (2005) have shown that miRNAs are implicated in the control of normal and tension wood formation in trees. Chen et al. (2015a, 2016a) have been using the tension wood system for high-throughput sequencing (RNA-seq and degradome sequencing) as means to identify and characterize miRNAs involved in wood formation. Expression analysis, miRNA and target predictions, as well as genetic association mapping, showed that miR156i, miR156j, miR159, miR164, miR166, miR477, miR396a and miR6445b are involved in the formation of tension wood and miR482a.2 was proposed as a tree-specific miRNA in tension wood (Lu et al. 2005; Chen et al. 2015a, 2016a). Although association mapping is a good strategy to associate allelic variants to complex traits, such as that of secondary growth in trees, it is still unclear how the genes identified contribute to the traits. Recently, association mapping and expression quantitative trait nucleotide (eQTN) mapping has been employed to lignin biosynthesis to understand the genetic networks and gene contribution underlying the wood properties (Quan et al. 2019).

8.5 miRNAs and the Regulation of Cork Cambium

Cork cambium (also named phellogen) is a lateral meristem concentric to the vascular cambium that forms a continuous ring-like file of stem cells that bidirectionally produces cork (or phellem) towards the surface of the stems and roots, and phelloderm inwardly. The group of phellem, phellogen and phelloderm tissues form the periderm. The epitome of the activity of cork cambium can be best observed in Cork oak tree stems (*Quercus suber* L.) that produce massive amounts of phellem cells. In a process that very much resembles the development of secondary xylem, the phellem cells undergo several differentiation steps in a sequential manner that start at the phellogen with active cell division, followed by cell expansion, deposition of suberin in the cell walls and ending in a programmed death of the phellem cells at maturity (Natividade 1950; Esau 1960; Graça and Pereira 2004; Soler et al. 2007; Inácio et al. 2018).

Few regulators of phellogen activity have been described so far (Miguel et al. 2016; Almeida et al. 2013a, b; Capote et al. 2018) but some studies report on global transcriptomic changes during cork formation (Teixeira et al. 2018; Boher et al. 2018; Vulavala et al. 2019). We have earlier found a *SHORT-ROOT-like* gene, *PtSHRL2B* involved in *Populus* phellogen activity (Miguel et al. 2016) and Almeida et al. (2013a, b) described *QsMYB1* (MYB84) is up-regulated in cork tissues and modulates phellogen action during environmental stresses in *Q. suber*. While miRNAs action in vascular cambium and secondary xylem have been the subject of numerous studies, there is a huge gap when it comes to the description of miRNAs in cork cambium activity and in periderm tissues. This lack of information is in part due to the lack of advanced genomic and genetic tools to study phellogen activity in cork oak. However, the recent release of the cork oak transcriptome and the first draft of cork oak genome (Pereira-Leal et al. 2014; Ramos et al. 2018), together with emergence of other more amenable model systems to study periderm such as the Arabidopsis hypocotyl (Wunderling et al. 2018; Campilho et al. 2020) or birch stem (Alonso-Serra et al. 2019), are expected to increase our knowledge about cork cambium formation dynamics in future work. A recent study in our lab (Célia Miguel Lab, ITQB-NOVA/iBET/BioISI-FCUL, Portugal) profiled for the first time the miRNAs expression patterns in leaf and cork tissues of Cork oak (Chaves et al. 2014). In this study, it was found that the conserved miRNA families miR167, miR165/166, miR396 and miR159 were highly expressed both in leaves and phellem tissues. MiR390 and miR168 were found highly expressed in phellem tissues when compared to leaves. Moreover, seven miRNA families (miR1140, miR158, miR1863, miR2916, miR479, miR5083 and miR530) were found phellem-specific (Chaves et al. 2014). Some of the miRNAs found have already been described implicated in secondary growth. As mentioned in the above sections, miR165/166 target the *HD-ZIP III* and are involved in vascular development, procambial and cambial stem cell maintenance and xylem differentiation; miR390 and miR160 target *ARF*-related genes and are involved in auxin responses in Arabidopsis root and in cambial dormancy release in Populus (Ding et al. 2014), miR530 targets *KNAT1* to promote xylem fibers differentiation in Arabidopsis hypocotyls and associates with wood formation and properties traits in Populus (Yang et al. 2015); miR159 that is up-regulated during vascular cambium activity and tension wood formation, and is known to regulate gibberellin signalling (Chen et al. 2016a). In birch (*Betula pendula*) stems, a transcriptomic study revealed that 17 transcription factors were specifically expressed in the phellogen and phelloderm fractions of cryo-sectioned tissues (Alonso-Serra et al. 2019). Among these, *KAN1*, indirectly regulated by miR165/166 in organ polarity (Kerstetter et al. 2001), *WUSCHEL RELATED HOMEOBOX 4* (*WOX4*), a known regulator of cell division in the vascular cambium (Hirakawa et al. 2010) and *AINTEGUMENTA* (*ANT*), regulator of cell proliferation in the vascular cambium during secondary growth of roots (Randal et al. 2015), were found highly expressed. Although not direct targets of miR165/166, *KAN* genes are intrinsically interconnected with *HD-ZIP III* in vascular patterning (Ilegems et al. 2010) and induction of miR165a is known to lead to downregulation of *WOX4* and *ANT* through *HD-ZIP III* genes in the vascular cambium (Smetana et al. 2019). It may well be that other miRNAs and non-conserved miRNAs play important roles in phellogen activity, taking into consideration that most datasets published so far on cambium activity and secondary growth focus solely on secondary xylem/phloem/wood formation processes. We might have overlooked miRNAs and transcriptional networks exclusively involved in cork formation and in the future will see the unraveling of novel mechanisms specifically related to periderm tissues.

8.6 Concluding Remarks

miRNAs are part of the genetic basis of complex processes such as secondary growth. Although secondary growth continues to be a challenge subject to study, novel model systems, such as the Arabidopsis hypocotyl, are emerging and bring advantages to the study of molecular mechanisms involved in secondary growth. There are several gaps in the understanding of miRNAs role during secondary growth. For instance, how the transition from primary to secondary growth is orchestrated and what miRNA-mediated mechanisms underlie such events is currently not known. It is plausible that mobile signals have a role in the transition from primary to secondary growth. Could the cell-cell communication during such transition be accomplished by mobile miRNA molecules? Currently, cork cambium is the least studied of all meristems in plants. It will be interesting to follow future work now that the genome of cork oak has been sequenced. Another aspect that will need further study is the coordination between cork cambium and vascular cambium activity. It is intuitive that communication must exist between the two meristems to coordinate during radial growth, but the genetic basis of this is completely unknown. The long list of novel non-conserved miRNAs from high throughput sequencing that appear specific to cambia (cork and vascular) is not deeply explored. Even though in the last couple of decades long lists of genes, including miRNAs, have been produced by high-throughput sequencing, experimental validation and mechanistic understanding of the regulatory networks implies that in addition to statistical modelling and prediction we pursue experimental functional characterization in the post-genomic era. Advances in genetic transformation, genomic technologies and computational methods can aid in a systems biology approach to understand the role of these miRNAs in secondary growth in the future.

Acknowledgements The authors would like to thank funding from Fundação para a Ciência e Tecnologia, in the form of CEEC/IND/00175/2017 contract to Ana Milhinhos, Ph.D. fellowship PD/BD/114359/2016 to Susana Lopes, GREEN-it (grant no. UID/Multi/04551/2013) and BioISI (grant no. UID/Multi/04046/2019). The authors apologise to authors whose work could not be included in this review due to space limitations.

References

Almeida T, Menendez E, Capote T et al (2013a) Molecular characterization of *Quercus suber* MYB1, a transcription factor up-regulated in cork tissues. J Plant Physiol 170:172–178

Almeida T, Pinto G, Correia B et al (2013b) *QsMYB1* expression is modulated in response to heat and drought stresses and during plant recovery in *Quercus suber*. Plant Physiol Biochem 73:274–281

Alonso-Serra J, Safronov O, Lim K et al (2019) Tissue-specific study across the stem reveals the chemistry and transcriptome dynamics of birch bark. New Phytol 222:1816–1831

Aukerman MJ, Sakai H (2003) Regulation of flowering time and floral organ identity by a MicroRNA and its *APETALA2-like* target genes. Plant Cell 15:2730–2741

Baima S, Nobili F, Sessa G et al (1995) The expression of the ATHB-8 homeobox gene is restricted to provascular cells in *Arabidopsis thaliana*. Development 121:4171–4182

Baima S, Possenti M, Matteucci A et al (2001) The Arabidopsis ATHB-8 HD-ZIP protein acts as a differentiation-promoting transcription factor of the vascular meristems. Plant Physiol 126:643–655

Barakat A, Wall PK, Diloreto S et al (2007) Conservation and divergence of microRNAs in Populus. BMC Genom 8:481

Bartel B, Bartel DP (2003) MicroRNAs: at the root of plant development? Plant Physiol 132:709–717

Benfey PN, Linstead PJ, Roberts K et al (1993) Root development in Arabidopsis: four mutants with dramatically altered root morphogenesis. Development 119:57–70

Boher P, Soler M, Sánchez A et al (2018) A comparative transcriptomic approach to understanding the formation of cork. Plant Mol Biol 96:103–118

Bossinger G, Spokevicius AV (2018) Sector analysis reveals patterns of cambium differentiation in poplar tree stems. J Exp Bot 69:4339–4348

Campilho A, Nieminen K, Ragni L (2020) The development of the periderm: the final frontier between a plant and its environment. Curr Opin Plant Biol 53:10–14

Capote T, Barbosa P, Usié A et al (2018) ChIP-Seq reveals that QsMYB1 directly targets genes involved in lignin and suberin biosynthesis pathways in cork oak (*Quercus suber*). BMC Plant Biol 18:198

Carlsbecker A, Lee JY, Roberts CJ et al (2010) Cell signalling by microRNA165/6 directs gene dose-dependent root cell fate. Nature 465:316–321

Chaves I, Lin Y-C, Pinto-Ricardo C et al (2014) miRNA profiling in leaf and cork tissues of *Quercus suber* reveals novel miRNAs and tissue-specific expression patterns. Tree Genet Genomes 10:721–737

Chen J, Chen B, Yang X et al (2015a) Association genetics in Populus reveals the interactions between Pt-miR397a and its target genes. Sci Rep 5:11672

Chen J, Quan M, Zhang D (2015b) Genome-wide identification of novel long non-coding RNAs in *Populus tomentosa* tension wood, opposite wood and normal wood xylem by RNA-seq. Planta 241:125

Chen B, Du Q, Chen J et al (2016a) Dissection of allelic interactions among Pto-miR257 and its targets and their effects on growth and wood properties in *Populus*. Heredity 117:73–83

Chen J, Xie J, Chen B et al (2016b) Genetic variations and miRNA–target interactions contribute to natural phenotypic variations in *Populus*. New Phytol 212:150–160

Couzigou JM, Combier JP (2016) Plant microRNAs: key regulators of root architecture and biotic interactions. New Phytol 212:22–35

Cui J, Lu W, Lu Z et al (2019) Identification and analysis of microRNAs in the SAM and leaves of *Populus tomentosa*. Forests 10:130

Dastidar MG, Mosiolek M, Bleckmann A et al (2016) Sensitive whole mount *in situ* localization of small RNAs in plants. Plant J 88:694–702

Digby J, Wareing PF (1966) The effect of applied growth hormones on cambial division and the differentiation of the cambial derivatives. Ann Bot 30:539–548

Ding Q, Zeng J, He XQ (2014) Deep sequencing on a genome-wide scale reveals diverse stage-specific microRNAs in cambium during dormancy-release induced by chilling in poplar. BMC Plant Biol 14:267

Dolan L, Janmaat K, Willemsen V (1993) Cellular organisation of the *Arabidopsis thaliana* root. Development 119:71–84

Du J, Miura E, Robischon M et al (2011) The Populus Class III HD ZIP Transcription Factor POPCORONA Affects Cell Differentiation during Secondary Growth of Woody Stems. PLoS ONE 6:e17458

Du Q, Avci U, Li S et al (2015) Activation of miR165b represses AtHB15 expression and induces pith secondary wall development in Arabidopsis. Plant J 83:388–400

Ebert MS, Sharp PA (2010) Emerging roles for natural microRNA sponges. Curr Biol 20:R858–R861

Emery JF, Floyd SK, Alvarez J et al (2003) Radial patterning of Arabidopsis shoots by class III HD-ZIP and KANADI genes. Curr Biol 13:1768–1774

Esau K (ed) (1960) Anatomy of seed plants. Wiley, New York, p 376

Evans LM, Slavov GT, Rodgers-Melnick R et al (2014) Population genomics of *Populus trichocarpa* identifies signatures of selection and adaptive trait associations. Nat Genet 46:1089–1096

Felipo-Benavent A, Úrbez C, Blanco-Touriñán N et al (2018) Regulation of xylem fiber differentiation by gibberellins through DELLA-KNAT1 interaction. Development 145:dev164962

Graça J, Pereira H (2004) The periderm development in *Quercus suber*. IAWA J 25:325–335

Guo C, Xu Y, Shi M et al (2017) Repression of miR156 by miR159 regulates the timing of the juvenile-to-adult transition in Arabidopsis. Plant Cell 29:1293–1304

He F, Xu C, Fu X et al (2018) The MicroRNA390/TRANS-ACTING SHORT INTERFERING RNA3 module mediates lateral root growth under salt stress via the auxin pathway. Plant Physiol 177:775–791

Hirakawa Y, Kondo Y, Fukuda H (2010) TDIF peptide signaling regulates vascular stem cell proliferation via the WOX4 homeobox gene in Arabidopsis. Plant Cell 22:2618–2629

Huijser P, Schmid M (2011) The control of developmental phase transitions in plants. Development 138:4117–4129

Ilegems M, Douet V, Meylan-Bettex M et al (2010) Interplay of auxin, KANADI and Class III HD-ZIP transcription factors in vascular tissue formation. Development 137:975–984

Inácio V, Martins MT, Graça J et al (2018) Cork oak young and traumatic periderms show PCD typical chromatin patterns but different chromatin-modifying genes expression. Front Plant Sci 9

Izhaki A, Bowman JL (2007) KANADI and class III HD-Zip gene families regulate embryo patterning and modulate auxin flow during embryogenesis in Arabidopsis. Plant Cell 19:495–508

Ji L, Liu X, Yan J et al (2011) ARGONAUTE10 and ARGONAUTE1 regulate the termination of floral stem cells through two microRNAs in Arabidopsis. PLoS Genet 7:e1001358

Kerstetter RA, Bollman K, Taylor RA et al (2001) KANADI regulates organ polarity in Arabidopsis. Nature 411:706–709

Kim J, Jung J-H, Reyes JL et al (2005) microRNA-directed cleavage of *ATHB15* mRNA regulates vascular development in Arabidopsis inflorescence stems. Plant J 42:84–94

Klevebring D, Street NR, Fahlgren N et al (2009) Genome-wide profiling of Populus small RNAs. BMC Genom 10:620

Ko J-H, Prassinos C, Han K-H (2006) Developmental and seasonal expression of PtaHB1, a Populus gene encoding a class III HD-Zip protein, is closely associated with secondary growth and inversely correlated with the level of microRNA (miR166). New Phytol 169:469–478

Kozomara A, Griffiths-Jones S (2014) miRBase: annotating high confidence microRNAs using deep sequencing data. Nucleic Acids Res 42(Database issue):D68–D73

Le Roy J, Blervacq AS, Créach A et al (2017) Spatial regulation of monolignol biosynthesis and laccase genes control developmental and stress-related lignin in flax. BMC Plant Biol 17:124

Li C, Lu S (2014) Molecular characterization of the SPL gene family in *Populus trichocarpa*. BMC Plant Biol 14:131

Liebsch D, Sunaryo W, Holmlund M et al (2014) Class I KNOX transcription factors promote differentiation of cambial derivatives into xylem fibers in the Arabidopsis hypocotyl. Development 141:4311–4319

Lu S, Sun Y-H, Shi R et al (2005) Novel and mechanical stress-responsive microRNAs in *Populus trichocarpa* that are absent from Arabidopsis. Plant Cell 17:2186–2203

Lu S, Li Q, Wei H et al (2013) Ptr-miR397a is a negative regulator of laccase genes affecting lignin content in *Populus trichocarpa*. Proc Natl Acad Sci USA 110:10848–10853

Mallory AC, Bartel DP, Bartel B (2005) MicroRNA-directed regulation of Arabidopsis AUXIN RESPONSE FACTOR17 is essential for proper development and modulates expression of early auxin response genes. Plant Cell 17:1360–1375

Mallory AC, Reinhart BJ, Jones-Rhoades MW et al (2004) MicroRNA control of PHABULOSA in leaf development: importance of pairing to the microRNA 5′ region. EMBO J 23:3356–3364

Marin-Gonzalez E, Suarez-Lopez P (2012) "And yet it moves": cell-to-cell and long-distance signaling by plant microRNAs. Plant Sci 196:18–30

Mauriat M, Moritz T (2009) Analyses of GA20ox- and GID1-over-expressing aspen suggest that gibberellins play two distinct roles in wood formation. Plant J 58:989–1003

McConnell JR, Barton MK (1998) Leaf polarity and meristem formation in Arabidopsis. Development 125:2935–2942

McConnell JR, Emery J, Eshed Y et al (2001) Role of PHABULOSA and PHAVOLUTA in determining radial patterning in shoots. Nature 411:709–713

Miguel A, Milhinhos A, Novak et al (2016) The SHORT-ROOT-like gene PtSHR2B is involved in Populus phellogen activity. J Exp Bot 67:1545–1555

Milhinhos A, Prestele J, Bollhoner B et al (2013) Thermospermine levels are controlled by an auxin-dependent feedback loop mechanism in Populus xylem. Plant J 75:685–698

Milhinhos A, Vera-Sirera F, Blanco-Touriñán N et al (2019) SOBIR1/EVR prevents precocious initiation of fiber differentiation during wood development through a mechanism involving BP and ERECTA. Proc Natl Acad Sci USA 116:18710–18716

Miyashima S, Koi S, Hashimoto T (2011) Non-cell-autonomous microRNA165 acts in a dose-dependent manner to regulate multiple differentiation status in the Arabidopsis root. Development 138:2303–2313

Miyashima S, Roszak P, Sevilem I et al (2019) Mobile PEAR transcription factors integrate hormone and miRNA cues to prime cambial growth. Nature 565:490–494

Natividade JV (ed) (1950) Subericultura. DGSFA, Lisboa

Ohashi-Ito K, Fukuda H (2003) HD-Zip III homeobox genes that include a novel member, ZeHB-13 (Zinnia)/ATHB-15 (Arabidopsis), are involved in procambium and xylem cell differentiation. Plant Cell Physiol 44:1350–1358

Otsuga D, DeGuzman B, Prigge MJ et al (2001) REVOLUTA regulates meristem initiation at lateral positions. Plant J. 25:223–236

Pereira-Leal JB, Abreu IA, Alabaça CS et al (2014) A comprehensive assessment of the transcriptome of cork oak (*Quercus suber*) through EST sequencing. BMC Genom 15:371

Porth I, Klápste J, McKown AD et al (2014) Extensive functional pleiotropy of REVOLUTA substantiated through forward genetics. Plant Physiol 164:548–554

Potkar R, Recla J, Busov V (2013) ptr-MIR169 is a posttranscriptional repressor of PtrHAP2 during vegetative bud dormancy period of aspen (*Populus tremuloides*) trees. Biochem Biophys Res Commun 431:512–518

Prigge MJ, Otsuga D, Alonso JM et al (2005) Class III homeodomain-leucine zipper gene family members have overlapping, antagonistic, and distinct roles in Arabidopsis development. Plant Cell 17:61–76

Puzey JR, Karger A, Axtell M et al (2012) Deep annotation of *Populus trichocarpa* microRNAs from diverse tissue sets. PLoS ONE 7:e33034

Qiu Z, Li X, Zhao Y et al (2015) Genome-wide analysis reveals dynamic changes in expression of microRNAs during vascular cambium development in Chinese fir, *Cunninghamia lanceolata*. J Exp Bot 66:3041–3054

Quan M, Wang Q, Phangthavong S et al (2016) Association studies in *Populus tomentosa* reveal the genetic interactions of Pto-MIR156c and its targets in wood formation. Front Plant Sci 7:1159

Quan M, Xiao L, Lu W et al (2018) Association genetics in Populus reveal the allelic interactions of Pto-MIR167a and its targets in wood formation. Front Plant Sci 9:744

Quan M, Du Q, Xiao L et al (2019) Genetic architecture underlying the lignin biosynthesis pathway involves noncoding RNAs and transcription factors for growth and wood properties in Populus. Plant Biotechnol J 17:302–315

Ragni L, Nieminen K, Pacheco-Villalobos D et al (2011) Mobile gibberellin directly stimulates arabidopsis hypocotyl xylem expansion. Plant Cell 23:1322–1336

Ramachandran P, Carlsbecker A, Etchells P (2017) Class III HD-ZIPs govern vascular cell fate: an HD view on patterning and differentiation. J Exp Bot 68:55–69

Ramos AM, Usié A, Barbosa P et al (2018) The draft genome sequence of cork oak. Sci Data 5:180069

Randall RS, Miyashima S, Blomster T et al (2015) AINTEGUMENTA and the D type cyclin CYCD3;1 regulate root secondary growth and respond to cytokinins. Biol Open 4:1229–1236

Ratcliffe OJ, Riechmann JL, Zhang JZ (2000) INTERFASCICULAR FIBERLESS1 is the same gene as REVOLUTA. Plant Cell 12:315–317

Reinhart BJ, Weinstein EG, Rhoades MW et al (2002) MicroRNAs in plants. Genes Dev 16:1616–1626

Rhoades MW, Reinhart BJ, Lim LP et al (2002) Prediction of plant microRNA targets. Cell 110:513–520

Ridoutt BG, Pharis RP, Sands R (1996) Fibre length and gibberellins A1 and A20 are decreased in *Eucalyptus globulus* by acylcyclohexanedione injected into stem. Physiol Plant 96:559–566

Robischon M, Du J, Miura E, Groover A (2011) The Populus class III HD ZIP, popREVOLUTA, influences cambium initiation and patterning of woody stems. Plant Physiol 155(3):1214–1225. https://doi.org/10.1104/pp.110.167007

Rodriguez RE, Ercoli MF, Debernardi JM et al (2015) MicroRNA miR396 regulates the switch between stem cells and transit-amplifying cells in Arabidopsis roots. Plant Cell 27:3354–3366

Sakaguchi J, Watanabe Y (2012) miR165/166 and the development of land plants. Dev Growth Differ 54:93–99

Scheres B, Wolkenfelt H, Willemsen V et al (1994) Embryonic origin of the Arabidopsis primary root and root meristem initials. Development 120:2475–2487

Shi W, Quan M, Du Q et al (2017) The interactions between the long non-coding RNA NERDL and its target gene affect wood formation in *Populus tomentosa*. Front Plant Sci 8:1035

Shi D, Lebovka I, López-Salmerón V et al (2019) Bifacial cambium stem cells generate xylem and phloem during radial plant growth. Development 146:dev171355

Smetana O, Mäkilä R, Lyu M et al (2019) High levels of auxin signalling define the stem-cell organizer of the vascular cambium. Nature 565:485–489

Soler M, Serra O, Molinas M et al (2007) A genomic approach to suberin biosynthesis and cork differentiation. Plant Physiol 144:419–431

Sundell D, Street NR, Kumar M et al (2017) AspWood: High-spatial-resolution rranscriptome profiles reveal uncharacterized modularity of wood formation in *Populus tremula*. Plant Cell 29:1585–1604

Teixeira RT, Fortes AM, Bai H et al (2018) Transcriptional profiling of cork oak phellogenic cells isolated by laser microdissection. Planta 247:317–338

Thamm A, Sanegre-Sans S, Paisley J et al (2019) A simple mathematical model of allometric exponential growth describes the early three-dimensional growth dynamics of secondary xylem in Arabidopsis roots. Roy Soc Open Sci 6:190126

Tian J, Chen J, Li B et al (2016) Association genetics in *Populus* reveals the interactions between *Pto*-miR160a and its target *Pto-ARF16*. Mol Genet Genomics 291:1069–1082

Todesco M, Balasubramanian S, Cao J et al (2012) Natural variation in biogenesis efficiency of individual *Arabidopsis thaliana* microRNAs. Curr Biol 22:166–170

Tuominen H, Puech L, Fink S et al (1997) A radial concentration gradient of indole-3-acetic acid is related to secondary xylem development in hybrid aspen. Plant Physiol 115:577–585

Uggla C, Moritz T, Sandberg G et al (1996) Auxin as a positional signal in pattern formation in plants. Proc Natl Acad Sci USA 93:9282–9286

Vulavala VK, Fogelman E, Faigenboim A et al (2019) The transcriptome of potato tuber phellogen reveals cellular functions of cork cambium and genes involved in periderm formation and maturation. Sci Rep 9:10216

Wang H, Avci U, Nakashima J et al (2010) Mutation of WRKY transcription factors initiates pith secondary wall formation and increases stem biomass in dicotyledonous plants. Proc Natl Acad Sci USA 107:22338–22343

Wang JW, Park MY, Wang LJ et al (2011) MicroRNA control of vegetative phase change in trees. PLoS Genet 2:e1002012

Wang C, Zhang S, Yu Y et al (2014) MiR397b regulates both lignin content and seed number in Arabidopsis via modulating a laccase involved in lignin biosynthesis. Plant Biotechnol J 12:1132–1142

Wareing PF (1958) Interaction between indole-acetic and gibberellic acid in cambial activity. Nature 181:1744–1745

Weigel D, Jurgens G (2002) Stem cells that make stems. Nature 415:751–754

Wu G, Poethig RS (2006) Temporal regulation of shoot development in *Arabidopsis thaliana* by miR156 and its target SPL3. Development 133:3539–3547

Wu G, Park MY, Conway SR et al (2009) The sequential action of miR156 and miR172 regulates developmental timing in Arabidopsis. Cell 138:750–759

Wunderling A, Ripper D, Barra-Jimenez A et al (2018) A molecular framework to study periderm formation in Arabidopsis. New Phytol 219:216–229

Xiao L, Quan M, Du Q et al (2017) Allelic Interactions among *Pto*-MIR475b and its four target genes potentially affect growth and wood properties in *Populus*. Front Plant Sci 8

Xu C, Shen Y, He F et al (2018) Auxin-mediated Aux/IAA-ARF-HB signaling cascade regulates secondary xylem development in *Populus*. New Phytol 222:752–767

Yang X, Du Q, Chen J et al (2015) Association mapping in Populus reveals the interaction between Pto-miR530a and its target Pto-KNAT1. Planta 242:77–95

Yoshida S, Barbier de Reuille P, Lane B et al (2014) Genetic control of plant development by overriding a geometric division rule. Dev Cell 29:75–87

Zhang Z, Zhang X (2012) Argonautes compete for miR165/166 to regulate shoot apical meristem development. Curr Opin Plant Biol 15:652–658

Zhao Y, Lin S, Qiu Z et al (2015) MicroRNA857 is involved in the regulation of secondary growth of vascular tissues in Arabidopsis. Plant Physiol 169:2539–2552

Zhong R, Ye ZH (1999) *IFL1*, a gene regulating interfascicular fiber differentiation in Arabidopsis encodes a homeodomain-leucine zipper protein. Plant Cell 11:2139–2152

Zhong R, Ye ZH (2004) *Amphivasal vascular bundle 1*, a gain-of-function mutation of the *IFL1/REV* gene, is associated with alterations in the polarity of leaves, stems and carpels. Plant Cell Physiol 45:369–385

Zhou G-K, Kubo M, Zhong R et al (2007) Overexpression of miR165 affects apical meristem formation, organ polarity establishment and vascular development in Arabidopsis. Plant Cell Physiol 48:391–404

Zhou Y, Honda M, Zhu H et al (2015) Spatiotemporal sequestration of miR165/166 by Arabidopsis argonaute10 promotes shoot apical meristem maintenance. Cell Rep 10:1819–1827

Zhu H, Hu F, Wang R et al (2011) Arabidopsis Argonaute10 specifically sequesters miR166/165 to regulate shoot apical meristem development. Cell 145:242–256

Zhu Y, Song D, Sun J (2013) PtrHB7, a class III HD-Zip gene, plays a critical role in regulation of vascular cambium differentiation in Populus. Mol Plant 6:1331–1343

Zhu Y, Song D, Xu P et al (2018) A HD-ZIP III gene, PtrHB4, is required for interfascicular cambium development in *Populus*. Plant Biotechnol J 16:808–817

Part III
microRNAs in Plant Abiotic and Biotic Stresses

Chapter 9
Regulation of High-Temperature Stress Response by Small RNAs

Henrik Mihály Szaker, Péter Gyula, György Szittya and Tibor Csorba

Abstract Temperature extremes constitute one of the most common environmental stresses that adversely affect the growth and development of plants. Transcriptional regulation of temperature stress responses, particularly involving protein-coding gene networks, has been intensively studied in recent years. High-throughput sequencing technologies enabled the detection of a great number of small RNAs that have been found to change during and following temperature stress. The precise molecular action of some of these has been elucidated in detail. In the present chapter, we summarize the current understanding of small RNA-mediated modulation of high-temperature stress-regulatory pathways including basal stress responses, acclimation, and thermo-memory. We gather evidence that suggests that small RNA network changes, involving multiple upregulated and downregulated small RNAs, balance the trade-off between growth/development and stress responses, in order to ensure successful adaptation. We highlight specific characteristics of small RNA-based temperature stress regulation in crop plants. Finally, we explore the perspectives of the use of small RNAs in breeding to improve stress tolerance, which may be relevant for agriculture in the near future.

Keywords Plants · Heat stress · RNA silencing · Thermomemory · Adaptation

9.1 Introduction

Temperature is a primary environmental factor that is continuously monitored by living organisms to adjust their growth and reproduction accordingly. Due to climate change, the global temperature is constantly increasing. Besides this, the frequency and amplitude of extreme temperature events are also elevated. Extreme temperatures exceeding the optimal range of a given organism have an adverse impact on almost every aspect of life including growth and reproduction. The extreme heat exposure damages lipid membranes, proteins, and nucleic acids, etc. Being sessile organisms,

---

H. M. Szaker · P. Gyula · G. Szittya · T. Csorba (✉)
National Agricultural Research and Innovation Center, Agricultural Biotechnology Institute (NARIC-ABC), Szent-Györgyi A. 4, Gödöllő 2100, Hungary
e-mail: csorba.tibor@abc.naik.hu

C. Miguel et al. (eds.), *Plant microRNAs*, Concepts and Strategies in Plant Sciences, https://doi.org/10.1007/978-3-030-35772-6_9

plants are especially exposed to extreme temperature fluctuations and heat stress (Luria et al. 2019; Mazzeo et al. 2018; Mittler et al. 2012; Ohama et al. 2017; Scharf et al. 2012). To mitigate the negative effects, heat stress response (HSR) pathways are initiated. HSR is a multilayered complex set of physiological and phenological alterations that help plants to cope with the extreme temperatures and sudden temperature changes and enables survival and reproduction (Mittler et al. 2012; Ohama et al. 2017; Scharf et al. 2012). HSR acts on epigenetic, transcriptional and post-transcriptional levels (Ohama et al. 2017).

9.2 Regulation of HSR

Plant heat stress response is initiated by a number of signaling mechanisms. One central role of heat stress signaling is the activation of Heat Stress Factors (HSFs). HSFs are pioneer transcription factors (TFs) that recognize and bind to tandem inverted repeats of nGAAn consensus sequence within the target gene loci to activate transcription (Maruyama et al. 2017). HSF family comprises multiple members (for example *Arabidopsis thaliana* has 21, *Brassica napus* 64, soybean 52, maize 30 members) (Scharf et al. 2012; Zhu et al. 2017). HSF members show high functional diversification and partial genetic redundancy. HSFs might be induced during other stresses or combined stress conditions as well. The master regulators of HSR are the Heat Stress Transcription factors HSFA1 family proteins (in *A. thaliana* HSFA1A, HSFA1B, and HSFA1D) (Liu and Charng 2013; Liu et al. 2011; Ohama et al. 2017; Yoshida et al. 2011). In the absence of these factors, induction of a large number of downstream HS-inducible genes is absent, and plants become HS-sensitive. Temperature-sensitivity of plants becomes apparent already at high ambient temperatures in *hsfa1*-mutant plants, suggesting that the HSF-mediated stress pathways are initiated at these temperatures (Cortijo et al. 2017). *HSFA2* is a direct target of HSFA1s and is a major enhancer of thermotolerance. HSFA1 and HSFA2 proteins form a heteromultimer TF complex (superactivator complex) and strongly induce expression of downstream HSFs, transcription factors, heat shock proteins (HSPs), non-chaperone proteins or non-coding RNAs (Chan-Schaminet et al. 2009; Scharf et al. 1998, 2012). If plants encounter mild temperature stress they become acclimated. This, so-called acquired thermotolerance, allows plants to survive later stronger heat stress exposures (Lin et al. 2014; Liu and Charng 2012). Acquired thermotolerance may be actively maintained for several days post-stress through a process called heat stress memory (Brzezinka et al. 2016; Charng et al. 2007; Lamke et al. 2016; Liu et al. 2018; Stief et al. 2014). Transcriptional memory allows a faster or stronger stress response to be initiated when the same stress is re-encountered. HSFA2 is a central component of thermomemory. Transient binding of HSFA2 induces deposition of Histone 3 lysine 4 dimethylation and trimethylation (H3K4me2, H3K4me3) at memory loci (Lamke et al. 2016). These histone modifications persist after active transcription was subsided, marking them as recently transcriptionally active. Upon recurring heat stress, transcription of these memory loci is hyperinduced, and thus enables a quicker and more efficient HSR. Besides HSFA2, a growing number of stress memory factors have been discovered (Brzezinka et al. 2018; Charng et al. 2006; Lin et al. 2014; Liu et al. 2018; Meiri and Breiman 2009). Thermomemory was shown to be transmitted transgenerationally (Liu et al. 2019). Micro RNAs (miRNA) and small interfering RNAs (siRNA) are intimately involved in high-temperature stress response and thermomemory.

9.3 Abiotic Stresses Alter sRNAs' Level

The central molecules of RNA silencing pathways are short non-coding regulatory RNAs (small RNAs, sRNA), with the size of 21–24 nucleotides (Axtell 2013; Borges and Martienssen 2015; Rogers and Chen 2013; Vaucheret 2006). There are several classes of sRNAs including miRNA and small interfering RNAs (trans-acting siRNAs or ta-siRNAs; heterochromatic siRNAs or het-siRNA, repeat-associated siRNAs or ra-siRNA, etc.). miRNAs and siRNAs differ in their biogenesis. miRNAs are encoded on distinct genomic loci; their precursors (pri-miRNA) are transcribed by RNA Polymerase II, capped, and polyadenylated. In *A. thaliana*, DICER-LIKE 1 (DCL1), together with SERRATE (SE) and HYPONASTIC LEAVES 1 (HYL1) forms the Microprocessor complex. Microprocessor complex processes pri-miRNAs in two subsequent steps to generate pre-miRNA then mature miRNAs. miRNAs are mostly 21-nt long; they are loaded into RNA-induced Silencing Complex (RISC) effector protein ARGONAUTE (AGO1 in *A. thaliana*) and guide sequence-specifically the cleavage or inhibit translation of complementary target mRNAs through a process called Post-Transcriptional Gene Silencing (PTGS) or RNA silencing. 22-nt-long (or in special cases, 21-nt-long) miRNAs not only cleave their targets but also trigger the biogenesis of secondary siRNAs. Secondary siRNA biogenesis requires the activity of DCLs, AGOs and RNA-DEPENDENT RNA POLYMERASES (RDRs). In *A. thaliana*, DCL4, AGO1 or AGO7, and RDR6 are the key enzymes involved in secondary siRNA biogenesis (Dalmay et al. 2000; Martinez de Alba et al. 2013). The secondary siRNA further represses target RNAs *in trans* (at PTGS level), therefore, are called trans-acting siRNAs. miRNAs and ta-siRNAs are involved in almost every aspect of plant growth and development.

The 24-nt-long siRNAs (het-siRNA, ra-siRNA) originate from long double-stranded RNA molecules and serve to silence specific regions of the genome through guiding DNA methylation (RNA-directed DNA methylation, RdDM) and subsequent histone modifications (Borges and Martienssen 2015). This affects gene expression at the transcriptional level; therefore, the process is called Transcriptional Gene Silencing (TGS). During TGS, stable repression of transcription is established at transposons, chromosomal repeats, transgenic inserts but also may occur at certain protein-coding genes. The main function of TGS is the preservation of genome stability and epigenetic regulation of certain genes.

To date, thousands of sRNA have been identified through bioinformatic predictions, cloning and deep sequencing, or hybridization methods both from model plants and numerous crops (Baksa et al. 2015; Barciszewska-Pacak et al. 2015; Beckers et al. 2017; Gyula et al. 2018; Kruszka et al. 2014; Liu et al. 2017; May et al. 2013; Mohorianu et al. 2013; Zhou et al. 2016).

Besides their role in growth, development and genome stability, sRNAs are also important components of stress responses. Environmental alteration of sRNAs was observed in a variety of species like *A. thaliana*, *Brassica rapa*, *Populus euphratica*, *Triticum aestivum, Oryza sativa*, etc. (Barciszewska-Pacak et al. 2015; Chen et al. 2012; Kumar et al. 2015; Mangrauthia et al. 2017; Sunkar and Zhu 2004; Xin et al. 2011; Yu et al. 2012). Different environmental conditions including heat, drought, salt, oxidative stress, hypoxia, cold, nutrient deprivation, UV-B, strong light, etc. caused severe alteration of sRNA levels or biosynthesis of novel sRNAs (see Table 9.1). Stress-responsiveness suggests regulatory roles during stress or recovery period. This is supported also by the differences between sRNAs' stress-response observed between stress-sensitive and -tolerant varieties or cultivars (Liu et al. 2017; Mangrauthia et al. 2017).

The fact that stress-induced changes have similar trends, occur in multiple species and correlate with the stress-sensitivity, points to a conserved mechanism involving sRNAs as active players. However, some sRNAs' stress response was found to be very specific. For example, miR159, miR160, miR164, miR166, miR319 or miR398 were oppositely regulated in different species (Kumar et al. 2015; Liu et al. 2017; Xin et al. 2010): miR156 family members are induced in response to HS in *A. thaliana, Brassica rapa,* and wheat (Stief et al. 2014; Xin et al. 2010; Yu et al. 2012), but repressed in cassava, or rice (Ballen-Taborda et al. 2013; Liu et al. 2017); miR164 was upregulated and miR164-target NASCENT POLYPEPTIDE-ASSOCIATED COMPLEX1 (NAC1) was suppressed in *Arabidopsis* (May et al. 2013); oppositely to this, tae-miR164 was downregulated and its target HSP17 upregulated in wheat (Kumar et al. 2015); miR164, miR160, miR319, and miR398 were upregulated in *Arabidopsis* but downregulated in wheat (Kumar et al. 2015). Some miRNAs possess tissue-specific changes characteristic to roots, leaves or flowers (Gyula et al. 2018; Liu et al. 2017; Mangrauthia et al. 2017). In specific cases, novel sRNA-target interactions have evolved from conserved miRNA and mRNA target components: for example, in sunflower, the HaWRKY6 transcription factor recently evolved to be the target of the highly conserved miR396 (Giacomelli et al. 2012). These findings warn that sRNA-based stress response regulation is complex. Detailed studying of unique sRNA-target pairs and characterization of their putative roles during HSR response in every specific case is required before generalization is made about their role in non-model plants.

A handful of sRNA::target pairs have been studied in detail. Accumulating evidence shows that sRNA-mediated regulation plays essential roles in plant responses to abiotic stresses, including high-temperature stress response (Cui et al. 2014; Guan et al. 2013; He et al. 2018; Ma et al. 2015; Stief et al. 2014) (see Table 9.1). sRNAs were shown to be required for adaptation to the mild elevation in ambient temperatures, some sRNAs are involved in the establishment of acute stress response, acquired thermo-tolerance or thermomemory in response to heat, while others indirectly alter development in the post-stress period. Importantly, changes in sRNA

Table 9.1 sRNA changes during high-temperature stress response in plants

sRNA	Known targets	Species	Direction of change	References
miR156	SPLS (SQUAMOSA PROMOTER BINDING PROTEIN LIKE)	*Arabidopsis thaliana*	Up	Stief et al. (2014)
		Brassica rapa	Up (miR156h, g)	Yu et al. (2012)
		Manihot esculenta	Down	Ballen-Taborda et al. (2013)
		Oryza sativa	Down	Sailaja et al. (2014), Liu et al. (2017)
		Solanum pimpinellifolium L.	Up	Zhou et al. (2016)
		Triticum aestivum L.	Up	Xin et al. (2011)
miR159	MYBS (MYB DNA-BINDING DOMAIN CONTAINING PROTEINS)	*Cucumis sativus*	Down (if +ABA or +Luffa rootstock)	Li et al. (2016a)
		Panicum virgatum	Up	Hivrale et al. (2016)
		Triticum aestivum L.	Up	Xin et al. (2011), Wang et al. (2012)
miR160	ARF10, ARF16, ARF17 (AUXIN RESPONSIVE FACTORS)	*Apium graveolens* L.	Up	Li et al. (2014a)
		Arabidopsis thaliana	Up	Lin et al. (2018), May et al. (2013)
		Helianthus annuus	Up (1,5h HS) Down (3-6h HS)	Ebrahimi Khaksefidi et al. (2015)
		Hordeum vulgare	Up (miR160a)	Kruszka et al. (2014)
		Oryza sativa	Up (shoot) Down (root)	Sailaja et al. (2014)
		Panicum virgatum	Up	Hivrale et al. (2016)

(continued)

Table 9.1 (continued)

sRNA	Known targets	Species	Direction of change	References
		Populus tomentosa	Down (miR160a, b, c)	Chen et al. (2012)
		Solanum pimpinellifolium L.	Up	Zhou et al. (2016)
		Triticum aestivum L.	Down	Kumar et al. (2015)
miR164	NAC1 (NASCENT POLYPEPTIDE-ASSOCIATED COMPLEX1), HSP17 (HEAT SHOCK PROTEIN 17)	*Apium graveolens* L.	Up	Li et al. (2014a)
		Arabidopsis thaliana	Up	May et al. (2013)
		Panicum virgatum	Up	Hivrale et al. (2016)
		Triticum aestivum L.	Down	Kumar et al. (2015)
miR166	PHV (PHAVOLUTA), REV (REVOLUTA), HOX9 (HOMEOBOX LEUCINE ZIPPER PROTEIN 9)	*Arabidopsis thaliana*	Up	May et al. (2013)
		Hordeum vulgare	Up (miR166a)	Kruszka et al. (2014)
		Panicum virgatum	Up	Hivrale et al. (2016)
		Triticum aestivum L.	Up (miR166d)	(Xin et al. 2011)
miR167	ARFS (AUXIN RESPONSIVE FACTORS), IAA-ALA RESISTANT3 (IAR3)	*Brassica rapa*	Up (endosperm, pollen)	Bilichak et al. (2015)
		Helianthus annuus	Up (1,5h HS), Down (3-6h HS)	Ebrahimi Khaksefidi et al. (2015)
		Hordeum vulgare	Up (miR167 h)	Kruszka et al. (2014)
		Oryza sativa	Down	Sailaja et al. (2014)
		Panicum virgatum	Up	Hivrale et al. (2016)
		Populus tomentosa	Down (miR167c, d)	Chen et al. (2012)

(continued)

Table 9.1 (continued)

sRNA	Known targets	Species	Direction of change	References
		Solanum pimpinellifolium L.	Up	Zhou et al. (2016)
		Triticum aestivum L.	Up	Xin et al. (2011)
miR168	AGO1 (ARGONAUTE 1)	*Apium graveolens* L.	Up	Li et al. (2014a)
		Brassica rapa	Up (endosperm)	Bilichak et al. (2015)
		Oryza sativa	Down	Sailaja et al. (2014)
		Panicum virgatum	Up	Hivrale et al. (2016)
		Populus tomentosa	Down (miR168a, b)	Chen et al. (2012)
		Triticum aestivum L.	Up	Xin et al. (2011)
miR169	NF-YS (NUCLEAR TRANSCRIPTION FACTOR Y)	*Arabidopsis thaliana*	Up	Guan et al. (2013)
		Oryza sativa	Up (shoot), Down (root)	Sailaja et al. (2014)
		Panicum virgatum	Down	Hivrale et al. (2016)
		Populus tomentosa	Up (miR169j) Down (miR169a, b, n)	Chen et al. (2012)
		Triticum aestivum L.	Up	Xin et al. (2011)
miR171	SCL6 (SCARECROW-LIKE6)	*Arabidopsis thaliana*	Up	Mahale et al. (2013)
		Panicum virgatum	Down	Hivrale et al. (2016)
		Populus tomentosa	Down	Chen et al. (2012)
		Populus trichocarpa	Down	Lu et al. (2008)
miR172	AP2 (APETALA2)	*Arabidopsis thaliana*	Down	May et al. (2013)

(continued)

Table 9.1 (continued)

sRNA	Known targets	Species	Direction of change	References
		Helianthus annuus	Down	Ebrahimi Khaksefidi et al. (2015)
		Oryza sativa	Down	Mangrauthia et al. (2017)
		Triticum aestivum L.	Down	Xin et al. (2011)
miR319	TCPS (TEOSINTE BRANCHED1, CYCLOIDEA, AND PROLIFERATING CELL NUCLEAR ANTIGEN BINDING FACTOR), MYBS (MYB DNA-BINDING DOMAIN CONTAINING PROTEINS)	*Arabidopsis thaliana*	Up	Li et al. (2014b)
		Oryza sativa	Up	Mangrauthia et al. (2017)
		Panicum virgatum	Up	Hivrale et al. (2016)
		Triticum aestivum L.	Down	Kumar et al. (2015)
miR390	ARFS (AUXIN RESPONSIVE FACTORS)	*Panicum virgatum*	Up	Hivrale et al. (2016)
		Brassica rapa	Up (endosperm, pollen)	Bilichak et al. (2015)
miR393	TIR1 (TRANSPORT INHIBITOR RESPONSE 1)	*Arabidopsis thaliana*	Down	Guan et al. (2013)
		Oryza sativa	Up	Mangrauthia et al. (2017)
		Panicum virgatum	Up	Hivrale et al. (2016)
		Solanum pimpinellifolium L.	Up	Zhou et al. (2016)
		Triticum aestivum L.	Up	Xin et al. (2011)
miR394	LCR (LEAF CURLING RESPONSIVENESS)	*Apium graveolens* L.	Up	Li et al. (2014a)

(continued)

Table 9.1 (continued)

sRNA	Known targets	Species	Direction of change	References
		Populus tomentosa	Down (miR394a, b)	Chen et al. (2012)
miR395	APSS (ATP SULFURYLASES), SULTR2;1 (SULFATE TRANSPORTER 2;1)	*Apium graveolens* L.	Up	Li et al. (2014a)
		Panicum virgatum	Up	Hivrale et al. (2016)
		Populus tomentosa	Down	Chen et al. (2012)
		Solanum pimpinellifolium L.	Up	Zhou et al. (2016)
		Triticum aestivum L.	Up	Kumar et al. (2015)
miR396	GRFS (GROWTH-REGULATING FACTOR), BHLH74 (BASIC HELIX-LOOP-HELIX 74), HAWRKY6	*Helianthus annuus*	Up (1,5h HS) Down (3-24h HS)	Giacomelli et al. (2012)
		Oryza sativa	Down	Liu et al. (2017), Mangrauthia et al. (2017)
		Panicum virgatum	Up	Hivrale et al. (2016)
miR397	LACCASES	*Arabidopsis thaliana*	Down	Mahale et al. (2013)
		Oryza sativa	Up	Jeong et al. (2011)
miR398	CSD1 (COPPER/ZINC SUPEROXID DISMUTASE1), CSD2, CCS (COPPER CHAPERONE FOR SOD), COX5 (SUBUNIT OF THE MITOCHONDRIAL CYTOCHROME C OXIDASE)	*Arabidopsis thaliana*	Up	Guan et al. (2013), Fang et al. (2019)
		Brassica rapa	Down (miR398a, b)	Yu et al. (2012)
		Helianthus annuus	Up	Ebrahimi Khaksefidi et al. (2015)
		Oryza sativa	Down	Sailaja et al. (2014)

(continued)

Table 9.1 (continued)

sRNA	Known targets	Species	Direction of change	References
		Panicum virgatum	Up	Hivrale et al. (2016)
		Populus tomentosa	Down (miR398a, b)	Chen et al. (2012)
		Triticum aestivum L.	Up	Xin et al. (2011)
		Zea mays	Up	Guan et al. (2013)
miR399	PHO2 (PHOSPHATE 2)	*Arabidopsis thaliana*	Down	Kim et al. (2011)
		Brassica rapa	Down	Yu et al. (2012)
		Oryza sativa	Down	Liu et al. (2017)
		Triticum aestivum L.	Up	Xin et al. (2011)
miR400	PPRS (PENTATRICOPEPTID REPEAT PROTEINS)	*Arabidopsis thaliana*	Down	Yan et al. (2012)
miR403	AGO2 (ARGONAUTE 2)	*Helianthus annuus*	Up	Ebrahimi Khaksefidi et al. (2015)
miR408	LACCASES, PLASTOCYANIN-LIKE FAMILY PROTEINS	*Apium graveolens* L.	Up	Li et al. (2014a)
		Oryza sativa	Up	Mangrauthia et al. (2017)
		Panicum virgatum	Down	Hivrale et al. (2016)
		Populus tomentosa	Down	Chen et al. (2012)
		Saccharina japonica	Down	Liu et al. (2013)
		Solanum pimpinellifolium L.	Up	Zhou et al. (2016)
		Triticum aestivum L.	Up	Kumar et al. (2015)

(continued)

Table 9.1 (continued)

sRNA	Known targets	Species	Direction of change	References
miR529	SPLS (SQUAMOSA PROMOTER BINDING PROTEIN LIKE)	*Panicum virgatum*	Up	Hivrale et al. (2016)
miR827	SPX (SYG1/PHO81/XPR1)-DOMAIN-CONTAINING GENES, NLA (NITROGEN LIMITATION ADAPTATION), PHT5 (PHOSPHATE TRANSPORTER 5)	*Panicum virgatum*	Up	Hivrale et al. (2016)
		Triticum aestivum L.	Up	Xin et al. (2011)
miR842	JACALIN LECTIN FAMILY MEMBERS	*Helianthus annuus*	Up	Ebrahimi Khaksefidi et al. (2015)
miR1884	–	*Oryza sativa*	Up (shoot) Down (root)	Sailaja et al. (2014)
miR1885b.3	–	*Brassica rapa*	Down	Yu et al. (2012)
miR2005	–	*Triticum aestivum* L.	Up	Xin et al. (2011)
miR5175	–	*Hordeum vulgare*	Up (miR5175a)	Kruszka et al. (2014)
miR5714	–	*Brassica rapa*	Up	Yu et al. (2012)
miR5716	–	*Brassica rapa*	Down	Yu et al. (2012)
miR5736	–	*Brassica rapa*	Up	Yu et al. (2012)
TAS1-derived ta-siRNAs	HTT1, HTT2 (HEAT-INDUCED TAS1 TARGET1 AND 2)	*Arabidopsis thaliana*	Up	Li et al. (2014b)
Ta-miR2002	–	*Triticum aestivum* L.	Up	Xin et al. (2011)
SiRNA 002061_0636_3054.1	–	*Triticum aestivum* L.	Down	Yao and Sun (2012)
SiRNA 005047_0654_1904.1	–	*Triticum aestivum* L.	Down	Yao and Sun (2012)
SiRNA 080621_1340_98.1	–	*Triticum aestivum* L.	Down	Yao and Sun (2012)

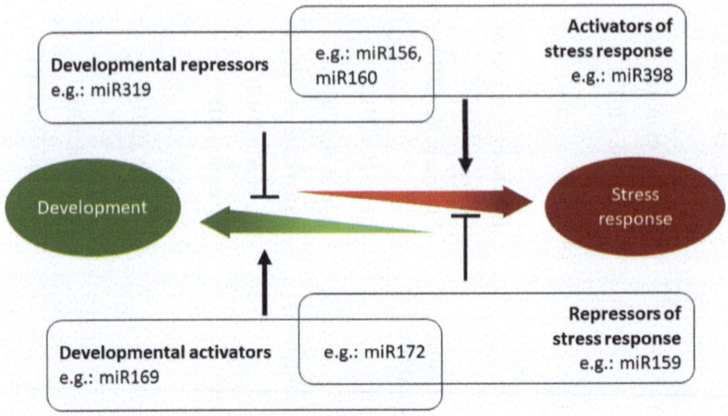

Fig. 9.1 Small RNAs act in a complex network to regulate the balance between developmental pathways and stress responses. Examples are given for each category (for activity of specific sRNAs, please see Table 9.1)

or/and target act in a complex, interconnected network that fine-tunes the balance between growth and stress response (Fig. 9.1).

9.4 sRNAs' Involvement in the Ambient Temperature Response

Plants substantially alter their developmental program upon changes in the ambient temperature (Quint et al. 2016; Wigge 2013). It was shown recently that changes in ambient temperature affect the steady-state levels of only a small fraction of the total sRNA loci (0.6%) in *A. thaliana* (Gyula et al. 2018). One of these temperature-responsive loci can add an extra regulatory layer to the auxin-dependent thermomorphogenesis by regulating epigenetically the expression of the auxin synthesis gene *YUCCA2* (*YUC2*). This finding also suggests that the temperature-dependency of these loci is not conferred by some common siRNA biogenesis components (i.e. PolIV/V, RDR2, DCL3) (Borges and Martienssen 2015), but some other, unknown *trans* factors.

Several *Arabidopsis* miRNAs were identified in microarray and northern blot experiments to be differentially expressed (miR156, miR163, miR169, miR172, miR398, and miR399) in response to ambient temperature changes (Lee et al. 2010). The gradient of miR156 and miR172 regulates the juvenile to adult phase transition and adult to reproductive phase transitions. miR156 is highly expressed during the juvenile stage and gradually decreasing its level as the plant ages (Poethig 2009). Targets of miR156 are the SQUAMOSA PROMOTER BINDING-LIKE (SPL) transcription factors. SPLs regulate a number of genes involved in floral time transition (Srikanth and Schmid 2011; Wang et al. 2009). Overexpression of miR156 in *Arabidopsis* causes altered vegetative phase transition and delayed flowering (Xing et al.

2013; Yu et al. 2010). The level of miR156 is high at low ambient temperatures (16 °C) and therefore could delay juvenile to adult phase transition or floral transition. The thermodynamic properties of mir156 precursor stem-loop structure can affect the cleavage efficiency of the mature miRNA/miRNA* duplex. Structural analysis studies showed that certain positions in the mir156a precursor sequence can particularly affect the cleavage efficiency, and in turn, the flowering time at different ambient temperatures (16 °C vs. 23 °C) (Kim et al. 2016). Therefore, the mir156 precursors can serve as a temperature sensor. Furthermore, warm temperature (23 °C) enhances the expression of miR172, which plays the opposite role to miR156. Overexpression of miR172 causes early flowering by down-regulating the AP2-domain transcription factors TARGET OF EAT1 (TOE1), TOE2, TOE3, SCHLAFMÜTZE (SMZ) and SCHNARCHZAPFEN (SNZ) (Aukerman and Sakai 2003; Chen 2004; Spanudakis and Jackson 2014).

Warm ambient temperature (27 °C) reduces the expression of miR169 (Gyula et al. 2018), which target mRNAs encoding subunits A of the NUCLEAR FACTOR Y (NF-Y) transcription factor complex. These TF form a conserved heterotrimeric TF complex that is composed of NF-Y subunit A (NF-YA), NF-YB and NF-YC subunits (Petroni et al. 2012). The miR169 family represses the NF-YA subunit, which affects the expression of the NF-Y -dependent genes, including the flowering time regulator *FLOWERING LOCUS T (FT),* the auxin synthesis gene *YUC2*, and could mediate the high ambient temperature-induced early flowering (Gyula et al. 2018; Siriwardana et al. 2016).

PHYTOCHROME B (PHYB) mediates red light and high ambient temperature signals through a subset of basic helix-loop-helix (b-HLH) transcription factors, the PHYTOCHROME INTERACTING FACTOR1 (PIF1), PIF3, PIF4 and PIF5 (Jung et al. 2016; Legris et al. 2016). PIF4 is a key regulator of the photo- and thermomorphogenesis. At high ambient temperature, the PIF4 alters the expression of hundreds of genes that result in morphological and physiological changes that are necessary for thermal adaptation. A recent study described a new, surprising function of this transcription factor (Sun et al. 2018). According to this, PIF4 not only regulates the transcription of some miRNA genes, but being the part of the Microprocessor complex, it affects the cleavage of the miRNA duplex from the miRNA-precursor in a red light-, and probably a temperature-regulated manner by promoting the degradation of DCL1 and HYL1 proteins. This moonlighting activity of PIF4 suggests that other bHLH proteins may also bind the DCL1 and HYL1 conveying other environmental or hormonal signals to the Microprocessor complex. Temperature affects miRNA processing at multiple levels; at low ambient temperature (16 °C), the miRNA processing machinery becomes more robust and it does not require the presence of DCL1 cofactors HYL1 and SE, for short base-to-loop processing of miRNA precursor (Re et al. 2019).

9.5 sRNAs Regulate Thermotolerance Response

Many sRNA (like miR156, miR160, miR164, miR166, miR169, miR171, miR319, miR398 or TAS1-derived ta-siRNAs) are also induced in response to heat (see Table 9.1). Heat-induced changes of miR160 contribute to thermotolerance through alteration of HSP expression. High-throughput sequencing and hybridization experiments have revealed that induction of miR160 occurs in various plant species (Kruszka et al. 2014; Kumar et al. 2015; Lin et al. 2018). HS induced miR160 transcriptionally, that caused repression of AUXIN RESPONSIVE FACTOR 10 (ARF10), ARF16 and ARF17 targets (Lin et al. 2018). Auxin decrease during heat also contributes to ARF10 and ARF16 repression through a miR160-independent pathway (Wang et al. 2005). Downregulation of ARFs was necessary for proper HSR: overexpression of miR160 improved seed germination and seedling survival in response to heat while miR160-mimicry plants became HS-sensitive (Lin et al. 2018). The HS-tolerant phenotype of *arf10*, *arf16*, and *arf17* mutant plants was similar to miR160-overexpressors. miR160's and ARFs' changes during HS alter the expression of multiple HSPs. Therefore, the intertwined activities of hormonal signals, protein homeostasis, and miR160-guided silencing contribute to basal thermotolerance in a positive manner. Some miRNAs might affect HS tolerance negatively. miR400 is a negative regulator of HS response in *Arabidopsis*. ath-miR400 is located within an intron and is co-transcribed with its host nascent RNA transcript. Under non-inductive conditions, the miR400-containing intron is spliced, and the miR400 maturated from this intron/precursor. Under HS conditions, alternative splicing of host genes led to partial intron retention and abolishment of miR400 processing (the level of the host gene was not affected). Overexpression of miR400 causes HS-sensitive phenotype, therefore, downregulation of miR400 is required for proper thermotolerance response (Yan et al. 2012). Alternative splicing, therefore, regulates HS response via the miRNA pathway.

miR159 has also a negative impact on HSR. It was shown that luffa (*Luffa cylindrica*) rootstock enhanced thermotolerance in cucumber (*Cucumis sativus*) plants (Li et al. 2016a). Luffa rootstock induced a significant downregulation of a subset of miRNAs in the shoot during HS, suggesting that miRNAs might participate in luffa rootstock-induced heat tolerance. Amongst these, miR156, miR159, miR169, and miR398 were linked to HS response previously (Guan et al. 2013; Kruszka et al. 2014; Xin et al. 2010). Luffa rootstock also induced enhanced abscisic acid (ABA) accumulation in the shoot. It is known that ABA plays a role in the regulation of several miRNAs, such as miR159. miR159 targets are the MYB transcription factors that are involved in abiotic stress responses. Foliar application of exogenous ABA suppressed miR159 and induced CsMYB29-like and CsGAMYB1 target levels (Li et al. 2016a). Transgenic tae-miR159-overexpressor rice, csa-miR159-overexpressor, and *myb33;myb65* double mutant *Arabidopsis* plants were shown to be HS-sensitive (Li et al. 2016a; Wang et al. 2012). AtHSP70 transcript and protein accumulation were impaired in miR159b overexpression lines suggesting that HSP70 may be directly or indirectly, negatively regulated by miR159 under natural conditions. miR159 downregulation, therefore, is required for efficient HSR. It was suggested that cucumber plants that originate in the temperate climate have evolved poor HSR compared to luffa plants. However, the role of the *Arabidopsis* miR159-MYB33/65 pathway in abiotic stress responses including heat was questioned by another study (Li et al. 2016b).

The decrease of ta-siRNAs is also required for proper HSR. It has been shown that targets of TAS1-derived ta-siRNA, named HEAT-INDUCED TAS1 TARGET1 (HTT1) and HTT2 regulate thermotolerance (Li et al. 2014b). HTT1 and HTT2 are strongly accumulated during HS due to the direct HSFA1s-mediated transcriptional activation. Parallel to this, HS inhibited the production of TAS1-derived ta-siRNAs, the negative regulators of HTT1 and HTT2 that contributed to the elevated level of HTT1 and HTT2 under HS conditions. Overexpression of TAS1a reduced expression of HTT genes and caused HS-sensitive phenotype while overexpression of HTT1 and HTT2 upregulated several HSF genes and improved thermotolerance of transgenic plants. HTT1 and HTT2 positively regulate HSR through transcriptional activation of HS-related targets within a transcriptional complex comprising HSP70-14, HSP40, and NF-Y complex subunit C2 (NF-YC2) factors.

9.6 Role of sRNAs in Acquired Thermotolerance and Heat Stress Memory

Stress events are recurring in nature. The encounter of nonlethal stress leads to the acquisition of thermotolerance (acquired thermotolerance or AT). AT functions to prime the plants for a quicker and/or stronger HSR during later, more severe stresses and to alter post-stress development of plants in order to be able to complete the life cycle before lethal stress (e.g. modulate growth rate, flowering time, senescence, etc.). *ago1* mutant plants are impaired in acquired thermotolerance suggesting the involvement of RNA silencing pathway also in this form of the stress response (Stief et al. 2014). AGO1 levels might be modulated in response to heat through an autoregulatory loop involving miR168 activity (Chen et al. 2012; Hivrale et al. 2016; Vaucheret et al. 2006; Xin et al. 2010).

miRNAs and ta-siRNA classes are both involved in AT. The miR156 family is highly conserved (Cui et al. 2014; Stief et al. 2014) and controls juvenile to adult transition. miR156 is strongly induced transcriptionally during heat stress and its activation persists for several days (Stief et al. 2014). miR156 family downregulates SPLs to promote sustained expression of heat-stress memory-related genes including *APX2*, *sHSP17.6A*, *HSFA2*, *HSA32*, *sHSP22*. The expression of MIR156 genes following HS is critical for acquired thermotolerance and heat stress memory. Epigenetic modifications like H3K4me2/3 are involved in transcriptional memory (Lamke et al. 2016).

The most deleterious impact of heat is caused during the reproductive phase of plant life. Flowers are the most heat-susceptible tissues: heat affects pollen viability, grain filling, and maturation. Through all these, HS compromises reproductive success (Liu et al. 2017; Stief et al. 2014). Flowering time, therefore, is regulated in a complex manner to align it to the most favorable environmental conditions. miR156 is a repressor, miR172 is an activator of flowering time through age pathway (Wang 2014). The decline of the miR156 level with aging provides a permissive environment for flowering through de-repression of SPLs that positively regulate floral activator genes including LEAFY, FRUITFULL, APETALA1, APETALA2 or FT (Kim et al. 2012; Wang et al. 2009; Yamaguchi et al. 2009). Bra-miR156 upregulation and BraSPL2 downregulation also occur in *B. rapa* (Yu et al. 2012). Heat (37–45 °C) not only induces miR156 but also represses miR172 in *Arabidopsis*, wheat, or *Helianthus annuus* (Ebrahimi Khaksefidi et al. 2015; May et al. 2013; Stief et al. 2014; Xin et al. 2010) while its targets are upregulated. Antagonistic changes of miR156-SPL and miR172-AP2 regulatory modules converge to record plant age as "time well spent" (excluding stress time) (Stief et al. 2014) and modulate flowering time accordingly.

miR319 also contributes to flowering time changes in response to HS. The miR319 family targets TEOSINTE BRANCHED/CYCLOIDEA/PCF domain proteins (TCPs). TCPs are positive regulators of flowering time: overexpression of miR319 and loss-of-function of the miR319 target TCP4 cause late flowering. miR319 is upregulated and *TCP* family genes are downregulated during HS (Hivrale et al. 2016). ta-siRNAs also affect flowering time regulation (Liu et al. 2019) (*see later*). In summary, the spectrum of sRNAome fine-tunes stress response both through positive and negative activities in order to mitigate the detrimental impact of heat and successfully accomplish reproduction.

9.7 sRNAs Are Required for Trans-Generational Stress Memory

Temperature affects the efficiency of RNA silencing (Szittya et al. 2003; Zhong et al. 2013). Moderate heat stress (30 °C) induces the epigenetic release of RNA silencing through inhibition of ta-siRNA biogenesis (Zhong et al. 2013). It was shown recently that this pathway requires the activity of HSFA2, REF6, and SGIP1 (Liu et al. 2019). HSFA2 transcriptionally activates *RELATIVE OF EARLY FLOWERING 6* (*REF6*), an H3K27-demethylase gene (Lu et al. 2011). REF6, in turn, de-represses *HSFA2*, through H3K27me3 demethylation. As a consequence, it is established a heritable feedback loop that positively affects both heat response and thermomemory. REF6-HSFA2 interplay activates SGS3-INTERACTING PROTEIN 1 (SGIP1), an E3 ubiquitin ligase. SGIP1 mediates degradation of SUPPRESSOR OF GENE SILENCING (SGS3) protein (Kumakura et al. 2009; Mourrain et al. 2000), the cofactor of RDR6. SGS3 degradation leads to the abolishment of ta-siRNA production in the heat-stressed plants. Low levels of SGS3 are epigenetically maintained also in the non-stressed progeny (Liu et al. 2019; Zhong et al. 2013). HS-induced transcriptional activation and absence of ta-siRNAs activity converge to release HTT5, which drives early flowering (Liu et al. 2019). The basis for trans-generational adaptation is the maternally transmitted H3K27me3 demethylation of the *HSFA2* locus.

Besides the PTGS pathway, RdDM-mediated TGS also plays a role in the defense against the detrimental effects of excessive heat. It has been shown that under heat stress, the Copia-type retrotransposon ONSEN is transcriptionally activated and produces extrachromosomal DNA copies (Ito et al. 2011). HS-induction of ONSEN requires HSE *cis* motifs and HSFA1s and HSFA2 *trans* factors (Pecinka et al. 2010). Activation of ONSEN was stimulated in siRNA biogenesis mutants: a high frequency of new ONSEN insertions was observed in the progeny of stressed siRNA-deficient plants. As retrotransposition was not observed in wild-type plants subjected to the same treatment (Ito et al. 2011), siRNA pathway has a crucial role in restricting stress-triggered retrotransposition ensuring genome stability during and following heat stress. Transient derepression of ONSEN may result occasionally in new retrotransposition events. Heat-induced trans-generational trait changes via ta-siRNA and het-siRNA pathway may offer adaptive advantage and flexibility amongst warmer environments.

Trans-generational thermomemory, however, can be associated with negative consequences: the release of HTT5 in absence of ta-siRNAs attenuated plants' immunity. Abiotic resistance might also affect growth rate amongst non-stress conditions or in response to different stresses. Therefore it is conceivable that unknown mechanisms hamper full inheritance of epigenetic traits. Chromatin regulators DECREASE IN DNA METHYLATION1 (DDM1) and MORPHEUS' MOLECULE1 (MOM1) were found to redundantly prevent transgenerational inheritance of environmentally induced traits (Iwasaki and Paszkowski 2014). In accordance with this idea, ta-siRNA biogenesis and basal level of thermoresistance were reset in later generations (Liu et al. 2019), providing flexibility for adaptation in the long term.

9.8 A Complex sRNA Network Contributes to HSR

The above examples (see also Table 9.1) illustrate that under elevated ambient temperature (27–29 °C) and heat stress (32–45 °C) conditions a number of sRNAs are changed to coordinate temperature response. The altered sRNAome causes changes in the mRNA target network that affects a large spectrum of cellular and physiological processes. Under optimal conditions, resources are directed to support growth and development. Under elevated ambient temperatures, stress responses are already initiated but at the same time, the flowering transition is accelerated to ensure reproductive success: the cost of this is that fewer seeds are produced. During severe stress, growth and development are rather stalled and resources are primarily mobilized to manage the negative impacts of stress. The trade-off between development and stress response is continuously fine-tuned based on the strength, type and duration of temperature changes encountered.

The miR156-SPL9-DFR pathway was suggested to coordinate the balance between growth and abiotic stress tolerance in *Arabidopsis* (Cui et al. 2014). Under optimal conditions, miR156 is suppressed to accelerate development, while under stress conditions it is induced to maintain juvenile state for a longer period. The mechanism is conserved also in rice. Downstream genes that take part in this include miR156-targets SPL9 and DIHYDROFLAVONOL-4-REDUCTASE (DFR) that modulate anthocyanin metabolism. miR156 is induced in response to multiple abiotic stresses including heat, salt or mannitol. SPL9 and SPL10 targets of miR156 promote transcription of miR172 that is required for the juvenile-to-adult transition. miR172 overexpression cause stress-sensitivity, further supporting the antagonistic relationship between development and stress response.

Auxin perception and signaling pathway also attenuate growth through miRNA activity. Expression of auxin-responsive genes is mediated by the auxin receptor TRANSPORT INHIBITOR RESPONSE 1 (TIR1) through ARFs release from auxin-mediated heterodimerization. miR393 negatively regulates TIR1, while miR160 and miR167 negatively regulate ARFs. Upon stress, miR160 is induced, it downregulates ARF10, ARF16, and ARF17 leading to suppression of ARF-induced gene expression and thereby growth attenuation (Sunkar et al. 2012). Upregulation of miR160 and parallel downregulation of ARF targets was also observed in barley and *H. annuus* (Ebrahimi Khaksefidi et al. 2015; Kruszka et al. 2014). In wheat, however, miR160 was downregulated in response to heat (Kumar 2014). Besides these, miR390 and miR393 also participates in auxin-mediated HSR (Hivrale et al. 2016; Vidal et al. 2010).

The miR166 target PHAVOLUTA (PHV), REVOLUTA (REV), and HOMEOBOX LEUCINE ZIPPER PROTEIN 9 (HOX9) and miR319 target TCP are involved in leaf development; miR171 targets SCARECROW-LIKE6s (SCL6s) and miR164 target NAC1 regulate shoot and root development; all these miRNAs are induced and their targets are repressed in response to heat (Mahale et al. 2013; Kruszka et al. 2014; Schommer et al. 2008), further underpinning the hypothesis that developmental pathways are being retarded temporarily to allow stress recovery.

In summary, combined sRNA changes contribute to the balance between growth and stress response and aim completion of life-cycle to establish the next generation (Fig. 9.1).

9.9 Combined Stress Pathways Are Regulated by sRNAs

In nature, abiotic stresses occur mostly in combination. Multiple miRNAs, including the above-mentioned miRNAs (miR156, miR160, miR167, and miR393) are upregulated or downregulated in response to heat, cold, salt, mannitol, UV-B, drought, hypoxia or oxidative stress in diverse species (Khraiwesh et al. 2012). This suggests that miRNAs are integrators of environmental stimuli under combined environmental stress conditions.

Exposure to high temperature causes oxidative stress as well. Signals generated during oxidative stress are absolutely required for an effective HSR (Dickinson et al. 2018). Subcellular energy organelles, the chloroplasts, drive a series of metabolic reprogramming (Wang et al. 2018) and initiate multiple retrograde signaling to regulate abiotic (light, heat, oxidative) stress responses (Dickinson et al. 2018; Fang et al. 2019). One of these pathways acts through miRNAs (Fang et al. 2019): plastid synthesized tocopherols and 3′-phosphoadenosine 5′-phosphate (PAP) inhibit nuclear exoribonucleases (XRN)-catalyzed pri-miRNA degradation. This promotes enhanced miRNA biogenesis, amongst others that of miR398 that confers thermotolerance to the plants. Efficient accumulation of miR398 also requires transcriptional induction via direct binding of HSFA1b and HSFA7b to its gene promoter region (Guan et al. 2013). As a consequence, the level of miR398 targets Cu/Zn SUPEROXIDE DISMUTASE 1 (CSD1), CSD2, and COPPER CHAPERONE FOR SUPEROXIDE DISMUTASE (CCS) mRNAs are downregulated in response to heat (Guan et al. 2013). Transgenic plants expressing miR398-resistant forms of CSD1, CSD2, and CCS are HS-sensitive and expression of numerous HSFs and HSPs during the heat is reduced. Oppositely to this, *csd1*, *csd2*, and *ccs* mutant plants are HS-resistant and expression of certain HSFs and HSPs is induced. It was suggested that heat-induced miR398 alters cellular redox status through negative regulation of CSD1, CSD2 and CCS targets. The altered redox environment is directly or indirectly sensed by HSFs (Volkov et al. 2006) and thermotolerance pathways are initiated through HSR-regulon. miR398 family members and their target genes are highly conserved across monocots and dicots. miR398 heat-induction was found to occur also in maize, *Brassica rapa* and *Populus tomentosa* (Guan et al. 2013; Kotak et al. 2007; Yu et al. 2012).

Another miRNA, miR408, also contributes to redox changes during different abiotic stress conditions. Expression of miR408 is significantly induced by abiotic stresses including heat (also cold, salt, oxidative, osmotic drought) in *Arabidopsis* and important crop plants. miR408 targets are the *PHYTOCYANIN* family genes that function as electron-transfer factors to regulate redox state (Hivrale et al. 2016; Kumar et al. 2015; Ma et al. 2015; Mangrauthia et al. 2017; Zhou et al. 2016). miR398 and miR408, therefore, may provide abiotic stress tolerance through regulation of redox status during heat or combined stress conditions. Manipulation of redox state through alteration of miR398, miR408 or their targets, therefore, may be a useful strategy to breed thermotolerant crops.

9.10 Breeding for Thermotolerance

In order to enhance HS tolerance in crop plants, it is necessary to better understand the molecular mechanisms that regulate the HSR and to develop new breeding strategies. In a large number of studies, the biological relevance of sRNA and target changes in

HSR has been demonstrated using sRNA or target mutants, miRNA-overexpressors, or target-mimicry lines. These findings spurred the idea of developing crops having altered miRNA or target level leading to improved agronomical performance, yield or nutritional value amongst abiotic (e.g. heat stress) conditions. This may be achieved through classical or molecular breeding. Artificial miRNAs (amiRNA) have been successfully employed to silence endogenous targets in model plants, crop plants and unicellular organisms (Alvarez et al. 2006; Khraiwesh et al. 2008; Molnar et al. 2009; Schwab et al. 2006; Warthmann et al. 2008). amiRNAs may be designed to have high specificity (avoid off-target effects). Oppositely to amiRNA technology that suppresses target levels, target-mimicry constructs lead to the abolishment of miRNA activity and consequently target-mRNA accumulation (Franco-Zorrilla et al. 2007). amiRNAs and target-mimicry constructs provide a flexible tool for downregulation of related genes or gene families. In contrast to miRNA-based methods, siRNA-based RNA silencing technology (RNA interference or RNAi) cannot be successfully employed for heat-tolerant plant breeding since these depend on the activity of RDR6/SGS3 complex (SGS3 protein is degraded under high-temperature conditions) (Liu et al. 2019) for amplification of gene silencing response. At the same time, it is important to consider the possible side-effect of the improved abiotic tolerance: improved heat-tolerance weakens immunity and might cause poor development under non-stress conditions (Liu et al. 2019).

9.11 Conclusions

There is an increased risk of crop yield reduction worldwide caused by global warming and extreme temperature fluctuations. Elucidation of temperature response, therefore, is becoming more important. Present evidence suggests that epigenetic, transcriptional, sRNA-mediated post-transcriptional and protein-level regulatory mechanisms are all intercalated to precisely and flexibly adjust the trade-off between development and stress response within and trans-generationally. Future research is needed for functional confirmation of HS-responsive sRNA/target changes and for the elucidation of the relationship of sRNA-mediated HSR and other regulatory networks. Complete understanding of the mechanism of sRNA action during HS, combined abiotic stresses or cross-protection between stresses may provide tools for breeding resilient crops.

Acknowledgements H.M.Sz. is a Ph.D. student at the Eötvös Loránd University, Budapest, Faculty of Natural Sciences; T.Cs. was supported by János Bolyai Research Scholarship of Hungarian Academy of Science and Hungarian National Research, Development and Innovation Office, K-115934 and K-129283. G.S. is funded by grants from Hungarian National Research, Development and Innovation Office, K-119701 and K-129171.

References

Alvarez JP, Pekker I, Goldshmidt A, Blum E, Amsellem Z, Eshed Y (2006) Endogenous and synthetic microRNAs stimulate simultaneous, efficient, and localized regulation of multiple targets in diverse species Plant Cell 18:1134–1151. https://doi.org/10.1105/tpc.105.040725

Aukerman MJ, Sakai H (2003) Regulation of flowering time and floral organ identity by a MicroRNA and its APETALA2-like target genes. Plant Cell 15:2730–2741. https://doi.org/10.1105/tpc.016238

Axtell MJ (2013) Classification and comparison of small RNAs from plants. Annu Rev Plant Biol 64:137–159. https://doi.org/10.1146/annurev-arplant-050312-120043

Baksa I et al (2015) Identification of Nicotiana benthamiana microRNAs and their targets using high throughput sequencing and degradome analysis BMC genomics 16:1025. https://doi.org/10.1186/s12864-015-2209-6

Ballen-Taborda C, Plata G, Ayling S, Rodriguez-Zapata F, Becerra Lopez-Lavalle LA, Duitama J, Tohme J (2013) Identification of Cassava MicroRNAs under abiotic stress. Int J Genomics 2013:857986. https://doi.org/10.1155/2013/857986

Barciszewska-Pacak M et al (2015) Arabidopsis microRNA expression regulation in a wide range of abiotic stress responses. Front Plant Sci 6:410. https://doi.org/10.3389/fpls.2015.00410

Beckers M, Mohorianu I, Stocks M, Applegate C, Dalmay T, Moulton V (2017) Comprehensive processing of high-throughput small RNA sequencing data including quality checking, normalization, and differential expression analysis using the UEA sRNA. Workbench RNA 23:823–835. https://doi.org/10.1261/rna.059360.116

Bilichak A, Ilnytskyy Y, Woycicki R, Kepeshchuk N, Fogen D, Kovalchuk I (2015) The elucidation of stress memory inheritance in Brassica rapa plants. Front Plant Sci 6:5. https://doi.org/10.3389/fpls.2015.00005

Borges F, Martienssen RA (2015) The expanding world of small RNAs in plants. Nat Rev Mol Cell Biol 16:727–741. https://doi.org/10.1038/nrm4085

Brzezinka K et al (2016) Arabidopsis FORGETTER1 mediates stress-induced chromatin memory through nucleosome remodeling. eLife 5. https://doi.org/10.7554/elife.17061

Brzezinka K, Altmann S, Baurle I (2018) BRUSHY1/TONSOKU/MGOUN3 is required for heat stress memory. Plant Cell Environ. https://doi.org/10.1111/pce.13365

Chan-Schaminet KY, Baniwal SK, Bublak D, Nover L, Scharf KD (2009) Specific interaction between tomato HsfA1 and HsfA2 creates hetero-oligomeric superactivator complexes for synergistic activation of heat stress gene expression. J Biol Chem 284:20848–20857. https://doi.org/10.1074/jbc.m109.007336

Charng YY, Liu HC, Liu NY, Hsu FC, Ko SS (2006) Arabidopsis Hsa32, a novel heat shock protein, is essential for acquired thermotolerance during long recovery after acclimation. Plant Physiol 140:1297–1305. https://doi.org/10.1104/pp.105.074898

Charng YY, Liu HC, Liu NY, Chi WT, Wang CN, Chang SH, Wang TT (2007) A heat-inducible transcription factor, HsfA2, is required for extension of acquired thermotolerance in Arabidopsis. Plant Physiol 143:251–262. https://doi.org/10.1104/pp.106.091322

Chen X (2004) A microRNA as a translational repressor of APETALA2 in Arabidopsis flower development. Science 303:2022–2025. https://doi.org/10.1126/science.1088060

Chen L, Ren Y, Zhang Y, Xu J, Sun F, Zhang Z, Wang Y (2012) Genome-wide identification and expression analysis of heat-responsive and novel microRNAs in *Populus tomentosa*. Gene 504:160–165. https://doi.org/10.1016/j.gene.2012.05.034

Cortijo S et al (2017) Transcriptional regulation of the ambient temperature response by H2A.Z nucleosomes and HSF1 transcription factors in Arabidopsis. Mol Plant 10:1258–1273. https://doi.org/10.1016/j.molp.2017.08.014

Cui LG, Shan JX, Shi M, Gao JP, Lin HX (2014) The miR156-SPL9-DFR pathway coordinates the relationship between development and abiotic stress tolerance in plants. Plant J Cell Mol Biol 80:1108–1117. https://doi.org/10.1111/tpj.12712

Dalmay T, Hamilton A, Rudd S, Angell S, Baulcombe DC (2000) An RNA-dependent RNA polymerase gene in Arabidopsis is required for posttranscriptional gene silencing mediated by a transgene but not by a virus. Cell 101:543–553

Dickinson PJ et al (2018) Chloroplast signaling gates thermotolerance in Arabidopsis. Cell Rep 22:1657–1665. https://doi.org/10.1016/j.celrep.2018.01.054

Ebrahimi Khaksefidi R et al (2015) Differential expression of seven conserved microRNAs in response to abiotic stress and their regulatory network in *Helianthus annuus*. Front Plant Sci 6:741. https://doi.org/10.3389/fpls.2015.00741

Fang X et al (2019) Chloroplast-to-nucleus signaling regulates microRNA biogenesis in Arabidopsis. Dev Cell 48(371–382):e374. https://doi.org/10.1016/j.devcel.2018.11.046

Franco-Zorrilla JM et al (2007) Target mimicry provides a new mechanism for regulation of microRNA activity. Nat Genet 39:1033–1037. https://doi.org/10.1038/ng2079

Giacomelli JI, Weigel D, Chan RL, Manavella PA (2012) Role of recently evolved miRNA regulation of sunflower HaWRKY6 in response to temperature damage. New Phytologist 195:766–773. https://doi.org/10.1111/j.1469-8137.2012.04259.x

Guan Q, Lu X, Zeng H, Zhang Y, Zhu J (2013) Heat stress induction of miR398 triggers a regulatory loop that is critical for thermotolerance in Arabidopsis. Plant J Cell Mol Biol 74:840–851. https://doi.org/10.1111/tpj.12169

Gyula P, Baksa I, Toth T, Mohorianu I, Dalmay T, Szittya G (2018) Ambient temperature regulates the expression of a small set of sRNAs influencing plant development through NF-YA2 and YUC2. Plant Cell Environ 41:2404–2417. https://doi.org/10.1111/pce.13355

He F, Xu C, Fu X, Shen Y, Guo L, Leng M, Luo K (2018) The microRNA390/TRANS ACTING SHORT INTERFERING RNA3 module mediates lateral root growth under salt stress via the auxin pathway. Plant Physiol. https://doi.org/10.1104/pp.17.01559

Hivrale V et al (2016) Characterization of drought- and heat-responsive microRNAs in switchgrass. Plant Sci Int J Experim Plant Biol 242:214–223. https://doi.org/10.1016/j.plantsci.2015.07.018

Ito H, Gaubert H, Bucher E, Mirouze M, Vaillant I, Paszkowski J (2011) An siRNA pathway prevents transgenerational retrotransposition in plants subjected to stress. Nature 472:115–119. https://doi.org/10.1038/nature09861

Iwasaki M, Paszkowski J (2014) Identification of genes preventing transgenerational transmission of stress-induced epigenetic states. Proc Natl Acad Sci USA 111:8547–8552. https://doi.org/10.1073/pnas.1402275111

Jeong DH, Park S, Zhai J, Gurazada SG, De Paoli E, Meyers BC, Green PJ (2011) Massive analysis of rice small RNAs: mechanistic implications of regulated microRNAs and variants for differential target RNA cleavage. Plant Cell 23:4185–4207. https://doi.org/10.1105/tpc.111.089045

Jung JH et al (2016) Phytochromes function as thermosensors in Arabidopsis. Science 354:886–889. https://doi.org/10.1126/science.aaf6005

Khraiwesh B, Ossowski S, Weigel D, Reski R, Frank W (2008) Specific gene silencing by artificial MicroRNAs in *Physcomitrella patens*: an alternative to targeted gene knockouts. Plant Physiol 148:684–693. https://doi.org/10.1104/pp.108.128025

Khraiwesh B, Zhu JK, Zhu J (2012) Role of miRNAs and siRNAs in biotic and abiotic stress responses of plants. Biochem Biophys Acta 1819:137–148. https://doi.org/10.1016/j.bbagrm.2011.05.001

Kim W, Ahn HJ, Chiou TJ, Ahn JH (2011) The role of the miR399-PHO2 module in the regulation of flowering time in response to different ambient temperatures in *Arabidopsis thaliana*. Mol Cells 32:83–88. https://doi.org/10.1007/s10059-011-1043-1

Kim JJ, Lee JH, Kim W, Jung HS, Huijser P, Ahn JH (2012) The microRNA156-SQUAMOSA PROMOTER BINDING PROTEIN-LIKE3 module regulates ambient temperature-responsive flowering via FLOWERING LOCUS T in Arabidopsis. Plant Physiol 159:461–478. https://doi.org/10.1104/pp.111.192369

Kim W, Kim HE, Jun AR, Jung MG, Jin S, Lee JH, Ahn JH (2016) Structural determinants of miR156a precursor processing in temperature-responsive flowering in Arabidopsis. J Experim Botany 67:4659–4670. https://doi.org/10.1093/jxb/erw248

Kotak S, Larkindale J, Lee U, von Koskull-Doring P, Vierling E, Scharf KD (2007) Complexity of the heat stress response in plants. Curr Opin Plant Biol 10:310–316. https://doi.org/10.1016/j.pbi.2007.04.011

Kruszka K et al (2014) Transcriptionally and post-transcriptionally regulated microRNAs in heat stress response in barley. J Exp Bot 65:6123–6135. https://doi.org/10.1093/jxb/eru353

Kumakura N, Takeda A, Fujioka Y, Motose H, Takano R, Watanabe Y (2009) SGS3 and RDR6 interact and colocalize in cytoplasmic SGS3/RDR6-bodies. FEBS Lett 583:1261–1266. https://doi.org/10.1016/j.febslet.2009.03.055

Kumar R (2014) Role of microRNAs in biotic and abiotic stress responses in crop plants. Appl Biochem Biotechnol 174:93–115. https://doi.org/10.1007/s12010-014-0914-2

Kumar RR et al (2015) Novel and conserved heat-responsive microRNAs in wheat (*Triticum aestivum* L.). Funct Integr Genom 15:323–348. https://doi.org/10.1007/s10142-014-0421-0

Lamke J, Brzezinka K, Altmann S, Baurle I (2016) A hit-and-run heat shock factor governs sustained histone methylation and transcriptional stress memory. EMBO J 35:162–175. https://doi.org/10.15252/embj.201592593

Lee H et al (2010) Genetic framework for flowering-time regulation by ambient temperature-responsive miRNAs in Arabidopsis. Nucl Acids Res 38:3081–3093. https://doi.org/10.1093/nar/gkp1240

Legris M et al (2016) Phytochrome B integrates light and temperature signals in Arabidopsis. Science 354:897–900. https://doi.org/10.1126/science.aaf5656

Li MY, Wang F, Xu ZS, Jiang Q, Ma J, Tan GF, Xiong AS (2014a) High throughput sequencing of two celery varieties small RNAs identifies microRNAs involved in temperature stress response. BMC genomics 15:242. https://doi.org/10.1186/1471-2164-15-242

Li S, Liu J, Liu Z, Li X, Wu F, He Y (2014b) HEAT-INDUCED TAS1 TARGET1 mediates thermotolerance via HEAT STRESS TRANSCRIPTION FACTOR A1a-directed pathways in Arabidopsis. Plant Cell 26:1764–1780 https://doi.org/10.1105/tpc.114.124883

Li H et al (2016a) Microarray and genetic analysis reveals that csa-miR159b plays a critical role in abscisic acid-mediated heat tolerance in grafted cucumber plants. Plant Cell Environ 39:1790–1804. https://doi.org/10.1111/pce.12745

Li Y, Alonso-Peral M, Wong G, Wang MB, Millar AA (2016b) Ubiquitous miR159 repression of MYB33/65 in Arabidopsis rosettes is robust and is not perturbed by a wide range of stresses. BMC Plant Biol 16:179. https://doi.org/10.1186/s12870-016-0867-4

Lin JS et al (2018) MicroRNA160 modulates plant development and heat shock protein gene expression to mediate heat tolerance in Arabidopsis. Front Plant Sci 9:68. https://doi.org/10.3389/fpls.2018.00068

Lin MY, Chai KH, Ko SS, Kuang LY, Lur HS, Charng YY (2014) A positive feedback loop between HEAT SHOCK PROTEIN101 and HEAT STRESS-ASSOCIATED 32-KD PROTEIN modulates long-term acquired thermotolerance illustrating diverse heat stress responses in rice varieties. Plant Physiol 164:2045–2053. https://doi.org/10.1104/pp.113.229609

Liu J et al (2019) An H3K27me3 demethylase-HSFA2 regulatory loop orchestrates transgenerational thermomemory in Arabidopsis. Cell Res. https://doi.org/10.1038/s41422-019-0145-8

Liu HC, Charng YY (2012) Acquired thermotolerance independent of heat shock factor A1 (HsfA1), the master regulator of the heat stress response. Plant Signal Behav 7:547–550. https://doi.org/10.4161/psb.19803

Liu HC, Charng YY (2013) Common and distinct functions of Arabidopsis class A1 and A2 heat shock factors in diverse abiotic stress responses and development. Plant Physiol 163:276–290. https://doi.org/10.1104/pp.113.221168

Liu HC, Liao HT, Charng YY (2011) The role of class A1 heat shock factors (HSFA1s) in response to heat and other stresses in Arabidopsis. Plant Cell Environ 34:738–751. https://doi.org/10.1111/j.1365-3040.2011.02278.x

Liu F, Wang W, Sun X, Liang Z, Wang F (2013) RNA-Seq revealed complex response to heat stress on transcriptomic level in Saccharina japonica (Laminariales, Phaeophyta). J Appl Phycol 26(3):1585–1596

Liu Q et al (2017) Integrating small RNA sequencing with QTL mapping for identification of miRNAs and their target genes associated with heat tolerance at the flowering stage in rice. Front Plant Sci 8:43. https://doi.org/10.3389/fpls.2017.00043

Liu HC, Lamke J, Lin SY, Hung MJ, Liu KM, Charng YY, Baurle I (2018) Distinct heat shock factors and chromatin modifications mediate the organ-autonomous transcriptional memory of heat stress. Plant J Cell Mol Biol. https://doi.org/10.1111/tpj.13958

Lu S, Sun YH, Chiang VL (2008) Stress-responsive microRNAs in Populus. Plant J Cell Mol Biol 55:131–151. https://doi.org/10.1111/j.1365-313X.2008.03497.x

Lu F, Cui X, Zhang S, Jenuwein T, Cao X (2011) Arabidopsis REF6 is a histone H3 lysine 27 demethylase. Nat Genet 43:715–719. https://doi.org/10.1038/ng.854

Luria G, Rutley N, Lazar I, Harper JF, Miller G (2019) Direct analysis of pollen fitness by flow cytometry: implications for pollen response to stress. Plant J Cell Mol Biol. https://doi.org/10.1111/tpj.14286

Ma C, Burd S, Lers A (2015) miR408 is involved in abiotic stress responses in Arabidopsis. Plant J Cell Mol Biol 84:169–187. https://doi.org/10.1111/tpj.12999

Mahale BM, Fakrudin B, Ghosh S, Krishnaraj PU (2013) LNA mediated in situ hybridization of miR171 and miR397a in leaf and ambient root tissues revealed expressional homogeneity in response to shoot heat shock in *Arabidopsis thaliana*. J Plant Biochem Biotechnol 23(1):93–103

Mangrauthia SK, Bhogireddy S, Agarwal S, Prasanth VV, Voleti SR, Neelamraju S, Subrahmanyam D (2017) Genome-wide changes in microRNA expression during short and prolonged heat stress and recovery in contrasting rice cultivars. J Exp Bot 68:2399–2412. https://doi.org/10.1093/jxb/erx111

Martinez de Alba AE, Elvira-Matelot E, Vaucheret H (2013) Gene silencing in plants: a diversity of pathways. Biochimica et biophysica acta 1829:1300–1308. https://doi.org/10.1016/j.bbagrm.2013.10.005

Maruyama K et al (2017) Design of an optimal promoter involved in the heat-induced transcriptional pathway in Arabidopsis, soybean, rice and maize. Plant J Cell Mol Biol 89:671–680. https://doi.org/10.1111/tpj.13420

May P, Liao W, Wu Y, Shuai B, McCombie WR, Zhang MQ, Liu QA (2013) The effects of carbon dioxide and temperature on microRNA expression in Arabidopsis development. Nat Commun 4:2145. https://doi.org/10.1038/ncomms3145

Mazzeo MF, Cacace G, Iovieno P, Massarelli I, Grillo S, Siciliano RA (2018) Response mechanisms induced by exposure to high temperature in anthers from thermo-tolerant and thermo-sensitive tomato plants: a proteomic perspective. PloS one 13:e0201027. https://doi.org/10.1371/journal.pone.0201027

Meiri D, Breiman A (2009) Arabidopsis ROF1 (FKBP62) modulates thermotolerance by interacting with HSP90.1 and affecting the accumulation of HsfA2-regulated sHSPs. Plant J Cell Mol Biol 59:387–399. https://doi.org/10.1111/j.1365-313x.2009.03878.x

Mittler R, Finka A, Goloubinoff P (2012) How do plants feel the heat? Trends Biochem Sci 37:118–125. https://doi.org/10.1016/j.tibs.2011.11.007

Mohorianu I, Stocks MB, Wood J, Dalmay T, Moulton V (2013) CoLIde: a bioinformatics tool for CO-expression-based small RNA loci identification using high-throughput sequencing data. RNA Biol 10:1221–1230. https://doi.org/10.4161/rna.25538

Molnar A et al (2009) Highly specific gene silencing by artificial microRNAs in the unicellular alga *Chlamydomonas reinhardtii*. Plant J Cell Mol Biol 58:165–174. https://doi.org/10.1111/j.1365-313X.2008.03767.x

Mourrain P et al (2000) Arabidopsis SGS2 and SGS3 genes are required for posttranscriptional gene silencing and natural virus resistance. Cell 101:533–542

Ohama N, Sato H, Shinozaki K, Yamaguchi-Shinozaki K (2017) Transcriptional regulatory network of plant heat stress response. Trends Plant Sci 22:53–65. https://doi.org/10.1016/j.tplants.2016.08.015

Pecinka A, Dinh HQ, Baubec T, Rosa M, Lettner N, Mittelsten Scheid O (2010) Epigenetic regulation of repetitive elements is attenuated by prolonged heat stress in Arabidopsis. Plant Cell 22:3118–3129. https://doi.org/10.1105/tpc.110.078493

Petroni K et al (2012) The promiscuous life of plant NUCLEAR FACTOR Y transcription factors. Plant Cell 24:4777–4792. https://doi.org/10.1105/tpc.112.105734

Poethig RS (2009) Small RNAs and developmental timing in plants. Curr Opin Genet Develop 19:374–378. https://doi.org/10.1016/j.gde.2009.06.001

Quint M, Delker C, Franklin KA, Wigge PA, Halliday KJ, van Zanten M (2016) Molecular and genetic control of plant thermomorphogenesis. Nat Plants 2:15190. https://doi.org/10.1038/nplants.2015.190

Re DA, Lang PLM, Yones C, Arce AL, Stegmayer G, Milone D, Manavella PA (2019) Alternative use of miRNA-biogenesis co-factors in plants at low temperatures. Development 146. https://doi.org/10.1242/dev.172932

Rogers K, Chen X (2013) Biogenesis, turnover, and mode of action of plant microRNAs. Plant Cell 25:2383–2399. https://doi.org/10.1105/tpc.113.113159

Sailaja B, Voleti SR, Subrahmanyam D, Sarla N, Vishnuprasanth V, Bhadana VP, Mangrauthia SK (2014) Prediction and expression analysis of miRNAs associated with heat stress in *Oryza sativa*. Rice Sci 21:3–12

Scharf KD, Heider H, Hohfeld I, Lyck R, Schmidt E, Nover L (1998) The tomato Hsf system: HsfA2 needs interaction with HsfA1 for efficient nuclear import and may be localized in cytoplasmic heat stress granules. Mol Cell Biol 18:2240–2251

Scharf KD, Berberich T, Ebersberger I, Nover L (2012) The plant heat stress transcription factor (Hsf) family: structure, function and evolution. Biochimica et biophysica acta 1819:104–119. https://doi.org/10.1016/j.bbagrm.2011.10.002

Schommer C et al (2008) Control of jasmonate biosynthesis and senescence by miR319 targets. PLOS Biol 6:e230. https://doi.org/10.1371/journal.pbio.0060230

Schwab R, Ossowski S, Riester M, Warthmann N, Weigel D (2006) Highly specific gene silencing by artificial microRNAs in Arabidopsis. Plant Cell 18:1121–1133. https://doi.org/10.1105/tpc.105.039834

Siriwardana CL, Gnesutta N, Kumimoto RW, Jones DS, Myers ZA, Mantovani R, Holt 3rd BF (2016) NUCLEAR FACTOR Y, Subunit A (NF-YA) Proteins positively regulate flowering and act through FLOWERING LOCUS T. PLOS Genet 12:e1006496. https://doi.org/10.1371/journal.pgen.1006496

Spanudakis E, Jackson S (2014) The role of microRNAs in the control of flowering time. J Exp Bot 65:365–380. https://doi.org/10.1093/jxb/ert453

Srikanth A, Schmid M (2011) Regulation of flowering time: all roads lead to Rome. Cellular Mol Life Sci (CMLS) 68:2013–2037. https://doi.org/10.1007/s00018-011-0673-y

Stief A, Altmann S, Hoffmann K, Pant BD, Scheible WR, Baurle I (2014) Arabidopsis miR156 regulates tolerance to recurring environmental stress through spl transcription factors. Plant Cell 26:1792–1807. https://doi.org/10.1105/tpc.114.123851

Sun Z, Li M, Zhou Y, Guo T, Liu Y, Zhang H, Fang Y (2018) Coordinated regulation of Arabidopsis microRNA biogenesis and red light signaling through Dicer-like 1 and phytochrome-interacting factor 4. PLOS Genet 14:e1007247. https://doi.org/10.1371/journal.pgen.1007247

Sunkar R, Zhu JK (2004) Novel and stress-regulated microRNAs and other small RNAs from Arabidopsis. Plant Cell 16:2001–2019. https://doi.org/10.1105/tpc.104.022830

Sunkar R, Li YF, Jagadeeswaran G (2012) Functions of microRNAs in plant stress responses. Trends Plant Sci 17:196–203. https://doi.org/10.1016/j.tplants.2012.01.010

Szittya G et al (2003) Low temperature inhibits RNA silencing-mediated defence by the control of siRNA generation. EMBO J 22:633–640. https://doi.org/10.1093/emboj/cdg74

Vaucheret H (2006) Post-transcriptional small RNA pathways in plants: mechanisms and regulations. Genes Dev 20:759–771. https://doi.org/10.1101/gad.1410506

Vaucheret H, Mallory AC, Bartel DP (2006) AGO1 homeostasis entails coexpression of MIR168 and AGO1 and preferential stabilization of miR168 by AGO1. Mol Cell 22:129–136. https://doi.org/10.1016/j.molcel.2006.03.011

Vidal EA, Araus V, Lu C, Parry G, Green PJ, Coruzzi GM, Gutierrez RA (2010) Nitrate-responsive miR393/AFB3 regulatory module controls root system architecture in *Arabidopsis thaliana*. Proc Nat Acad Sci USA 107:4477–4482. https://doi.org/10.1073/pnas.0909571107

Volkov RA, Panchuk II, Mullineaux PM, Schoffl F (2006) Heat stress-induced H_2O_2 is required for effective expression of heat shock genes in Arabidopsis. Plant Mol Biol 61:733–746. https://doi.org/10.1007/s11103-006-0045-4

Wang JW (2014) Regulation of flowering time by the miR156-mediated age pathway. J Experim Botany 65:4723–4730. https://doi.org/10.1093/jxb/eru246

Wang JW, Wang LJ, Mao YB, Cai WJ, Xue HW, Chen XY (2005) Control of root cap formation by MicroRNA-targeted auxin response factors in Arabidopsis. Plant Cell 17:2204–2216. https://doi.org/10.1105/tpc.105.033076

Wang JW, Czech B, Weigel D (2009) miR156-regulated SPL transcription factors define an endogenous flowering pathway in *Arabidopsis thaliana*. Cell 138:738–749. https://doi.org/10.1016/j.cell.2009.06.014

Wang Y, Sun F, Cao H, Peng H, Ni Z, Sun Q, Yao Y (2012) TamiR159 directed wheat TaGAMYB cleavage and its involvement in anther development and heat response. PloS One 7:e48445. https://doi.org/10.1371/journal.pone.0048445

Wang QL, Chen JH, He NY, Guo FQ (2018) Metabolic reprogramming in chloroplasts under heat stress in plants. Int J Mol Sci 19. https://doi.org/10.3390/ijms19030849

Warthmann N, Chen H, Ossowski S, Weigel D, Herve P (2008) Highly specific gene silencing by artificial miRNAs in rice. PloS One 3:e1829. https://doi.org/10.1371/journal.pone.0001829

Wigge PA (2013) Ambient temperature signalling in plants. Curr Opin Plant Biol 16:661–666. https://doi.org/10.1016/j.pbi.2013.08.004

Xin M, Wang Y, Yao Y, Xie C, Peng H, Ni Z, Sun Q (2010) Diverse set of microRNAs are responsive to powdery mildew infection and heat stress in wheat (*Triticum aestivum* L.). BMC Plant Biol 10:123. https://doi.org/10.1186/1471-2229-10-123

Xin M et al (2011) Identification and characterization of wheat long non-protein coding RNAs responsive to powdery mildew infection and heat stress by using microarray analysis and SBS sequencing. BMC Plant Biol 11:61. https://doi.org/10.1186/1471-2229-11-61

Xing S, Salinas M, Garcia-Molina A, Hohmann S, Berndtgen R, Huijser P (2013) SPL8 and miR156-targeted SPL genes redundantly regulate Arabidopsis gynoecium differential patterning. Plant J Cell Mol Biol 75:566–577. https://doi.org/10.1111/tpj.12221

Yamaguchi A, Wu MF, Yang L, Wu G, Poethig RS, Wagner D (2009) The microRNA-regulated SBP-Box transcription factor SPL3 is a direct upstream activator of LEAFY, FRUITFULL, and APETALA1. Develop Cell 17:268–278. https://doi.org/10.1016/j.devcel.2009.06.007

Yan K et al (2012) Stress-induced alternative splicing provides a mechanism for the regulation of microRNA processing in *Arabidopsis thaliana*. Mol Cell 48:521–531. https://doi.org/10.1016/j.molcel.2012.08.032

Yao Y, Sun Q (2012) Exploration of small non coding RNAs in wheat (*Triticum aestivum* L.). Plant Mol Biol 80:67–73. https://doi.org/10.1007/s11103-011-9835-4

Yoshida T et al (2011) Arabidopsis HsfA1 transcription factors function as the main positive regulators in heat shock-responsive gene expression. Mol Genetics Genomics (MGG) 286:321–332. https://doi.org/10.1007/s00438-011-0647-7

Yu N, Cai WJ, Wang S, Shan CM, Wang LJ, Chen XY (2010) Temporal control of trichome distribution by microRNA156-targeted SPL genes in *Arabidopsis thaliana*. Plant Cell 22:2322–2335. https://doi.org/10.1105/tpc.109.072579

Yu X et al (2012) Identification of conserved and novel microRNAs that are responsive to heat stress in Brassica rapa. J Exp Bot 63:1025–1038. https://doi.org/10.1093/jxb/err337

Zhong SH et al (2013) Warm temperatures induce transgenerational epigenetic release of RNA silencing by inhibiting siRNA biogenesis in Arabidopsis. Proc Natl Acad Sci USA 110:9171–9176. https://doi.org/10.1073/pnas.1219655110

Zhou R, Wang Q, Jiang F, Cao X, Sun M, Liu M, Wu Z (2016) Identification of miRNAs and their targets in wild tomato at moderately and acutely elevated temperatures by high-throughput sequencing and degradome analysis. Sci Rep 6:33777. https://doi.org/10.1038/srep33777

Zhu X, Huang C, Zhang L, Liu H, Yu J, Hu Z, Hua W (2017) Systematic analysis of Hsf family genes in the *Brassica napus* genome reveals novel responses to heat, drought and high CO_2 stresses. Front Plant Sci 8:1174. https://doi.org/10.3389/fpls.2017.01174

Chapter 10
Role of microRNAs in Plant–Fungus Interactions

Mireia Bundó, Sonia Campo and Blanca San Segundo

Abstract MicroRNAs (miRNAs) are small noncoding RNAs involved in post-transcriptional regulation of gene expression in eukaryotes. Given the rapid advancement of sequencing techniques, there has been a dramatic increase in the number of described miRNAs in plants. miRNAs act as master regulators of diverse plant developmental processes and adaptation to environmental stress. Evidence has also emerged on miRNAs involved in antifungal and antibacterial resistance by regulating ETI and PTI responses. Small RNAs, including miRNAs, also mediate cross-kingdom regulation of gene expression in host/pathogen interactions where specific host-derived miRNAs can be transmitted to the pathogen to downregulate genes essential for pathogen virulence. Thus, plants might have adapted cross-kingdom RNA interference mechanisms as part of their battery of immune responses used to arrest pathogen infection. However, although a large number of miRNAs are known to be regulated during pathogen infection, the biological role of the majority of these pathogen-regulated miRNAs has not yet been determined. In this review, we focus on miRNAs for which functional evidence exists on their involvement in the plant response to pathogen infection, mainly fungal pathogens. Elucidation of the function of miRNAs in plant immunity holds great potential for developing novel strategies to improve disease resistance in plants.

Keywords Disease resistance · Fungal pathogen · Immunity · MicroRNA (miRNA) · Plant defense

---

M. Bundó · S. Campo · B. San Segundo
Centre for Research in Agricultural Genomics (CRAG), CSIC-IRTA-UAB-UB, Campus Universitat Autònoma de Barcelona (UAB), Bellaterra (Cerdanyola del Vallés), Barcelona, Spain

B. San Segundo (✉)
Consejo Superior de Investigaciones Científicas (CSIC), Barcelona, Spain
e-mail: blanca.sansegundo@cragenomica.es

C. Miguel et al. (eds.), *Plant microRNAs*, Concepts and Strategies in Plant Sciences, https://doi.org/10.1007/978-3-030-35772-6_10

10.1 Plant miRNAs: Biogenesis and Origin

Small RNAs are short non-coding RNAs that guide gene silencing in most eukaryotes (Baulcombe 2004; Vaucheret 2006; Bologna and Voinnet 2014; Borges and Martienssen 2015). In plants, endogenous small RNAs can be classified into two major categories: short-interfering RNAs (siRNAs) and microRNAs (miRNAs) which are distinguished by their mode of biogenesis and mechanism of action. siRNAs derive from perfectly paired, long double-stranded RNA precursors that are produced either by the activity of an RNA-dependent RNA polymerase (RDR) or from antisense transcripts (Axtell 2013). siRNAs can be further classified into several sub-classes of siRNAs, such as phased siRNAs (phasiRNAs) and heterochromatic siRNAs (hc-siRNAs) (Arikit et al. 2013; Axtell 2013). In contrast, miRNAs derive from single-stranded RNA precursors with imperfectly base-paired hairpin structures. MiRNAs act as post-transcriptional regulators of gene expression by triggering cleavage or translational inhibition of target transcripts (Llave et al. 2002; Brodersen et al. 2008).

Plant miRNAs are transcribed by RNA polymerase II from *MIR* genes as long precursor transcripts (primary miRNAs or pri-miRNAs) that adopt partial self-complementarity, allowing the precursor to form a stem-loop structure. The pri-miRNAs are processed in two steps by a RNase III DICER-like, typically DCL1, to produce a pre-miRNA (precursor miRNA) and then a double-stranded duplex, the miRNA-5p/miRNA-3p duplex (previously named miRNA/miRNA*). Although DCL1 is the main DICER-like protein involved in miRNA biogenesis, DCL3 and DCL4 are also capable of processing miRNA precursor transcripts (Rajagopalan et al. 2006; Vazquez et al. 2008; Cuperus et al. 2011). In plants, the generation of the miRNA duplexes occurs within specialized compartments in the nucleus, called Dicing-bodies or D-bodies. Following miRNA precursor processing, the 3′ ends of miRNA duplexes are methylated by Hua Enhancer 1 (HEN1) that protect the mature miRNAs from the action of exonucleases. The functional strand of the duplex is then loaded into an ARGONAUTE (AGO) protein to form the miRNA-induced silencing complex (RISC) where the miRNA recognizes target transcripts by sequence complementarity. The 5′ end of the miRNA strand largely determines the choice of AGO proteins (Mi et al. 2008). In Arabidopsis, most miRNAscarry 5′ Uridine at their 5′ ends and are usually bound by AGO1. However, miRNAs can be loaded into other members of the AGO family. Readers interested in the details of biogenesis of miRNAs are referred to specialized reviews (Rogers and Chen 2013; Arikit et al. 2013; Budak and Akpinar 2015; Yu et al. 2017; Wang et al. 2019).

Plant miRNAs can be grouped into distinct families, each family comprising distinct precursor miRNA sequences generating similar, if not identical, mature miRNAs that often share the same physiological function. Traditionally, it has been considered that only one of the two strands of the miRNA-5p/miRNA-3p duplex is functional. However, increasing evidence supports that both strands of the duplex can be active (Liu et al. 2017). For instance, both strands of the Arabidopsis miR393 duplex play a role in immunity against bacterial infection. Here, miR393-5p guide cleavage of auxin receptor transcripts (TIR1, AFB2, AFB3), whereas miR393-3p cleave transcripts encoding a SNARE protein involved in exocytosis of antimicrobial proteins(Navarro et al. 2006; Zhang et al. 2011).

During the past years, the adoption of deep sequencing of small RNA transcriptomes has increased exponentially the number of annotated miRNAs in miRBase, the central on line repository for miRNAs (http://www.mirbase.org). Currently, up to 48,860 plant mature miRNAs are registered in miRBase (release 22.1) (Kozomara et al. 2019). The original guidelines for annotation of miRNAs in miRBase initially involved a combination of criteria, including evidence on their expression (detection by RNA gel blot, sequencing of small RNA cDNA libraries), prediction of a hairpin precursor structure containing the mature miRNA, conservation among species, and reduction of accumulation in a *DCL* mutant (Ambros et al. 2003). Nevertheless, these criteria have proven to be insufficient for reliable miRNA annotation and improvements of the initial criteria came out in 2008 (i.e. by including detection of sequences representing both strands of the miRNA duplex derived from a particular precursor structure) (Meyers et al. 2008). More recently, criteria for the confident annotation of plant miRNAs suitable for the era of 'big data' have been published (Axtell and Meyers 2018). The updated criteria require next-generation sequencing of small RNAs and replication (i.e. annotation of novel miRNAs will require support from independent small RNA-seq libraries). Other updated criteria refer to precursor length (e.g. fold-back structures longer than 300 nucleotides should be avoided), mature miRNA length which should not be below 20 nucleotides or above 24 nucleotides), mismatched positions in the miRNA duplex, and 5′ and 3′ positional variants in small RNA sequences of the miRNA duplex (Axtell and Meyers 2018).

Currently (release 22.1), *Medicago truncatula* and *Glycine max* are the plant species with the highest number of mature miRNA species registered in miRBase (756 miRNAs for both), followed by *Oryza sativa* (738 miRNAs). An examination of the history of annotations in miRBase, however, reveals that the number of new sequences registered in the last years does not substantially increase (Fig. 10.1). Now, the challenge is to identify species-specific miRNAs, and miRNAs showing tissue-specific or conditional expression patterns (e.g. developmental- or stress-regulated miRNAs). Even though many miRNAs from different plant species are registered in miRBase, the biological function of most plant miRNAs remains unknown.

MIR genes have been proposed to originate by inverted duplication of pre-existing gene sequences (Allen et al. 2004; Rajagopalan et al. 2006; Fahlgren et al. 2010). An inverted duplication of a gene, when transcribed, would produce a hairpin structure with high complementarity (a 'proto-miRNA'). During evolution, the accumulation of mutations in the genome would reduce complementarity in the hairpin structure, but retaining complementarity in the region where the mature miRNA locates. The successive accumulation of mutations would also provoke a shift in DCL usage to generate canonical miRNAs(ancient miRNAs, typically 21 nucleotides in length) which are preferentially processed by DCL1 (Vazquez et al. 2008). Consistent with this notion, young, recently evolved miRNAs are mostly processed by DCL3 or DCL4 and show extended complementarity in their precursor structure. Also, ancient *MIR* genes are generally conserved in plant species, whereas young *MIR* genes are less

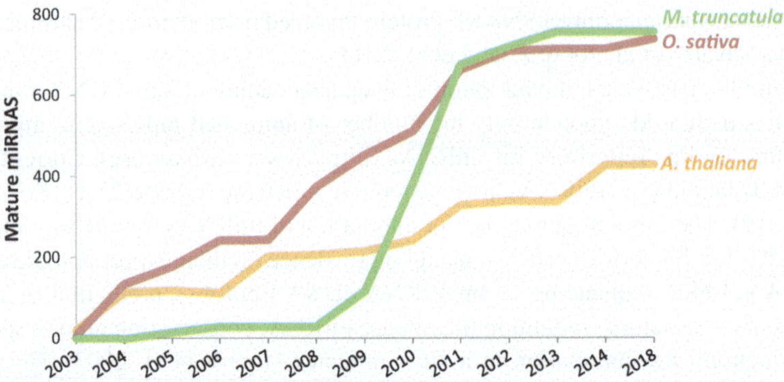

Fig. 10.1 History of annotations in miRBase of *M. truncatula*, *Oryza sativa*, and *Arabidopsis thaliana* miRNAs. The miRBase Tracker tool was used to track miRNA annotations in the miRBase database (http://www.mirbasetracker.org) (Van Peer et al. 2014)

conserved and appear to be species-specific (Cuperus et al. 2011; Qin et al. 2014). Other mechanisms that have been proposed to explain the origin of plant miRNAs include the evolution from inverted-repeat transposable elements that create hairpin RNAs resembling 'proto-miRNAs', or from random hairpin structures present in the genome (De Felippes et al. 2008; Axtell et al. 2011; Nozawa et al. 2012).

During evolution, many *MIR* genes have been gained and lost (Nozawa et al. 2012). The evolution and diversification of plant miRNAs can be driven by polyploidization (whole genome duplication), segmental duplications of subchromosomal regions, or tandem duplications (local duplications that involve only a few genes). Members of a miRNA family might have experienced gene duplication events, and some of the duplicated *MIR* genes might have diverged functionally, while others might have become pseudogenes.

Most plant *MIR* genes are found dispersed over the genome and can be defined as intergenic and intragenic *MIR* genes. *MIR* genes in intergenic regions are transcribed as independent transcriptional units under the control of their own regulatory sequences (promoter, terminator). Contrary to this, a number of plant miRNAs are known to derive from introns of host genes (intronic miRNAs, or mirtrons) that are transcribed as a common precursor transcript (Zhu et al. 2008; Meng and Shao 2012). Mirtrons do not arise from the canonical miRNA biogenesis pathway but from an alternative pathway through RNA splicing of host genes. The expression of intronic miRNAs and the host gene, from which they derive, is expected to be co-regulated.

Contrary to animal genomes, where *MIR* loci are frequently found in clusters which can be independently transcribed or simultaneously transcribed into polycistronic transcripts, only a few polycistronic miRNAs have been described in plants, most of them in Arabidopsis and rice (Boualem et al. 2008; Merchan et al. 2009; Barik et al. 2014; Baldrich et al. 2016; Salvador-Guirao et al. 2018b). Polycistronic miRNAs fall into two categories: homologous polycistronic miRNAs if the precursor structure contains more than one copy of the same miRNA family, and non-homologous polycistronic miRNAs in which the precursor contains unrelated miRNAs (e.g. miRNAs belonging to different miRNA families). Polycistronic transcription of multiple miRNAs would allow a coordinated expression of miRNAs potentially regulating functionally related or unrelated biological processes in a coordinated manner. An example of homologous polycistronic miRNAs is miR166k-166h, containing two miR166 family members (miR166k and miR166h). The miR166 family is highly conserved in plants and has conserved target genes, the Class II homeodomain-leucine zipper (HD-ZIP III) transcription factors involved in diverse developmental processes (Itoh et al. 2008). Additionally, the polycistronic miR166k-166h has been shown to function as a positive regulator in rice immunity via post-transcriptional control of *ethylene-insensitive 2* (*EIN2*) (Salvador-Guirao et al. 2018b). Here, the miR166k species encoded by the miR166k-166h polycistron was found to mediate cleavage of *EIN2*, a central ethylene signal transduction component that mediates plant immune responses (Zipfel 2013).

10.2 Function of miRNAs in Plants

It has long been recognized that miRNAs act as master regulators of plant development and adaptation to environmental stress. Most of the miRNAs initially identified target transcription factor genes and are highly conserved across the plant kingdom, supporting that these miRNAs have undergone positive selection for their biological role. Also, conserved miRNAs usually have conserved target genes that play essential functions in diverse physiological and developmental processes, including leaf and root development, organ polarity, phase change and flowering (Palatnik et al. 2003; Chen 2004; Mallory et al. 2004; Rubio-Somoza and Weigel 2011). MiRNAs involved in the control of developmental processes are usually represented by large gene families. MiRNAs are also involved in phenotypic plasticity triggered by various environmental stimuli, such as light, temperature and nutrients. There are many excellent articles covering different areas of miRNA function in plant development and adaptation to environmental stress (Chen 2009; Rubio-Somoza and Weigel 2011; D'Ario et al. 2017; Song et al. 2019).

Diversification of miRNA functions might also occur in conserved plant miRNAs, and miRNAs can evolve to recognize novel targets (i.e. in a miRNA family, a particular family member might have a non-conserved target for this family) (Debernardi et al 2012; Salvador-Guirao et al. 2018b). Additionally, plants produce species-specific miRNAs that show spatio/temporal expression patterns and have target genes that are involved in diverse functions. Unlike conserved miRNAs, which tend to be expressed at high levels, non-conserved miRNAs are expressed at very low levels and are often processed imprecisely. Despite this, certain non-conserved miRNAs might be expressed at a high level in a specific tissue, or induced under a particular condition.

Certain miRNAs are capable of moving from cell-to-cell and systemically, thus, representing important signaling molecules in controlling responses at specific functional domains, as well as at the whole plant level. The long-distance mobility of miRNAs in plants is well demonstrated by the transport of miR399 from shoot to root to modulate phosphate homeostasis in Arabidopsis (Lin et al. 2008). In other studies, susceptibility to rhizobial infection in *Lotus japonicas* was reported to be controlled by shoot-to-root trafficking of miR2111 (Tsikou et al. 2018). Plasmodesmata, the membrane-lined channels that provide cytoplasmic connection between adjacent cells, are the route that miRNAs use for cell-to-cell movement (Sevilem et al. 2013). For instance, the symplastic movement of miR165 via plasmodesmanta is well documented (Vatén et al. 2011). Recently, a novel polarized gating mechanism at intercellular connection points has been reported in Arabidopsis which might explain intercellular miRNA mobility to create functional domains (Skopelitis et al 2018). Although the mechanisms underlying miRNA mobility have not been elucidated, it appears that they are independent of those controlling protein movements (Skopelitis et al. 2018).

Finally, although many studies demonstrated alterations in the accumulation of miRNAs in response to pathogen infection in different plant species, our understanding of the functional roles of miRNAs in plant immunity is far less than that in plant development. In this review, we summarize the current knowledge on miRNAs with a known function in plant–pathogen interactions, with emphasis on the interaction with fungal pathogens.

10.3 Mechanisms Regulating Plant Immunity

In nature, plants are constantly exposed to a range of microbial pathogens with different lifestyles and modes of attack, including fungal, bacterial and viral pathogens. To prevent pathogen infection, plants must recognize the invader and activate fast and effective defense mechanisms. The activation of a successful plant defense response is initiated by the recognition of conserved microbial molecular signatures, collectively named pathogen-associated molecular patterns (PAMPs) or microbe-associated molecular patterns (MAMPs) by host receptors (Pathogen Recognition Receptors, or PRRs) (Fig. 10.2). PRRs localize on the plasma membrane and fall into the receptor-like kinases (RLK) and receptor-like protein (RLPs) family. Binding of PAMPs to PRRs induces PAMP-triggered immunity (PTI) which is associated to a profound and dynamic reprograming of gene expression that provides protection against most pathogens (Jones and Dangl 2006; Couto and Zipfel 2016; Cook et al. 2015). In addition to PAMPs/MAMPs which are derived from microorganisms, some endogenous molecules activate the innate immune system due to damage caused by pathogen infection. These molecules are referred to as Damage-Associated Molecular Patterns (DAMPs) (Choi and Klessig 2016). Perception of DAMPs triggers signaling pathways similar to PAMPs/effectors for defense activation.

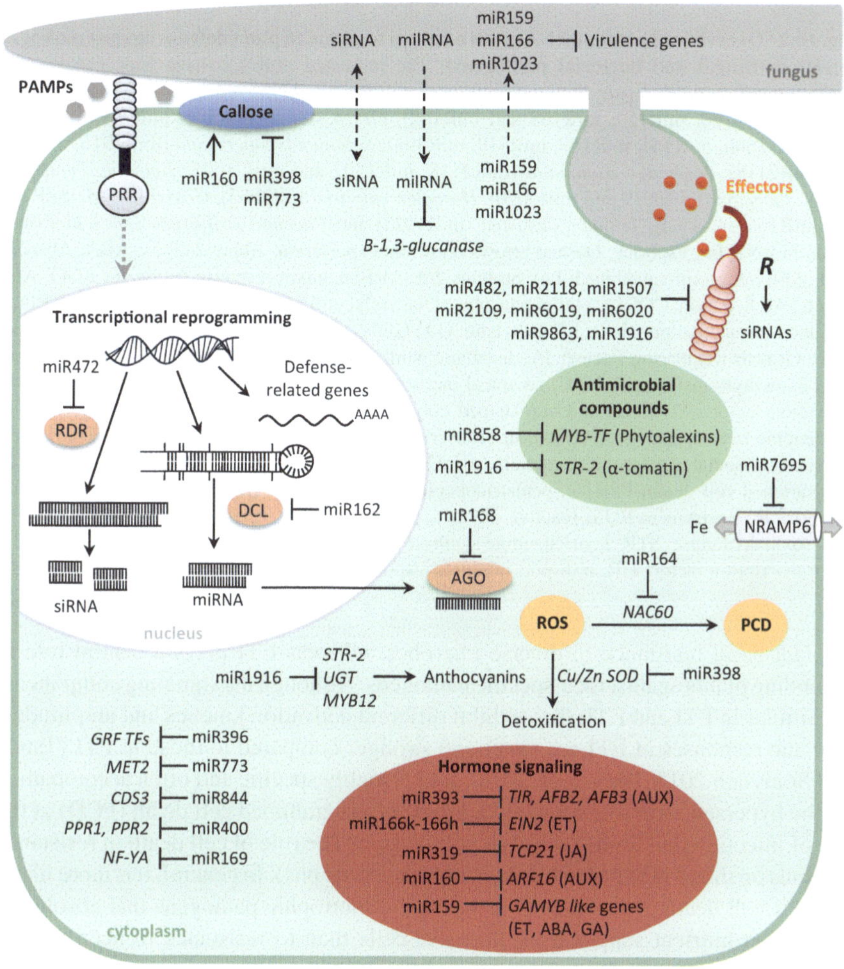

Pathogens adapted to their host plants can deliver virulence effector proteins into plant cells that target key PTI components and inhibit plant defense (Boller and Felix 2009; Boller and He 2009). Plants that are not able to detect these effectors are susceptible to the pathogen, resulting in effector-triggered susceptibility (ETS). However, during evolution, plants have evolved another layer of immunity in which receptors known as resistance (R) proteins detect specific effector proteins secreted by pathogens, the so-called effector-triggered immunity (ETI) (Fig. 10.2). In ETI, plants employ a diverse intracellular system of NBS-LRR (nucleotide-binding–leucine-rich repeat) receptors that recognize pathogens either directly or indirectly by the presence or activity of translocated pathogen effectors (Monteiro and Nishimura 2018). This recognition triggers a rapid and effective host defense. In general, PTI contributes

◀**Fig. 10.2 Overview of plant miRNAs with a known function in plant defense against pathogen infection (fungal and bacterial pathogens)**. The indicated miRNAs have been functionally characterized in *Arabidopsis thaliana* (miR393, miR396, miR773, miR858, miR844, miR400, miR398, miR160, miR159, miR168 and miR162), *Oryza sativa* (miR396, miR7695, miR160, miR166k-166h, miR164, miR169, miR319, miR398), *Solanum lycopersicum* (miR1916, miR482 and miR2118), *Medicago truncatula* (miR2118, miR1507 and miR2109), *Nicotiana benthamiana/N. tabacum* (miR6019 and miR6020), *Hordeum vulgare* (miR9863), *Gossypium* sp. (miR159 and miR166 exported), *Triticum aestivum* (miR1023) and *Puccinia striiformis f.sp. tritici* (milRNA, miRNA-like) species. Dashed arrows represent exportation. Abbreviations: ABA, Abscisic acid; AFB2/3, auxin-signaling F-box proteins 2/3; ARF16, auxin response factor 16; AGO, Argonaute; AUX, Auxin; CDS3, Cytidinephosphate Diacylglycerol Synthase 3;DCL, Dicer-Like; EIN2, Ethylene-insensitive protein 2; ET, Ethylene; GA, Gibberellins; GAMYB, Gibberellin myb gene; GRF, Growth-Regulating Factor; JA, Jasmonic acid; MYB, myeloblastosis transcription factor; MET2, methyltransferase 2; NAC, no apical meristem (NAM), Arabidopsis thaliana transcription activation factor (ATAF1/2) and cup-shaped cotyledon (CUC2); NBS-LRR, nucleotide-binding site leucine-rich repeat; NF-YA, Nuclear Transcription Factor Y Subunit Alpha NRAMP6, Natural resistance-associated macrophage protein 6); PAMP, Pathogen-associated Molecular Pattern; PCD, Programmed cell death; PPR1/2, pentatricopeptide repeat protein 1/2; PRR, Pattern recognition receptor; R, Resistance; ROS, reactive oxygen species; siRNAs, small interfering RNAs; SOD, Superoxid dismutase; STR-2, strictosidine synthase 2; TCP21, teosinte branched/cycloidea/pcf 21; TF, transcription factor; TIR, transport inhibitor response protein; UGT, UDP-glycosyltransferase

to plant basal resistance to diverse microbes, whereas ETI plays a central role in defending plants against race-specific pathogens. Although the signaling components are similar in PTI and ETI, they exhibit different activation kinetics and amplitudes: immune responses in ETI are faster and stronger compared to those in PTI (Tsuda and Somssich 2015; Peng et al. 2018). ETI is highly specific and often accompanied by the hypersensitive response (HR), a type of programmed cell death (PCD) at the site of infection that limits pathogen progression. The role of cell death in resistance depends on the type of pathogen (biotrophs, necrotrophs). In general, it is more likely that HR cell death contributes to resistance to biotrophic pathogens that absolutely depend on nutrient supply from host live cells than to resistance to necrotrophic pathogens.

During PTI and ETI, plants develop a wide range of defense mechanisms such as generation of reactive oxygen species (ROS), reinforcement of the cell wall and callose deposition, activation of protein phosphorylation cascades, production of antimicrobial compounds (e.g. phytoalexins), and induction of *Pathogenesis-Related* (*PR*) gene expression, among others. The production of ROS is one of the earliest plant responses to pathogen infection. Increased ROS levels contribute to cell wall strengthening via cross-linking of cell wall components, defense signaling and induction of HR (Torres et al. 2006). As opposed to having a defensive role during plant–pathogen interactions, excessive production of ROS might cause oxidative damage and ultimately cell death. Thus, proper ROS level is maintained by the balance between ROS production and ROS scavenging. Maintenance of ROS balance can also be perturbed by the presence of high levels of iron (Fe^{2+}) that would tip the normal cellular redox balance into oxidative stress owing to the enhanced production of hydroxyl radicals via the Fenton reaction (Demidchik 2015). Therefore, maintaining levels of cellular iron under control is paramount for keeping ROS within the redox biology range.

Phytohormones are regulators of plant growth and development and also play an essential role in the regulation of plant immune responses. Salicylic acid (SA), jasmonic acid (JA), ethylene (ET), and abscisic acid (ABA) are known to mediate defense responses to pathogens in plants (Denancé et al. 2013; Berens et al. 2017). Brassinosteroids and strigolactones have been shown to interact with the SA, JA, and ET signaling networks. In this way, synergistic and antagonistic interactions between hormone signaling pathways allow to fine-tune responses to different microbial pathogens. Hormone cross-talk between plant immune responses and developmental processes also helps plants to maintain a critical balance between growth and defense response.

10.4 miRNAs in Plant Immunity

Plant immune responses to bacterial and fungal pathogens, both PTI and ETI defense responses, have been historically considered as protein-based defense mechanisms, whereas RNA-based mechanisms largely regulate plant–virus interactions. Also, most studies on plant defense mechanisms focused on transcriptional reprograming of gene expression, and less is known about post-transcriptional processes controlling disease resistance. In recent years, however, increasing evidence emerged supporting that host endogenous small RNAs, siRNAs and miRNAs, are important players in processes associated to plant defense responses to bacterial and fungal pathogens (Katiyar-Agarwal and Jin 2010; Peláez and Sanchez 2013; Seo et al. 2013; Staiger et al. 2013; Gupta et al. 2014; Huang et al. 2016; Brant and Budak 2018; Islam et al. 2018).

During pathogen infection, important alterations in miRNA transcriptome are observed in many plant species which has led to the general view that plants utilize miRNAs as part of their arsenal to cope with the wide array of microbial pathogens they encounter. However, although many miRNAs have been shown to be pathogen-responsive in different pathosystems, the biological role of most of them remains elusive. The contribution of only a small number of miRNAs in antibacterial and antifungal immunity has been demonstrated. These miRNAs might function during PTI and/or ETI responses, as illustrated in Fig. 10.2. In this respect, studies in the dicotyledonous model plant *A. thaliana* during interaction with the bacterial pathogen *Pseudomonas syringae* allowed the identification of a series of miRNAs involved in the plant defense response (Staiger et al. 2013; Weiberg et al. 2014; Islam et al. 2018). Treatment with the flagellin-derived elicitor peptide flg22 from *P. syringae* causes an increase in miR393 accumulation which down-regulate the expression of auxin receptors which, in turn, enhances resistance to *P. syringae* (Navarro et al. 2006). Not only auxin signaling (regulated by miR393) but other hormone signaling pathways involved in plant immunity are known to be under miRNA regulation (Curaba et al. 2014). Clearly, miRNAs controlling hormone signaling might contribute to plant immunity either directly or indirectly via cross-talk between defense-related hormones.

Plant miRNAs might have a positive or negative effect on the regulation of immune responses, depending on the miRNA target gene that they regulate (e.g. negative or positive regulators of defense responses). Transgenic plants with altered expression of distinct miRNAs have been shown to exhibit a phenotype of disease resistance or susceptibility to infection by fungal pathogens, supporting that these miRNAs participate in defense responses.

In Arabidopsis, miR396, miR773 and miR858 negatively regulate PTI responses to infection by fungal pathogens, both necrotrophic (*Plectosphaerella cucumerina*) and hemibiotrophic (*Fusarium oxysporum*, *Colletotrichum higginsianum*) pathogens (Soto-Suarez et al. 2017; Salvador-Guirao et al. 2018a; Camargo-Ramírez et al. 2018). In particular, miR396, a negative regulator of immune responses in Arabidopsis, targets GROWTH-REGULATING FACTOR (GRF) family members. Interference with miR396 activity by the miRNA target mimic technology results in stronger PTI responses, or defense priming, and hence enhances resistance to fungal infection. Contrary to this, miR396 overexpression increases susceptibility to infection by fungal pathogens (Soto-Suarez et al. 2017). Overexpression of miR396 in rice enhances susceptibility to *M. oryzae* infection (Chandran et al. 2019). As for miR773, this miRNA targets METHYLTRANSFEREASE 2 (MET2). Interference with miR773 activity and overexpression results in resistance or susceptibility to fungal infection, respectively (Salvador-Guirao et al. 2018a). In other studies, the Arabidopsis miR858 (targeting flavonoid-specific MYB transcription factors) was found to modulate the accumulation of phenylpropanoid compounds with antifungal activity, including flavonoids, this particular miRNA functioning as a negative regulator of immune responses (Camargo-Ramírez et al. 2018). Interference with miR858 activity by target mimics (*MIM858* plants) results in disease resistance which was associated with a stronger induction of ET-mediated defense responses (Camargo-Ramírez et al. 2018). Contrary to this, *MIR858* overexpression enhances susceptibility to infection by fungal pathogens.

Other miRNAs that are negative regulators of the defense response of Arabidopsis plants to pathogen infection are miR844 and miR400, targeting cytidinephosphate diacylglycerol synthase 3 and a pentatricopeptide repeat protein gene (PRR), respectively. The expression of these miRNAs is downregulated in Arabidopsis during infection by the necrotrophic fungus *B. cinerea*, leading to enhanced expression of the corresponding target genes (Lee et al. 2015; Park et al. 2014). Transgenic Arabidopsis plants overexpressing miR844 displayed much more severe disease symptoms than wild-type plants, whereas a loss-of-function of *MIR844* mutant showed enhanced disease resistance (Lee et al. 2015). Similarly, miR400-overexpressing plants (as well as *ppr* mutants) were more susceptible to *P. syringae* and *B. cinerea* than wild-type plants (Park et al. 2014). In other studies, miR1916 was described to regulate the expression of target genes involved in the production of secondary metabolites in tomato (α-tomatine and anthocyanin), and its overexpression enhances susceptibility to *B. cinerea,* supporting that miR1916 is involved in defensive responses of tomato to pathogen infection (Chen et al. 2018).

MiR398 directs cleavage of transcripts encoding Cu/Zn superoxide dismutases (Cu/Zn SODs), acopper chaperone for SOD, and a subunit of cytochrome c oxidase (COX5b-1). Consistent with this function, a role for miR398 in protecting plants against oxidative stress caused during pathogen infection (e.g. fungal and bacterial pathogens) and abiotic stress has been proposed (Sunkar et al. 2006; Lu et al. 2010). Intriguingly, whereas miR398 overexpression in Arabidopsis compromises resistance against the bacterial pathogen *P. syringae,* its overexpression in rice enhances resistance to the fungal pathogen *Magnaporthe oryzae* (Li et al. 2010, 2014, 2019). These findings suggest that a particular miRNAs might function as a positive or a negative regulator of immune responses depending on the host plant, the type of pathogen, and/or the pathogen lifestyle (necrotrophic, biotrophic, or hemibiotrophic pathogens).

In addition to PTI, miRNAs are also important regulators of gene expression in ETI responses. Several miRNAs have been shown to target the transcripts of *R* genes encoding NBS-LRR proteins, the largest group of plant *R* genes, and trigger the production of secondary siRNAs, or phasiRNAs (Fig. 10.2). A miRNA superfamily comprising miR482 and miR2118 simultaneously silence multiple *R* genes involved in antibacterial and antiviral defense in tomato plants (Shivaprasad et al. 2012; Ouyang et al. 2014). In *M. truncatula*, miR2118, miR1507 and miR2109 miRNAs were also shown to target *NBS-LRR* genes and to produce phasiRNAs (Zhai et al 2011). It was proposed that miRNA-mediated regulation of *R* genes is a conserved mechanism that prevents *R*-triggered autoimmunity in the absence of the pathogen, while allowing *R* gene expression during pathogen infection. Regulation of the tobacco TIR-NB-LRR (Toll and Interleukin-1 receptor-NB-LRR) immune receptor by miR6019/miR6020 was reported, whereas miR9863 regulates *Mla* alleles encoding coiled-coil-NB-LRR (CC-NB-LRR) immune receptors in barley (Li et al. 2012; Liu et al. 2014). Moreover, the Arabidopsis miR472-RDR6 silencing pathway was found to modulate PTI and ETI through the control of disease resistance genes during infection with the bacterial pathogen *P. syringae* (Boccara et al. 2014).

Rice is the primary source of food for more than half of the world's population. Rice production is, however, severely compromised by the rice blast disease caused by the fungal pathogen *Magnaporthe oryzae* (Wilson and Talbot 2009). A plethora of rice miRNAs are regulated during infection of rice plants by *M. oryzae* or treatment with elicitors prepared from this fungus, pointing to a critical role of miRNAs in rice immunity (Campo et al. 2013; Li et al. 2014, 2016; Baldrich et al. 2015). However, only a few of these pathogen-regulated rice miRNAs have been functionally characterized (Fig. 10.2). They include positive (miR7695, miR160, miR398, and miR166k-166h) and negative (miR164a, miR169 and miR319) regulators of the rice defense response against *M. oryzae* infection (Campo et al. 2013; Li et al. 2014, 2017; Zhang et al. 2018; Wang et al. 2018; Salvador-Guirao et al. 2018b). Overexpression of miR160 or miR398 results in increased H_2O_2 accumulation at the infection site and up-regulation of defense-related gene expression, thus, promoting resistance to blast disease (Li et al. 2014). On the other hand, transgenic rice lines overexpressing miR169 are more susceptible to *M. oryzae* infection which is associated to a reduced expression of defense genes (Li et al. 2017).

The rice miR7695 represents a young, recently evolved miRNA that experienced natural and domestication selection events during evolution (Campo et al. 2013). MiR7695 occurs in *japonica*, but not in *indica*, subspecies from the *O. sativa* genus and its precursor is processed by DCL4. Moreover, miR7695 overexpression confers resistance to *M. oryzae* infection, supporting a positive role for miR7695 in blast resistance. It has been reported that miR7695 negatively regulates an alternatively spliced transcript of the *Nramp6* (*Natural resistance-associated macrophage pathogen 6*) gene encoding an iron and manganese transporter (Campo et al. 2013; Peris-Peris et al. 2017) (Fig. 10.2). A miR7695-mediated regulation of *Nramp6* during *M. oryzae* infection might then provide a flexible mechanism for the control of iron homeostasis which, in turn, would be important in disease resistance.

Different miRNAs might exert their regulatory role in pathogen resistance by affecting a particular stress-related process in an opposite manner. For instance, Arabidopsis miR160 positively regulates PAMP-induced callose deposition, while miR398 and miR773 negatively regulate callose deposition and disease resistance to bacteria (Li et al. 2010) (Fig. 10.2). Furthermore, a differential expression pattern of miRNA family members might be linked to different functions depending on the plant species and the invading pathogen.

In addition to controlling plant developmental programs, miRNAs are also key regulators in multiple hormones signalling pathways and cross-talk between defense-related hormones, including ET, JA, ABA or auxin signaling (Curaba et al. 2014). For instance, JA biosynthesis is regulated by miR319 (Zhang et al. 2016a, b) whereas miR159 constitutes a major connection among ET, ABA and GA in controlling developmental processes and programmed cell death (Curaba et al. 2014). Combined small RNA and degradome sequencing of rice tissues treated with *M. oryzae* elicitors revealed several miRNA-mediated regulatory networks in rice immune responses, including regulation of hormone signaling pathways (e.g. ethylene and auxin pathways) (Baldrich et al. 2015). Under this scenario, interactions between pathogen-regulated miRNAs and miRNA target genes involved in different hormonal regulatory networks would allow a coordinated regulation of diverse physiological processes for the regulation of immune responses. Despite this wealth of knowledge, there are still many unanswered questions regarding miRNA function in plant immunity. Because miRNAs provide quantitative regulation of gene expression, rather than on-off regulation, these riboregulators would help fine-tune host gene expression in reprogramming developmental programs and defense responses in a coordinated manner. This will avoid the fitness costs associated with the constitutive expression of host defense responses, thus, optimizing the growth-defense balance.

It is worth mentioning here that the functional analysis of *MIR* genes has been historically hampered due to the lack of an effective approach to silence *MIR* gene expression. The small size of *MIR* genes makes it difficult to produce T-DNA insertional knock-out mutants. To counter this limitation, alternative approaches have been developed aiming to sequester miRNAs to repress their activity (the target mimicry strategy), using miRNA-resistant target constructs in transgenic plants, and more recently, using targeted genome editing technologies, such as the CRISPR/Cas9 (clustered regulatory interspaced short palindromic repeats/Cas9) technology. Although the CRISPR/Cas9 system has been successfully used for the production of disease-resistant plants by targeting protein-coding genes (Wang et al. 2014, 2016a, b; Nekrasov et al. 2017; Chen et al. 2019), the use of CRISPR/Cas9 technology in editing miRNAs in plants is still in its infancy. The current scenario is that the use of the CRISPR/Cas9 technology will have important applications for targeted mutagenesis of *MIR* genes playing a role in disease resistance in plants. For more comprehensive, detailed information on technologies to delineate the function of miRNAs, we refer to Chap. 2.

Not only miRNAs but also components of the miRNA biogenesis pathway might play a role in disease resistance in plants. For instance, silencing of *OsDCL1* (a major component of the miRNA biogenesis pathway) has been shown to confer resistance to *M. oryzae*, whereas *OsDCL1* activation enhances susceptibility to infection by *M. oryzae* (hemibiotroph) and *Fusarium fukikuroi* (necrotroph) (Zhang et al. 2015; Salvador-Guirao et al. 2019). Susceptibility to pathogen infection in the rice *dcl1* activation mutants was associated with a weaker induction of defense gene expression and failure to produce diterpenoid phytoalexins during pathogen infection (Salvador-Guirao et al. 2019). In addition to *M. oryzae* infection, treatment with *M. oryzae* elicitors also regulates *OsDCL1* expression, suggesting that *OsDCL1* plays a role in PTI. Even though the most obvious trend that one would expect from *OsDCL1* activation is an increase in miRNA accumulation, alterations in the miRNA transcriptome caused by *OsDCL1* activation included both up-regulation and down-regulation of rice miRNAs. This observation suggests that factors other than processing of miRNA precursors by DCL1 might be involved in regulating miRNA accumulation in rice plants. Intriguingly, whereas *DCL1* appears to function as a negative regulator in rice immunity, this gene acts as a positive regulator of immune responses in Arabidopsis. Supporting this notion, susceptibility to infection by fungal (*B. cinerea*) and bacterial (*P. syringae*) pathogens have been observed in Arabidopsis *dcl1* mutants (defective in miRNA biogenesis) (Navarro et al. 2008; Seo et al. 2013; Weiberg et al. 2013). On the other hand, the Arabidopsis *dcl4* mutant displays enhanced susceptibility to the vascular wilt fungal pathogen *Verticillium dahliae*, the causal agent of the verticillium wilt disease, pointing to a positive role of DCL4 in the host defense response (Ellendorff et al. 2009).

Other components of the machinery for small RNA biogenesis and functioning might also contribute to defense against *Verticillium* wilt disease, such as RDR6 (previously named as SGS2) and AGO proteins (Ellendorff et al. 2009). In particular, AGO1 (the core component of the RISC complex involved in miRNA functioning) is known to be regulated by miR168, and *MIR168* was reported to be transcriptionally activated in response to treatment with fungal elicitors in Arabidopsis (Baldrich et al. 2014). It has been proposed that AGO1 contributes to PAMP-induced disease resistance in Arabidopsis by positively regulating callose deposition and defense gene expression (Navarro et al. 2008; Li et al. 2010). Together, this piece of evidence supports the notion that, in addition to miRNAs, components of the small RNA biogenesis and function pathways might contribute to plant immunity.

10.5 Cross-Kingdom Regulation of Gene Expression by miRNAs and Other Small RNAs in Plant–Pathogen Interactions

Small RNAs have been found to be mobile within organisms of different kingdoms triggering gene silencing in the interacting organism, the so-called cross-kingdom RNA interference (RNAi) (Knip et al. 2014). This phenomenon has been observed in both plants and animals interacting with pathogenic, symbiotic and parasitic organisms. For instance, small RNAs can move from humans to the protozoan parasite *Plasmodium falciparum*, bacteria to nematodes, plants to nematodes, plants to insects, or parasitic plants to plants (Mao et al. 2007; LaMonte et al. 2012; Liu et al. 2012; Knip et al. 2014; Shahid et al. 2018).

Cross-kingdom regulation of gene expression by small RNAs has been also demonstrated in plant–fungus interactions. Increasing evidence supports that fungal small RNAs can traffic from a fungal pathogen to its host as a strategy to cause disease (Wang et al. 2016a, b, 2018; Hua et al. 2018). Particularly, the fungus *B. cinerea* delivers small RNAs (*Bc-sRNAs*) into Arabidopsis and tomato cells where the fungal small RNAs interfere with host immune responses (Weiberg et al. 2013). The *B. cinerea* small RNAs are generated from transposable elements by fungal DCL1/2. When transferred to the host plant cells, they bind to the plant AGO1 and hijack the plant RNA interference machinery to selectively silence host defense genes (Weiberg et al. 2013). Then, small RNAs from *B. cinerea* function as pathogen effectors during infection. The observation that pathogen-encoded small RNAs are delivered into plant cells to suppress plant immunity adds another layer of complexity in our understanding of the molecular mechanisms underlying plant immunity (Weiberg et al 2013, 2014). In other studies, effectors from the oomycete plant pathogen *Phytophthora sojae* were found to suppress RNA silencing in the host plant by inhibiting the biogenesis of small RNAs (Qiao et al. 2013).

Conversely, plants can also transfer small RNAs into pathogens to attenuate their pathogenicity. Transfer of small RNAs from Arabidopsis and tomato plants expressing double-stranded RNAs (source of siRNAs) that target fungal genes (e.g. *Bc-DCL1* and *Bc-DCL2*) led to a reduction of disease symptoms caused by *B. cinerea* infection (Wang et al. 2016a, b). Recent studies have shown that *Phytophtora* infection increases the production of Arabidopsis siRNAs, i.e. siRNAs from a pentatrico peptide-repeat protein (PPR)-encoding gene loci, which contribute to defense against *Phytophtora* by silencing fungal target genes (Hou et al. 2019). As a counter-defense mechanism, a *Phytophthora* effector (PSR2) blocks this host-induced gene silencing by suppressing the biogenesis of these antimicrobial siRNAs. These findings indicate that hosts and pathogens can be engaged in an arms race centered on cross-kingdom RNAi-based immunity.

Delivery of artificially designed small RNA from plants to pathogen for cross-kingdom RNAi, the so-called host-induced gene silencing (HIGS) has proven to be a valuable tool for silencing target genes of invading pathogens and protection against fungal infection (Wang et al. 2017a, b; Qi et al. 2019). For instance, transgenic tomato and Arabidopsis plants expressing RNAi constructs targeting transcripts of genes that are essential for *Verticillium dahliae* pathogenicity are protected from *Verticillium* wilt disease, a devastating disease in many crops (Song and Thomma 2018).

Not only small RNAs but also miRNAs have been shown to function in cross-kingdom regulation of gene expression in plant–pathogen interactions (Wang et al. 2018) (Fig. 10.2). During infection with the fungus *V. daliae*, cotton plants accumulate miR159 and miR166 which are exported to the fungal cells to silence genes that are essential for fungal virulence (Zhang et al. 2016a, b). A wheat miRNA, miR1023, suppress the invasion of the fungal pathogen *Fusarium graminearum* by targeting and silencing an *alpha/beta hydrolase* gene in *F. graminearum* (Jiao and Peng 2018). On the other hand, a miRNA-like (milRNA) from the fungus *Puccinia striiformis* f. sp. *tritici*, the causal agent of wheat stripe rust, was found to act as a pathogen effector to suppress innate immunity in wheat plants by targeting the a β-1,3-glucanase gene (a *PR2* family member) (Wang et al. 2017a, b). The parasitic plant *Cuscuta campestris* has been shown to transfer miRNAs that hijack the host's own silencing machinery and reprogram Arabidopsis gene expression to the parasitic's advantage (Shahid et al. 2018). Several targets of the *C. campestris* miRNAs are linked to defense signaling, such as auxin receptors (TIR1, AFB2, AFB3), and the receptor-like cytoplasmic kinase BIK1, a component of the Arabidopsis signaling pathway that mediates pathogen resistance. Altogether, these observations suggest that plants might have developed defense strategies to inhibit pathogen infection in which host small RNAs, siRNAs and miRNAs, move into the fungal cells to regulate virulence genes. However, compared to small RNAs, cross-kingdom regulation of gene expression by plant miRNAs contributing to disease resistance has been less explored.

It is generally accepted that small RNAs trafficking between plants and pathogens takes place through extracellular vesicles, or exosomes (Kim et al. 2017; Lefebvre and Lécuyer 2017; Cai et al. 2018; Hou et al. 2019). Along with this, plants might have adapted exosome-mediated cross-kingdom RNA interference mechanisms as part of their battery of immune responses used to arrest pathogen invasion.

Pathogens are also capable of taking up small RNAs from the environment. Wang et al. (2016a, b) demonstrated a reduction in disease symptoms caused by *B. cinerea* by directly spraying small RNAs on plant surfaces, such as fruits (tomato, strawberry, grape), flower petals (rose), or vegetables (onion epidermis, lettuce). Spray application of a long non-coding double-stranded RNA targeting *Fusarium graminearum* cytochrome P450 genes was found to be effective to reduce disease development in *F. graminearum*-inoculated barley leaves (Koch et al. 2016). Exploiting this knowledge might allow the development of novel strategies involving the use of miRNAs as 'fungicides' for crop protection in an environmentally friendly way.

10.6 Final Remarks

Diseases caused by fungal pathogens continuously threaten crop production worldwide. Despite the great effort that has been made during the last years on the identification of miRNAs in plants, most of this research focused on miRNAs accumulating in different plant tissues and/or developmental stages. The specific function of numerous miRNAs involved in the control of developmental processes has been uncovered, but these miRNAs represent a relatively small fraction of known miRNAs. It is also noticeable that, although extensive data support that a substantial fraction of plant miRNAs is responsive to infection by fungal pathogens, their role in disease resistance remains largely unknown, and our understanding of miRNA-mediated processes underlying plant immunity is far from complete. Such studies tend to be complicated, at least in part, due to the fact that fungal pathogens have different lifestyles and strategies of pathogenesis, with necrotrophs gaining nutrients from rapidly killed tissue and biotrophs gaining nutrients from living host tissue (or in the case of hemibiotrophs, from living tissue that dies in a later stage of pathogenesis). As more functions of miRNAs are uncovered, it will be possible to distinguish between the contribution of miRNAs in specific pathosystems, and those that have a more general function in different host/pathogen interactions. The full characterization of the miRNA population in a different plant species will also provide clues for understanding the function and evolution of plant miRNAs in the context of disease resistance. Beyond double-stranded break generation for the production of *MIR* gene knockouts, the use of CRISPR/Cas9 variants might allow base editing (cytosine base-edition for conversion of cytosine in the DNA to uracil which is then replaced by thymine during DNA replication), or transcriptional regulation of *MIR* gene expression (i.e. using modified Cas9 nucleases fused to effector domains for activation or repression of *MIR* expression).

On the other hand, from what we have learned so far, a miRNA might have a function not only at the cell and/or tissue level but also at the whole-plant level. The new era of single-cell sequencing will certainly provide interesting insights into our research on miRNA function in plants.

Clearly, *MIR* genes can then be regarded as the next generation targets for genome editing, with evident applications for improvement of disease resistance in plant species of agronomical interest. However, to fully exploit the potential of genome editing technologies in crop protection (e.g. the CRISPR/Cas9 system; Chap. 2), a better knowledge of miRNAs controlling plant immune responses is needed. Since fungal diseases are major threats to important crops worldwide, unraveling the miRNA-mediated mechanisms underlying pathogen resistance in plants will help in designing novel strategies to improve disease resistance in plants.

Acknowledgements This work was supported by the Ministerio de Ciencia, Innovación y Universidades–Agencia Estatal de Investigación/FEDER (BIO2015-67212-R, RTI2018-101275-B-100, and BIO2017-92113-EXP). We also acknowledge support from the CERCA Programme ('Generalitat de Catalunya'), and MINECO ('Severo Ochoa Programme for Centres of Excellence in R&D' 2016-2019, SEV-2015-0533).

References

Allen E, Xie Z, Gustafson AM et al (2004) Evolution of microRNA genes by inverted duplication of target gene sequences in *Arabidopsis thaliana*. Nat Genet 36:1282–1290

Ambros V, Bartel B, Bartel DP et al (2003) A uniform system for microRNA annotation. RNA9:277–279

Arikit S, Zhai J, Meyers BC (2013) Biogenesis and function of rice small RNAs from non-coding RNA precursors. Curr Opin Plant Biol 16:170–179

Axtell MJ (2013) Classification and comparison of small RNAs from plants. Annu Rev Plant Biol 64:137–159

Axtell MJ, Meyers BC (2018) Revisiting criteria for plant microRNA annotation in the era of big data. Plant Cell 30:272–284

Axtell MJ, Westholm JO, Lai EC (2011) Vive la différence: biogenesis and evolution of microRNAs in plants and animals. Genome Biol 12:221–234

Baldrich P, Kakar K, Siré C et al (2014) Small RNA profiling reveals regulation of *Arabidopsis* miR168 and heterochromatic siRNA415 in response to fungal elicitors. BMC Genom 15:1083–1099

Baldrich P, Campo S, Wu MT et al (2015) MicroRNA-mediated regulation of gene expression in the response of rice plants to fungal elicitors. RNA Biol 12:847–863

Baldrich P, Hsing YIC, San Segundo B (2016) Genome-Wide analysis of polycistronic microRNAs in cultivated and wild rice. Genome Biol Evol 8:1104–1114

Barik S, SarkarDas S, Singh A et al (2014) Phylogenetic analysis reveals conservation and diversification of micro RNA166 genes among diverse plant species. Genomics 103:114–121

Baulcombe D (2004) RNA silencing in plants. Nature 431:356–363

Berens ML, Berry HM, Mine A et al (2017) Evolution of hormone signaling networks in plant defense. Annu Rev Phytopathol 55:401–425

Boccara M, Sarazin A, Thiébeauld O et al (2014) The *Arabidopsis* miR472-RDR6 silencing pathway modulates PAMP- and Effector-Triggered Immunity through the post-transcriptional control of disease resistance genes. PLoS Pathog 11:e1004814

Boller T, Felix G (2009) A renaissance of elicitors: perception of microbe-associated molecular patterns and danger signals by pattern-recognition receptors. Annu Rev Plant Biol 60:379–406

Boller T, He SY (2009) Innate immunity in plants: an arms race between pattern recognition receptors in plants and effectors in microbial pathogens. Science 324(5928):742–744

Bologna NG, Voinnet O (2014) The diversity, biogenesis, and activities of endogenous silencing small RNAs in *Arabidopsis*. Annu Rev Plant Biol 65:473–503

Borges F, Martienssen RA (2015) The expanding world of small RNAs in plants. Nat Rev Mol Cell Biol 16:727–741

Boualem A, Laporte P, Jovanovic M et al (2008) MicroRNA166 controls root and nodule development in *Medicago truncatula*. Plant J. 54:876–887

Brant EJ, Budak H (2018) Plant small non-coding RNAs and their roles in biotic stresses. Front Plant Sci 9:1038

Brodersen P, Sakvarelidze-Achard L, Bruun-Rasmussen M et al (2008) Widespread translational inhibition by plant miRNAs and siRNAs. Science 320:1185–1190

Budak H, Akpinar BA (2015) Plant miRNAs: biogenesis, organization and origins. Funct Integr Genomics 15:523–531

Cai Q, Qiao L, Wang M et al (2018) Plants send small RNAs in extracellular vesicles to fungal pathogen to silence virulence genes. Science 360:1126–1129

Camargo-Ramírez R, Val-Torregrosa B, San Segundo B (2018) MiR858-mediated regulation of flavonoid-specific MYB transcription factor genes controls resistance to pathogen infection in *Arabidopsis*. Plant Cell Physiol 59:190–204

Campo S, Peris-Peris C, Siré C et al (2013) Identification of a novel microRNA (miRNA) from rice that targets an alternatively spliced transcript of the *Nramp6 (Natural resistance-associated macrophage protein 6)* gene involved in pathogen resistance. New Phytol 199:212–227

Chandran V, Wang H, Gao F et al (2019) miR396-OsGRFs module balances growth and rice blast disease-resistance. Front Plant Sci 9:1999

Chen X (2004) A microRNA as a translational repressor of APETALA2 in *Arabidopsis* flower development. Science 303:2022–2025

Chen X (2009) Small RNAs and their roles in plant development. Annu Rev Cell Dev Biol 25:21–44

Chen L, Meng J, He XL et al (2018) *Solanum lycopersicum* microRNA1916 targets multiple target genes and negatively regulates the immune response in tomato. Plant Cell Environ 42:1393–1407

Chen K, Wang Y, Zhang R et al (2019) CRISPR/Cas genome editing and precision plant breeding in agriculture. Annu Rev Plant Biol 70:1–31

Choi HW, Klessig DF (2016) DAMPs, MAMPs, and NAMPs in plant innate immunity. BMC Plant Biol 16:232–242

Cook DE, Mesarich CH, Thomma BP (2015) Understanding plant immunity as a surveillance system to detect invasion. Annu Rev Phytopathol 53:541–563

Couto D, Zipfel C (2016) Regulation of pattern recognition receptor signalling in plants. Nat Rev Immunol 16:537–552

Cuperus JT, Fahlgren N, Carrington JC (2011) Evolution and functional diversification of *MIRNA* genes. Plant Cell 23:431–442

Curaba J, Singh MB, Bhalla PL (2014) miRNAs in the crosstalk between phytohormone signalling pathways. J Exp Bot 65:1425–1438

D'Ario M, Griffiths-Jones S, Kim M (2017) Small RNAs: big impact on plant development. Trends Plant Sci 22:1056–1068

De Felippes FF, Schneeberger K, Dezulian T et al (2008) Evolution of *Arabidopsis thaliana* microRNAs from random sequences. RNA14:2455–2459

Debernardi JM, Rodriguez RE, Mecchia MA et al (2012) Functional specialization of the plant miR396 regulatory network through distinct microRNA-target interactions. PLoS Genet 8:e1002419

Demidchik V (2015) Mechanisms of oxidative stress in plants: From classical chemistry to cell biology. Environ Exp Bot 109:212–228

Denancé N, Sánchez-Vallet A, Goffner D, Molina A (2013) Disease resistance or growth: the role of plant hormones in balancing immune responses and fitness costs. Front Plant Sci 4:155

Ellendorff U, Fradin EF, De Jonge R, Thomma BPHJ (2009) RNA silencing is required for *Arabidopsis* defence against *Verticillium* wilt disease. J Exp Bot 60:591–602

Fahlgren N, Jogdeo S, Kasschau KD et al (2010) MicroRNA gene evolution in *Arabidopsis lyrata* and *Arabidopsis thaliana*. Plant Cell 22:1074–1089

Gupta OP, Sharma P, Gupta RK, Sharma I (2014) Current status on role of miRNAs during plant–fungus interaction. Physiol Mol Plant Pathol 85:1–7

Hou Y, Zhai Y, Feng L et al (2019) A *Phytophthora* effector suppresses trans-kingdom RNAi to promote disease susceptibility. Cell Host Microbe 25:153–165

Hua C, Zhao JH, Guo HS (2018) Trans-kingdom RNA silencing in plant–fungal pathogen interactions. Mol Plant 11:235–244

Huang J, Yang M, Lu L, Zhang X (2016) Diverse functions of small RNAs in different plant–pathogen communications. Front Microbiol 7:1552

Islam W, Qasim M, Noman A et al (2018) Plant microRNAs: front line players against invading pathogens. Microb Pathog 118:9–17

Itoh JI, Hibara KI, Sato Y, Nagato Y (2008) Developmental role and auxin responsiveness of Class III Homeodomain Leucine Zipper gene family members in rice. Plant Physiol 147:1960–1975

Jiao J, Peng D (2018) Wheat microRNA1023 suppresses invasion of *Fusarium graminearum* via targeting and silencing FGSG_03101. J Plant Interact 13:514–521

Jones JDG, Dangl JL (2006) The plant immune system. Nature 444:323–329

Katiyar-Agarwal S, Jin H (2010) Role of small RNAs in host-microbe interactions. Annu Rev Phytopathol 48:225–246

Kim KM, Abdelmohsen K, Mustapic M et al (2017) RNA in extracellular vesicles. Wiley Interdiscip Rev RNA 8:1–10

Knip M, Constantin ME, Thordal-Christensen H (2014) Trans-kingdom cross-talk: small RNAs on the move. PLoS Genet 10:e1004602

Koch A, Biedenkopf D, Furch A et al (2016) An RNAi-based control of *Fusarium graminearum* infections through spraying of long dsRNAs involves a plant passage and is controlled by the fungal silencing machinery. PLoS Pathog 12:e1005901

Kozomara A, Birgaoanu M, Griffiths-Jones S (2019) MiRBase: from microRNA sequences to function. Nucleic Acids Res 47:155–162

LaMonte G, Philip N, Reardon J et al (2012) Translocation of sickle cell erythrocyte microRNAs into *Plasmodium falciparum* inhibits parasite translation and contributes to malaria resistance. Cell Host Microbe 12:187–199

Lee HJ, Park YJ, Kwak KJ et al (2015) MicroRNA844-guided downregulation of *Cytidinephosphate Diacylglycerol Synthase3 (CDS3)* mRNA affects the response of *Arabidopsis thaliana* to bacteria and fungi. Mol Plant–Microbe Interact 28:892–900

Lefebvre FA, Lécuyer E (2017) Small luggage for a long journey: transfer of vesicle-enclosed small RNA in interspecies communication. Front Microbiol 8:337

Li Y, Zhang Q, Zhang J et al (2010) Identification of microRNAs involved in pathogen-associated molecular pattern-triggered plant innate immunity. Plant Physiol 152:2222–2231

Li F, Pignatta D, Bendix C et al (2012) MicroRNA regulation of plant innate immune receptors. Proc Natl Acad Sci USA 109:1790–1795

Li Y, Lu YG, Shi Y et al (2014) Multiple rice microRNAs are involved in immunity against the blast fungus *Magnaporthe oryzae*. Plant Physiol 164:1077–1092

Li ZY, Xia J, Chen Z et al (2016) Large-scale rewiring of innate immunity circuitry and microRNA regulation during initial rice blast infection. Sci Rep 6:25493

Li Y, Zhao SL, Li JL et al (2017) *Osa-miR169* negatively regulates rice immunity against the blast fungus *Magnaporthe oryzae*. Front Plant Sci 8:2

Li Y, Cao XL, Zhu Y et al (2019) *Osa-miR398b* boosts H_2O_2 production and rice blast disease-resistance via multiple superoxide dismutases. New Phytol 222:1507–1522

Lin S, Chiang SF, Lin W et al (2008) Regulatory network of microRNA399 and PHO2 by systemic signaling. Plant Physiol 147:732–746

Liu H, Wang X, Wang HD et al (2012) *Escherichia coli* noncoding RNAs can affect gene expression and physiology of *Caenorhabditis elegans*. Nat Commun 3:1073

Liu J, Cheng X, Liu D et al (2014) The miR9863 family regulates distinct *Mla*alleles in barley to attenuate NLR receptor-triggered disease resistance and cell-death signaling. PLoS Genet 10:e1004755

Liu W, Meng J, Cui J, Luan Y (2017) Characterization and function of microRNA΄s in plants. Front Plant Sci 8:2200

Llave C, Xie Z, Kasschau KD, Carrington JC (2002) Cleavage of Scarecrow-like mRNA targets directed by a class of Arabidopsis miRNA. Science 297:2053–2056

Lu Y, Feng Z, Bian L et al (2010) miR398 regulation in rice of the responses to abiotic and biotic stresses depends on *CSD1* and *CSD2* expression. Funct Plant Biol 38:44–53

Mallory AC, Dugas DV, Bartel DP, Bartel B (2004) MicroRNA regulation of NAC-domain targets is required for proper formation and separation of adjacent embryonic, vegetative, and floral organs. Curr Biol 14:1035–1046

Mao YB, Cai WJ, Wang JW et al (2007) Silencing a cotton bollworm *P450 monooxygenase* gene by plant-mediated RNAi impairs larval tolerance of gossypol. Nat Biotechnol 25:1307–1313

Meng Y, Shao C (2012) Large-scale identification of mirtrons in arabidopsis and rice. PLoS ONE 7:e31163

Merchan F, Boualem A, Crespi M, Frugier F (2009) Plant polycistronic precursors containing non-homologous microRNAs target transcripts encoding functionally related proteins. Genome Biol 10:R136

Meyers BC, Axtell MJ, Bartel B et al (2008) Criteria for annotation of plant microRNAs. Plant Cell 20:3186–3190

Mi S, Cai T, Hu Y et al (2008) Sorting of small RNAs into Arabidopsis Argonaute complexes is directed by the 5′ terminal nucleotide. Cell 133:116–127

Monteiro F, Nishimura MT (2018) Structural, functional, and genomic diversity of plant NLR proteins: an evolved resource for rational engineering of plant immunity. Annu Rev Phytopathol 56:243–267

Navarro L, Dunoyer P, Jay F et al (2006) A plant miRNA contributes to antibacterial resistance by repressing auxin signaling. Science 312:436–439

Navarro L, Jay F, Nomura K et al (2008) Suppression of the microRNA pathway by bacterial effector proteins. Science 321:964–967

Nekrasov V, Wang C, Win J et al (2017) Rapid generation of a transgene-free powdery mildew resistant tomato by genome deletion. Sci Rep 7:482

Nozawa M, Miura S, Nei M (2012) Origins and evolution of microRNA genes in plant species. Genome Biol Evol 4:230–239

Ouyang S, Park G, Atamian HS et al (2014) MicroRNAs suppress NB domain genes in tomato that confer resistance to *Fusarium oxysporum*. PLoS Pathog 10:e1004464

Palatnik JF, Allen E, Wu X et al (2003) Control of leaf morphogenesis by microRNAs. Nature 425:257–263

Park YJ, Lee HJ, Kwak KJ et al (2014) MicroRNA400-guided cleavage of *Pentatricopeptide Repeat Protein* mRNAs renders *Arabidopsis thaliana* more susceptible to pathogenic bacteria and fungi. Plant Cell Physiol 55:1660–1668

Peláez P, Sanchez F (2013) Small RNAs in plant defense responses during viral and bacterial interactions: similarities and differences. Front Plant Sci 4:343

Peng Y, van Wersch R, Zhang Y (2018) Convergent and divergent signaling in PAMP-triggered immunity and effector-triggered immunity. Mol Plant–Microbe Interact 31:403–409

Peris-Peris C, Serra-Cardona A, Sánchez-Sanuy F et al (2017) Two NRAMP6 isoforms function as iron and manganese transporters and contribute to disease resistance in rice. Mol Plant-Microbe Interact 30(5):385–398

Qi T, Guo J, Peng H et al (2019) Host-induced gene silencing: a powerful strategy to control diseases of wheat and barley. Int J Mol Sci 20:206

Qiao Y, Liu L, Xiong Q et al (2013) Oomycete pathogens encode RNA silencing suppressors. Nat Genet 45:330–333

Qin Z, Li C, Mao L, Wu L (2014) Novel insights from non-conserved microRNAs in plants. Front Plant Sci 5:586

Rajagopalan R, Vaucheret H, Trejo J, Bartel DP (2006) A diverse and evolutionarily fluid set of microRNAs in *Arabidopsis thaliana*. Genes Dev 20:3407–3425

Rogers K, Chen X (2013) Biogenesis, turnover, and mode of action of plant microRNAs. Plant Cell 25:2383–2399

Rubio-Somoza I, Weigel D (2011) MicroRNA networks and developmental plasticity in plants. Trends Plant Sci 16:258–264

Salvador-Guirao R, Baldrich P, Weigel D et al (2018a) The microRNA miR773 is involved in the *Arabidopsis*immune response to fungal pathogens. Mol Plant–Microbe Interact 31:249–259

Salvador-Guirao R, Hsing Y, San Segundo B (2018b) The polycistronic miR166k-166h positively regulates rice immunity via post-transcriptional control of *EIN2*. Front Plant Sci 9:337

Salvador-Guirao R, Baldrich P, Tomiyama S et al (2019) *Osdcl1a* activation impairs phytoalexin biosynthesis and compromises disease resistance in rice. Ann Bot 123:79–93

Seo JK, Wu J, Lii Y et al (2013) Contribution of small RNA pathway components in plant immunity. Mol Plant–Microbe Interact 26:617–625

Sevilem I, Miyashima S, Helariutta Y (2013) Cell-to-cell communication via plasmodesmata in vascular plants. Cell Adhes Migr 7:27–32

Shahid S, Kim G, Johnson NR et al (2018) MicroRNAs from the parasitic plant *Cuscuta campestris* target host messenger RNAs. Nature 553:82–85

Shivaprasad PV, Chen HM, Patel K et al (2012) A microRNA superfamily regulates nucleotide binding site-leucine-rich repeats and other mRNAs. Plant Cell 24:859–874

Skopelitis DS, Hill K, Klesen S et al (2018) Gating of miRNA movement at defined cell-cell interfaces governs their impact as positional signals. Nat Commun 9:3107

Song Y, Thomma BPHJ (2018) Host-induced gene silencing compromises *Verticillium* wilt in tomato and Arabidopsis. Mol Plant Pathol 19:77–89

Song X, Li Y, Cao X, Qi Y (2019) MicroRNAs and their regulatory roles in plant–environment interactions. Annu Rev Plant Biol 70:489–525

Soto-Suarez M, Baldrich P, Weigel D et al (2017) The *Arabidopsis* miR396 mediates pathogen-associated molecular pattern-triggered immune responses against fungal pathogens. Sci Rep 7:44898

Staiger D, Korneli C, Lummer M, Navarro L (2013) Emerging role for RNA-based regulation in plant immunity. New Phytol 197:394–404

Sunkar R, Kapoor A, Zhu JK (2006) Posttranscriptional induction of two Cu/Zn superoxide dismutase genes in Arabidopsis is mediated by downregulation of miR398 and important for oxidative stress tolerance. Plant Cell 18:2051–2065

Torres MA, Jones JDG, Dangl JL (2006) Reactive oxygen species signaling in response to pathogens. Plant Physiol 141:373–378

Tsikou D, Yan Z, Holt DB et al (2018) Systemic control of legume susceptibility to rhizobial infection by a mobile microRNA. Science 362:233–236

Tsuda K, Somssich IE (2015) Transcriptional networks in plant immunity. New Phytol 206:932–947

Van Peer G, Lefever S, Anckaert J et al (2014) miRBase tracker: keeping track of microRNA annotation changes. Database (Oxford), vol 2014

Vatén A, Dettmer J, Wu S et al (2011) Callose biosynthesis regulates symplastic trafficking during root development. Dev Cell 21:1144–1155

Vaucheret H (2006) Post-transcriptional small RNA pathways in plants: mechanisms and regulations. Genes Dev 20:759–771

Vazquez F, Blevins T, Ailhas J et al (2008) Evolution of Arabidopsis *MIR* genes generates novel microRNA classes. Nucleic Acids Res 36:6429–6438

Wang Y, Cheng X, Shan Q et al (2014) Simultaneous editing of three homoeoalleles in hexaploid bread wheat confers heritable resistance to powdery mildew. Nat Biotechnol 32:947–951

Wang M, Weiberg A, Lin FM et al (2016a) Bidirectional cross-kingdom RNAi and fungal uptake of external RNAs confer plant protection. Nat Plants 2:16151

Wang F, Wang C, Liu P et al (2016b) Enhanced rice blast resistance by CRISPR/Cas9-targeted mutagenesis of the ERF transcription factor gene *OsERF922*. PLoS ONE 11:e0154027

Wang B, Sun Y, Song N et al (2017a) *Puccinia striiformis* f. sp. *tritici* microRNA-like RNA 1 (*Pst-milR1*), an important pathogenicity factor of *Pst*, impairs wheat resistance to *Pst* by suppressing the wheat *pathogenesis-related* 2 gene. New Phytol 215:338–350

Wang M, Thomas N, Jin H (2017b) Cross-kingdom RNA trafficking and environmental RNAi for powerful innovative pre- and post-harvest plant protection. Curr Opin Plant Biol 38:133–141

Wang Z, Xia Y, Lin S et al (2018) *Osa-miR164a* targets *OsNAC60* and negatively regulates rice immunity against the blast fungus *Magnaporthe oryzae*. Plant J 95:584–597

Wang J, Mei J, Ren G (2019) Plant microRNAs: biogenesis, homeostasis, and degradation. Front Plant Sci 10:360

Weiberg A, Wang M, Lin FM et al (2013) Fungal small RNAs suppress plant immunity by hijacking host RNA interference pathways. Science 342:118–123

Weiberg A, Wang M, Bellinger M, Jin H (2014) Small RNAs: a new paradigm in plant–microbe interactions. Annu Rev Phytopathol 52:495–516

Wilson RA, Talbot NJ (2009) Under pressure: investigating the biology of plant infection by *Magnaporthe oryzae*. Nat Rev Microbiol 7:185–195

Yu Y, Jia T, Chen X (2017) The 'how' and 'where' of plant microRNAs. New Phytol 216:1002–1017

Zhai J, Jeong DH, de Paoli E et al (2011) MicroRNAs as master regulators of the plant *NB-LRR* defense gene family via the production of phased, trans-acting siRNAs. Genes Dev 25:2540–2553

Zhang X, Zhao H, Gao S et al (2011) Arabidopsis Argonaute 2 regulates innate immunity via miRNA393*-mediated silencing of a Golgi-localized SNARE gene, MEMB12. Mol Cell 42:356–366

Zhang D, Liu M, Tang M et al (2015) Repression of microRNA biogenesis by silencing of *OsDCL1* activates the basal resistance to *Magnaporthe oryzae* in rice. Plant Sci 237:24–32

Zhang C, Ding Z, Wu K et al (2016a) Suppression of jasmonic acid-mediated defense by viral-inducible microRNA319 facilitates virus infection in rice. Mol Plant 9:1302–1314

Zhang T, Zhao YL, Zhao JH et al (2016b) Cotton plants export microRNAs to inhibit virulence gene expression in a fungal pathogen. Nat Plants. 2:16153

Zhang X, Bao Y, Shan D et al (2018) *Magnaporthe oryzae*induces the expression of a microRNA to suppress the immune response in rice. Plant Physiol 177:352–368

Zhu QH, Spriggs A, Matthew L et al (2008) A diverse set of microRNAs and microRNA-like small RNAs in developing rice grains. Genome Res 18:1456–1465

Zipfel C (2013) Combined roles of ethylene and endogenous peptides in regulating plant immunity and growth. Proc Natl Acad Sci USA 110:5748–5749

Chapter 11
Plant microRNAs Regulate Innate Immunity Through Diverse Mechanisms

Feng Li and Yudan Wang

Abstract Plant miRNAs are a class of small RNAs of 20–24 nucleotides that play broad roles in gene regulation by directing Argonaute proteins to target RNAs based on sequence complementarity. Plant innate immunity consists of two-layered defense mechanisms. In the first line of defense, pathogen molecular pattern triggered immunity (PTI) recognizes conserved pathogen molecules and confers broad-spectrum resistance. Virulent pathogens deliver effectors into plant cells to suppress PTI for successful infection. During plant-pathogen coevolution, the second line of defense evolved, which detects pathogen effectors and triggers a strong but race-specific defense reaction, usually accompanied by rapid programmed cell death, which is termed effector-triggered immunity (ETI). Antiviral RNA silencing is a form of PTI that directly targets viral pathogens. In this chapter, we summarize current knowledge of miRNA-mediated regulation of genes involving antiviral RNA silencing, PTI and ETI, and analyze their conservation in different plant lineages. Future directions for research on elucidating miRNA-mediated regulation of plant immunity are also discussed.

Keywords miRNAs · phasiRNAs · PTI · ETI · RNA silencing · NLR · Disease resistance

11.1 Introduction

miRNAs are short regulatory RNAs that are 20–24 nucleotides (nt) long. They are encoded by noncoding RNA genes and their primary (pri-)miRNA transcripts are transcribed by RNA polymerase II. Pri-miRNAs are capped, spliced, and polyadenylated like mRNAs, but unlike mRNAs, which are exported to cytoplasm for translation, pri-miRNAs are transported to the D-body in the nucleus for further processing (Yu et al. 2017). In the D-body, pri-miRNAs are cleaved by the dicing complex

F. Li (✉) · Y. Wang
Key Laboratory of Horticultural Plant Biology (MOE), College of Horticulture and Forestry Sciences, Huazhong Agricultural University,
Wuhan, China
e-mail: chdlifeng@mail.hzau.edu.cn

C. Miguel et al. (eds.), *Plant microRNAs*, Concepts and Strategies in Plant Sciences, https://doi.org/10.1007/978-3-030-35772-6_11

consisting of Dicer-like 1 (DCL1), Hyponastic leaf 1 (HYL1), and Serrate (SE) in two successive cleavage reactions, releasing pre-miRNA, and a miRNA/miRNA* duplex, respectively (Yu et al. 2017). Most conserved miRNAs, such as miR156, miR160, and miR172, are generated by DCL1, but there are also some miRNAs produced by other DCLs; for example, miR6026, miR1862, and miR822 are produced by DCL2, DCL3, and DCL4, respectively (Rajagopalan et al. 2006; Wang et al. 2018b; Wu et al. 2010). In certain cases, two different DCLs are involved in the two successive cleavage steps, such as rice miRNA miR1850.2, which is produced by sequential action of DCL1 and DCL3 (Wu et al. 2010). Once released by DCL cleavage, the miRNA/miRNA* duplex is then methylated by RNA methyltransferase Hua Enhancer 1 (HEN1) to protect it from degradation by exonucleases (Yu et al. 2017). Finally, the duplex is exported to the cytoplasm and forms the RNA-induced silencing complex (RISC) with various Argonaute proteins. AGO1-containing RISCs function in the cytoplasm by targeting mRNA for degradation or translation inhibition, while AGO4-containing RISCs enter the nucleus and function by directing homologous DNA methylation and associated histone methylation (Fang and Qi 2016).

In addition to directing transcriptional silencing and post-transcriptional silencing of target genes, some miRNAs can also trigger phased siRNA (phasiRNA) synthesis from their cleaved target transcript. In Arabidopsis, three mechanisms have been described for phasiRNA synthesis. The TAS3-miR390 phasiRNA system is highly conserved in plants and plays an important role in plant development (Xia et al. 2017). The TAS3 transcript has two miR390 binding sites located at the 5′ and 3′ ends. AGO7-miR390 RISC cleaves the TAS3 transcript at the 3′ binding site, tethers to the 5′ noncleavable miR390 binding site, and recruits Suppressor of Gene Silencing 3 (SGS3) and RNA-dependent RNA Polymerase 6 (RDR6). RDR6 transcribes the 5′ cleavage product into double-stranded (ds)RNA, which is further successively processed into 21-nt small interfering (si)RNAs by DCL4. The distance between the siRNA 5′ end and the 3′ miR390 cleavage site is in multiples of 21-nts, inphase with the miR390 cleavage site. PhasiRNAs are loaded into AGO1 to silence target genes expressed from other loci. Thus, phasiRNAs were initially named trans-acting siRNAs (tasiRNA) (Allen et al. 2005; Axtell et al. 2006). A second phasiRNA system is triggered by 22-nt miRNAs or siRNAs, such as the TAS2-miR173 system first described in Arabidopsis. The AGO1-miR173 RISC cleaves the TAS2 transcript at the only miR173 binding site and recruits SGS3/RDR6 to the 3′ cleavage product to generate dsRNA for phasiRNA synthesis (Allen et al. 2005; Chen et al. 2010; Cuperus et al. 2010). A third phasiRNA system is triggered by 21-nt miRNAs with a 22-nt passenger strand, usually with an asymmetric bulge in the passenger strand (Manavella et al. 2012). The last two phasiRNA systems are less conserved but are widely found, with different types of trigger miRNAs and phasiRNA transcripts and a broad range of targets.

miRNAs play broad roles in plants, including developmental regulation and response to biotic and abiotic stresses (Li et al. 2017). Plant innate immunity plays a central role in the plant response to biotic stress and consists of many interconnected functional modules, such as antiviral RNA silencing (Szittya and Burgyan 2013), pathogen-associated molecular pattern (PAMP)-triggered immunity (PTI) and effector-triggered immunity (ETI) (Chisholm et al. 2006). Recent studies show that miRNA-mediated regulation is deeply involved in all these plant innate immunity modules and plays important role in plant responses to different pathogens.

11.2 Regulation of Antiviral RNA Silencing by miRNAs

RNA silencing is a natural antiviral defense mechanism operating in diverse organisms (Guo et al. 2019). In plants, the antiviral RNA silencing pathway has been extensively studied and found to be conserved in different plant species (Carbonell and Carrington, 2015; Szittya and Burgyan 2013). Most of the studies focused on the interaction between model plant Arabidopsis and positive single-stranded RNA viruses, such as cucumber mosaic virus, tobacco rattle virus, and turnip mosaic virus. It was proposed that during viral infection, the viral replicating dsRNA intermediates are processed by DCL4, DCL2, and DCL3, generating 21-, 22-, and 24-nt long siRNAs, respectively (Fig. 11.1). For defense against many RNA viruses, DCL4, and DCL2 work redundantly, while for geminiviruses, DCL3 plays a major role (Deleris et al. 2006; Diaz-Pendon et al. 2007; Raja et al. 2014). The 21- and 22-nt siRNAs selectively interact with AGO1 and AGO2, while 24-nt siRNAs interact with AGO4 to form RISCs (Fig. 11.1). During RISC formation, different AGO proteins show bias for the 5′ end of their associated viral siRNAs; for example, AGO1 prefers viral siRNAs with 5′ U, while AGO2 and AGO4 prefer viral siRNAs with 5′ A (Wang et al. 2011). The AGO1- and AGO2-RISC function by directly binding and cleaving viral RNAs based on sequence complementarity between their associated viral siRNA and viral RNAs (Fig. 11.1). The AGO4-RISC functions in defense against DNA viruses via binding to viral RNAs attached to the viral DNA genome and recruiting DNA and histone modification enzymes to direct transcriptional gene silencing (Fig. 11.1) (Raja et al. 2014). During plant-virus coevolution, many viruses evolved suppressors for RNA silencing (VSR) to counteract antiviral defense with diverse mechanisms (Li and Ding, 2006). Due to many shared components between antiviral silencing pathways and miRNA pathways and similar chemical properties between viral siRNAs and miRNAs, VSRs can also inhibit miRNA function, which underlies certain virus-induced symptoms in plants (Kasschau et al. 2003).

11.2.1 Regulation of Silencing Initiators by miRNAs

DCL2 is regulated by miR1515, miR1507, and miR6026 in *Glycine max*, *Medicago truncatula*, and *Solanum lycopersicum*, respectively; all three miRNAs are 22-nt long and are capable of triggering phasiRNA production from DCL2 transcripts (Fig. 11.1) (Wang et al. 2018b; Zhai et al. 2011). Downregulation of miR1515a by short tandem target mimic (STTM) technology led to an increased transcript level of *DCL2* and reduced titer of soybean mosaic virus in *G. max* (Bao et al. 2018).

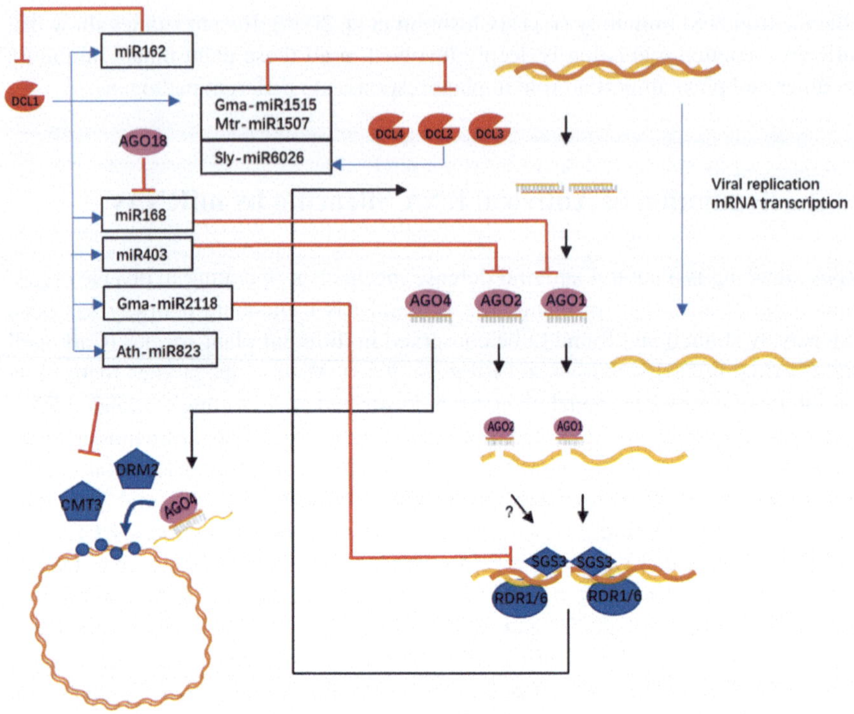

Fig. 11.1 miRNA-mediated regulation of antiviral RNA silencing. Red ovals with a notch represent different Dicer-like proteins as indicated. Purple ovals represent different Argonaute proteins. The double-stranded line and circle represent replication intermediates for RNA and DNA viruses, respectively. Black arrows represent siRNA biogenesis involving antiviral defense. Blue arrows represent miRNA biogenesis pathways. Red lines connect miRNAs and their targets in antiviral RNA silencing pathways or sequestration of miR168 by AGO18

In an Arabidopsis *dcl4* mutant, DCL2 processed DCL4-substrate dsRNA into 22-nt siRNA, which causes widespread unwanted gene silencing and developmental defects (Bouche et al. 2006; Wu et al. 2017b), indicating that DCL2 expression must be controlled for proper development. In tomato, the DCL2 subfamily undergoes gene duplication and has four members. DCL2b is essential for conferring resistance to tomato mosaic virus (TMV) (Wang et al. 2018a), indicating that DCL2 plays a more important role in tomato than in Arabidopsis in antiviral defense. It is interesting that tomato DCL2 is required for miR6026 biogenesis (Wang et al. 2018b), which constitutes a negative feedback regulation of DCL2 to balance its contribution to viral resistance and threats to development. In Arabidopsis, expression of DCL4 and DCL3 is elevated in a *dcl1* mutant but no miRNA directly targets these two DCLs, which suggests that miRNAs may indirectly regulate them by repressing their transcription factors (Qu et al. 2008).

11.2.2 Plant-Virus Arms Race Centered on the Silencing Effector AGO1

The *AGO1* gene is regulated by conserved miR168 and a silent mutation in the miR168 binding site of the *AGO1* gene, resulting in over accumulation of the AGO1 transcript and developmental defects, such as spoon-shaped, curled or twisted leaves resembling those of *dcl1*, *hyl1* and *hen1* mutants, suggesting miR168-mediated negative feedback regulation of AGO1 is essential for normal plant development (Vaucheret et al. 2004). Promoter analysis showed that Arabidopsis *MIR168* and *AGO1* were coexpressed, suggesting that they might be under similar transcriptional regulation (Vaucheret et al. 2006).

Due to the central role of AGO1 in anti-viral silencing, the miR168-AGO1 regulatory module is exploited by plant and viral pathogens to fine-tune anti-viral immunity. Infection of Arabidopsis and other plants by many viruses caused induction of miR168; however, AGO1 mRNA was also induced (Varallyay et al. 2010). In the case of Cymbidium ringspot virus (CymRSV) infection in *Nicotiana benthamiana* plants, the wild type and P19-defective virus-induced similar levels of AGO1 mRNA, but the wild type virus caused much stronger induction of miR168 compared to the P19-defective virus; as a result, the AGO1 protein level was significantly increased in P19-defective virus-infected cells but remained similar or reduced in wild type virus-infected cells (Varallyay et al. 2010). Induction of AGO1 mRNA transcription by CymRSV and other viruses could be considered a plant defense response, while further induction of MIR168 by the P19 suppressor is a new mechanism to counteract RNA silencing in addition to its siRNA sequestration activity. Intriguingly, it seemed that the two different counter silencing mechanisms of P19 were well-coordinated during evolution because P19 has very low affinity to miR168, thus allowing miR168 to repress AGO1 translation during wild type virus infection (Iki et al. 2018; Pertermann et al. 2018; Varallyay et al. 2010). This scheme is likely a general mechanism in shaping plant-virus interactions, as VSRs from TMV, TCV, TEV, and CMV also behaved like P19 (Varallyay and Havelda 2013). In rice, rice stripe virus (RSV) infection and transgenic expression of NS3 also resulted in *MIR168* and *AGO1* induction (Zheng et al. 2017). However, the rice AGO1 protein level was increased upon viral infection(Wu et al. 2015). Further study showed that increase in AGO1 protein level depended on rice-specific AGO18, which was induced by virus infection and specifically binds miR168, thus increasing its repression of AGO1 expression (Wu et al. 2015). Thus, miR168 plays an important role in both development and RNA silencing-based anti-viral immunity through regulation of AGO1. AGO2 is negatively regulated by miR403, which is conserved in eudicots (Chavez Montes et al. 2014). Since miR403 mainly associates with AGO1, and many VSRs inhibit AGO1 protein accumulation by inducing miR168 (Mi et al. 2008; Varallyay and Havelda 2013; Varallyay et al. 2010), it is expected that AGO2 expression will increase during infection by these viruses. Indeed, this was the case in *N. benthamiana* infected by CymRSV (Pertermann et al. 2018).

In summary, AGO1 provides conserved antiviral silencing functions in both monocots and dicots. As a counter defense, the conserved miR168-AGO1 module was exploited for effector-triggered susceptibility by viruses in both monocots and dicots. During coevolution, distinct counter-counter-defense mechanisms were developed in the two plant lineages: monocots evolved AGO18 to protect AGO1, while dicots evolved the miR403-AGO2 module as a backup for AGO1.

11.2.3 Regulation of Systemic Antiviral RNA Silencing by miRNAs

In Arabidopsis infected by CMV, viral siRNA accumulation largely depends on *RNA-dependent RNA polymerase 1* (*RDR1*); while the *rdr1* single mutant clearly enhanced CMV RNA accumulation, the *rdr1* and *rdr6* double mutation led to much higher viral RNA levels (Wang et al. 2010). These results suggest that primary viral siRNAs derived from replicating viral dsRNA could not restrict a virus efficiently and that secondary viral siRNAs generated from RDR1 or RDR6 synthesized dsRNA are required for robust immunity against a virus. SGS3 plays a key role in RDR6-dependent secondary siRNA synthesis and antiviral RNA silencing, and mutation in either gene enhanced potato virus X accumulation and abolished transgene silencing (Mourrain et al. 2000), which revealed that transgene silencing and antiviral silencing share a similar mechanism. In Arabidopsis, DCL2-dependent 22-nt transgene siRNAs trigger RDR6-dependent secondary siRNA synthesis, which is required for systemic silencing of the GFP transgene in grafting scions (Taochy et al. 2017). Viral infection induces systemic RNA silencing, which is important in restricting virus spread, and many viruses evolved VSR specifically to inhibit systemic RNA silencing (Xie and Guo, 2006). Thus, DCL2-RDR6-SGS3-mediated systemic silencing constitutes an important layer of anti-viral RNA silencing immunity. In addition to the regulation of DCL2 by miR1515, miR1507, and miR6026 that was discussed before, SGS3 in *G. max* is targeted by miR2118 (Zhai et al. 2011). It is expected that systemic RNA silencing against virus spread would be enhanced when VSR inhibits DCL2 targeting miRNAs and SGS3 targeting miR2118.

11.2.4 Regulation of RdDM-Based Immunity by miR823

DNA viruses are targeted by both posttranscriptional gene silencing that is mediated by DCL2, DCL4-dependent primary and RDR1, RDR6-dependent secondary siRNAs, and transcriptional gene silencing (TGS) (Pooggin 2013). TGS against DNA viruses depends on AGO4-DRM2 mediated de novo DNA methylation and CMT3 mediated maintenance methylation, which likely function by targeting the viral mini-chromosomes and restrict viral gene expression (Pooggin 2013). In Arabidopsis, miR823 regulates CMT3 and is induced by *Fusarium oxysporum* elicitors (Baldrich et al. 2014), which points to potential synergism between fungal and DNA virus infections.

As DCL1 is also subjected to negative feedback regulation by miR162 (Xie et al. 2003), regulation of miR162 by different types of stress adds an additional layer of regulation to the miRNA-mediated fine-tuning of antiviral RNA silencing by controlling *DCL1* expression, which generates most of the miRNAs regulating antiviral RNA silencing components (Fig. 11.1).

11.3 Regulation of Pattern-Triggered Immunity and Effector-Triggered Immunity by miRNAs

In addition to antiviral RNA silencing-based immunity, studies of plant–bacterial and plant–fungal interactions have unveiled a general two-layered defense mechanism in plants: Pathogen-associated molecular pattern (PAMP)-triggered immunity (PTI) and Effector-triggered immunity (ETI). PTI is elicited by the pattern recognition receptor (PRR) on the cell membrane upon recognition of conserved PAMPs (Couto and Zipfel 2016; Tang et al. 2017). PRRs include receptor-like kinase (RLK) and receptor-like protein (RLP), both of which contain variable ectodomain (ECD), a single-pass membrane domain; and while RLK has a cytoplasmic kinase domain, RLP lacks it. In the presence of PAMPs, RLK or RLP associate with a coreceptor, which is structurally like RLK but contains a shorter LRR domain, and phosphorylates it. The receptor complex then phosphorylates a receptor-like cytoplasmic kinase (RLCK), which in turn phosphorylates mitogen-activated protein kinase kinasekinase (MAPKKK) to trigger the MAPK cascade (Fig. 11.2), or phosphorylates calcium channels to induce calcium influx, or phosphorylates RbohD to induce an oxidative burst (Tang et al. 2017). PTI effectively protects plants from infection by most microbes, while pathogenic microbes deliver a plethora of effectors that inhibit PTI signaling with diverse mechanisms (Tang et al. 2017). During their coevolution with pathogens, plants evolved nucleotide-binding domain and leucine-rich repeat domain-containing protein (NB-LRR, or NLR)-based immunity, which recognizes pathogen effectors and triggers a rapid and strong resistance reaction, usually accompanied by the hypersensitive response (HR), a form of programmed cell death (Chisholm et al. 2006).

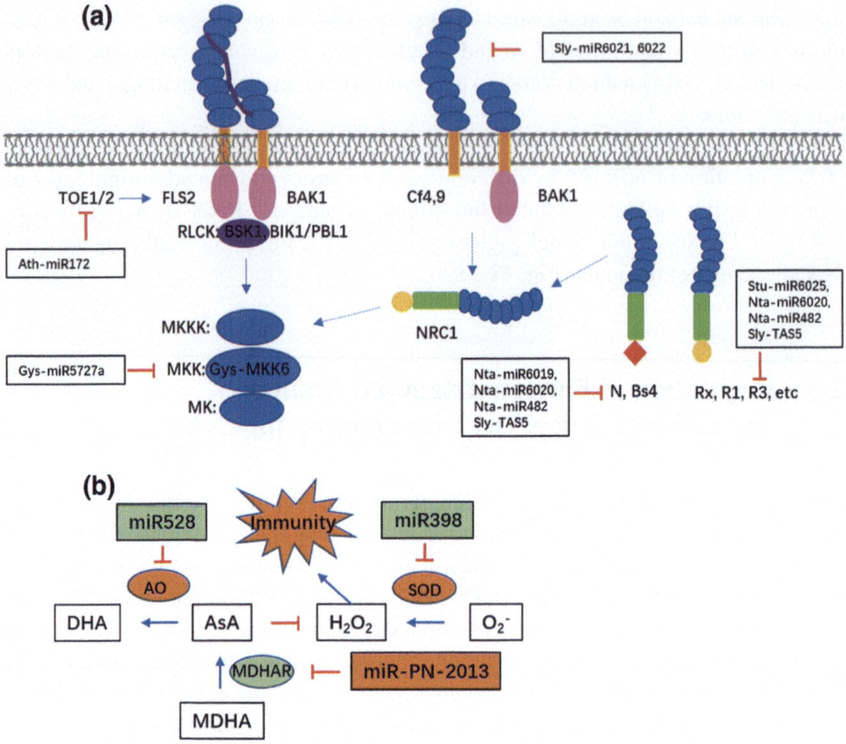

Fig. 11.2 miRNA-mediated regulation of PTI and ETI in plants. **a** MiRNAs directly target immune receptors and the MAPK signaling cascade and indirectly target FLS2 through regulation of transcription factors TOE1/2. **b** MiRNAs regulate ROS production in the immune response by targeting various enzymes of ROS metabolism. DHA, dehydroascorbate; AsA, L-ascorbicacid; AO, ascorbate oxidase; SOD, superoxidedismutase; MDHAR, monodehydroascorbatereductase; MDHA, monodehydroascorbate. The miRNAs and enzymes in the brown background promote H_2O_2 production and immunity, while those in the green background reduce H_2O_2 production and immunity

11.3.1 Direct Regulation of Immune Receptors by miRNAs and siRNAs

Plant *NLRs* form a large gene family in plant genomes, play important roles in protecting plants from diverse pathogens and are widely used in breeding disease-resistant crops (Deng et al. 2018a). However, a large number of *NLRs* also pose a serious threat to plant development, and *NLR* ectopic expression underlies hybrid necrosis in many cases (Deng et al. 2018a). Bioinformatics and molecular biological studies revealed that the tobacco mosaic virus resistance gene *N* is targeted by miR6019 and miR6020, which form a cluster in *Nicotiana* species, while miR6020 exists as a single miRNA in tomato (Fig. 11.2) (Deng et al. 2018b; Li et al. 2012b). The RLP type immune receptors Cf4 and Cf9 are also targeted by miRNAs, such as miR6021 and miR6022 in tomato (Fig. 11.2a) (Li et al. 2012b). The conserved role of miRNAs in *NLR* regulation was further confirmed by discovery of miR482/miR2118-mediated *NLR* regulation in different plant families (Li et al. 2012b; Shivaprasad et al. 2012; Zhai et al. 2011).

In addition to direct targeting by miRNAs, *NLRs* are also targeted by phasiRNAs (Fei et al. 2013). Two types of phasiRNA transcripts have been reported to participate in *NLR* regulation. Many NLR coding transcripts that are targeted by 22-nt miRNAs produce phasiRNAs, presumably targeting themselves and their closely related family members; for example, tobacco mosaic virus resistance gene *N* is targeted by miR6019-triggered phasiRNAs, and a 21-nt version of miR6019 that cannot trigger phasiRNA synthesis shows less efficient silencing of *N* (Deng et al. 2018b; Li et al. 2012b). The other source for *NLR* targeting phasiRNA is long noncoding (lnc)RNAs, such as TAS5 in tomato (Fig. 11.2a). TAS5 consists of fragments of *NLR* sequences from both *TNL* and *CNL* type genes and is cleaved by sly-miR482d/2118b to generate phasiRNAs targeting many *TNL* and *CNL* genes (Canto-Pastor et al. 2019; Li et al. 2012a).

Discrete homology between tobacco *MIR6019,MIR6020* and the *N* gene, and between tomato *MIR6022* and the *Cf9* gene, led to the hypothesis that *NLR* or *RLP* targeting miRNAs may evolve through duplication of these immune receptor genes (Li et al. 2012b). A study of four monocot and nine eudicot genomes found that approximately 67 and 63% of clustered *NLRs* and 33 and 37% of singleton *NLRs* are targeted by miRNAs, respectively, in eudicots and monocots (Zhang et al. 2014). In a later study across 75 plant genomes, similar observations were made, where the highly duplicated *NLR* family frequently generates new miRNAs to target their members, while heterogeneous *NLRs* in *Poaceae* and *Brassicaceae* are rarely targeted by miRNAs (Zhang et al. 2016). Furthermore, it was found that in the tomato *TAS5*, the phasiRNA generating regions consisted of multiple *NLR* fragments, suggesting that *TAS5* also evolved through *NLR* duplication. Thus, *NLR*-targeting miRNAs and phasiRNA genes arose from and coevolved with their target *NLRs*.

11.3.2 miRNAs Regulate Broad-Spectrum Resistance by Tuning NLR Expression

According to the classic model, *NLR*-mediated ETI targets race-specific pathogens, while there was also early evidence suggesting that NLR activation can cause nonrace-specific resistance. The gain-of-function mutation in the Arabidopsis *TNL* gene, *Suppressor of npr1-1, constitutive 1* (*SNC1*), resulted in constitutive activation of resistance signaling and resistance to both *Pseudomonas syringae* and *Peronosporaparasitica*(Zhang et al. 2003). Overexpression of the RPP1A TIR-NB-ARC domains in stable Arabidopsis transgenic lines led to constitutive resistance against *Hyaloeronosporaparasitica* and *P. syringae* (Weaver et al. 2006). It was shown recently that activation of *SNC1* inhibits miRNA and phasiRNA synthesis (Cai et al.

2018). The nuclear SNC1 represses transcription of miRNAs and phasiRNA loci, possibly through the transcriptional corepressor TPR1, leading to general upregulation of a majority of *NLR* genes in the Arabidopsis genome (Cai et al. 2018). These results suggested that there was a positive feedback circuit between *NLR* and miRNAs/phasiRNAs determining the *NLR* expression level, which influences general immunity in plants.

It is expected that the modulation of those miRNA levels could affect non-race-specific plant immunity, which was shown to be the case in several recent reports. In Arabidopsis, overexpression of the *NLR*-targeting 22-nt miR472 resulted in drastically increased accumulation of phasiRNAs from multiple *CNL* transcripts. Mutation in *RDR6* or *MIR472* led to increased expression of *CNL* genes and enhanced PTI response, while overexpression of miR472 reduced the PTI response (Boccara et al. 2014). These results suggested that miR472 can regulate the broad spectrum PTI response by tuning *NLR* expression in Arabidopsis through the phasiRNA pathway. Ouyang and colleagues compared the miRNA expression profile between tomato cultivars that are susceptible and resistant to *Fusarium oxysporum*. They found that miR482f and miR5300 were repressed in resistant tomatoes, while four targets of these two miRNAs encoding *NLR* proteins were upregulated in resistant tomatoes during infection. Knockdown of these *NLR* genes by VIGS confirmed their role in *F. oxysporum*resistance. However, these *NLR* genes are not *I-2*, the only known resistance (*R*) gene for *F. oxysporum* (Ouyang et al. 2014). These data suggest that downregulation of miR482f and miR5300 contributed resistance to *F. oxysporum*by releasing *NLR* silencing. More recently, two groups of researchers showed that knocking-down miR482b, miR482bce or miR2118b in stable transgenic tomato lines using STTM enhanced tomato resistance to *Phytopthorainfestans* and *P. syringae* (Canto-Pastor et al. 2019; Jiang et al. 2018). In soybean, transient expression of STTM against *NLR*-targeting miR1507a/c, miR482a or miR1515a increased resistance to *Soybean mosaic virus* (*SMV*) (Bao et al. 2018), suggesting that knocking-down these *NLR*-targeting miRNAs increased broad-spectrum resistance against *SMV*. In poplars, overexpression of miR472a reduced the PTI response elicited by flg22 and resistance against the biotrophic pathogen *Colletotrichumgloeosporioides*while knocking-down miR472a by STTM enhanced these immune responses (Su et al. 2018). However, for necrotrophic pathogens, miR472a played the opposite role: miR472a-OE plants showed enhanced resistance while miR472a-STTM plants showed enhanced susceptibility to *Cytospora chrysosperma* (Su et al. 2018). All these findings are in line with the mechanism that *NLR*-targeting miRNAs function as a negative regulator of broad-spectrum resistance against biotrophic pathogens by repressing overall *NLR* expression levels.

The ability of miRNA to regulate broad-spectrum resistance also helps explain early findings in which overexpression of viral suppressor of RNA silencing, P1/HC-Pro, caused enhanced broad-spectrum resistance in transgenic tobacco (Pruss et al. 2004). This study showed that transgenic expression of P1/HC-Pro in *Nicotiana tabacum cv* Xanthi NN resulted in enhanced gene-for-gene resistance against TMV mediated by the N gene, as evidenced by fewer and smaller necrotic lesions on inoculated leaf. The transgenic plants also showed stronger resistance against non-*R* gene-targeted viral, fungal, and bacterial pathogens, which was partially dependent on salicylic acid (SA) signaling (Pruss et al. 2004). These results were consistent with the scenario that P1/HC-Pro inhibited miRNA and phasiRNAs function and led to overexpression of *NLR* genes, which in turn led to SA dependent resistance. These results parallel findings in Arabidopsis, where an *rdr6* mutant showed enhanced *NLR* expression and SA-dependent resistance against *P. syringae* (*Pto*DC3000) (Boccara et al. 2014).

11.3.3 Developmental Regulation of PTI and ETI by miRNAs

In maize, it was shown that resistance to *Pucciniasorghi*, the common maize rust fungi, gradually increased as plants grew and became most apparent after the onset of reproductive growth of the plants, which is nonrace-specific and thus possibly a form of PTI (Headrick and Pataky 1987). This age-dependent resistance is modulated by *Corngrass1,* which prolongs the juvenile-vegetative phase and delays resistance to common rust (Abedon and Tracy 1996). Genetic analysis showed that the *Corngrass* phenotype was caused by overexpression of miR156b/c due to the insertion of a STONER element into the *MIR156b/c* promoter, and it can be phenocopied by insertion of multiple 35S promoter elements or overexpression of miR156 driven by the *ubiquitin* promoter (Chuck et al. 2007). These results suggest that miR156 negatively regulates immunity in maize during plant development. However, the molecular link between miR156 and age-dependent immunity remains unknown. The race-specific resistance against the bacterial blast pathogen *Xanthomonasoryzae* pv. *Oryzae* (Xoo) conferred by the rice *Xa21* and *Xa3/26* genes was also developmentally controlled (Cao et al. 2007; Century et al. 1999). Further analysis showed that gradually increased resistance was largely determined by increasing expression levels of *Xa21* and *Xa3/26* during plant growth (Zhao et al. 2009). Thus, both PTI and ETI are developmentally regulated in monocots.

In tobacco and tomato, the *NLR* gene *N* conferred incomplete resistance against TMVin the young seedling stage and full resistance in the fully expanded leaf stage (Deng et al. 2018b). The age-dependent *N*-mediated resistance is correlated with an increasing *N* expression level, which is controlled by gradually decreased expression of miR6019 and miR6020 during plant growth (Deng et al. 2018b). Overexpression of miR6019 impairs *N*-mediated TMVresistance in mature plants (Deng et al. 2018b). This work provided direct evidence for the role of miRNA in the regulation of ETI during plant development. However, the developmental regulation of miR6019 and miR6020 remains to be elucidated. In *Nicotiana benthamiana*, *RECEPTOR-LIKE PROTEIN REQUIRED FOR CSP22 RESPONSIVENESS* (*NbCSPR*) mediates PTI against bacteria by binding to csp22 peptide derived from bacterial cold shock protein. *NbCSPR*-mediated resistance is evident in 6-week-old plants but not in younger 4-week-old plants and is induced by flagellin (Saur et al. 2016), suggesting cooperation between *NbCSPR* and *FLAGELLIN-SENSING2* (*FLS2*). Interestingly, it was recently shown that transcription of *FLS2* is regulated by the miR172-TOE1/2 module during early seedling growth. MiR172b positively regulates *FLS2*-mediated PTI by downregulation of *FLS2* transcription repressor TOE1/2 (Zou et al. 2018). Thus, miRNAs participate in developmental regulation of both PTI and ETI in dicots through direct or indirect regulation of immune receptors.

11.3.4 miRNAs Regulate Signaling Transduction in Disease Resistance

In addition to regulating PTI and ETI at immune receptor levels, miRNAs also regulate downstream resistance signaling pathways. In cotton, MKK6 plays an essential role in resistance against *Fusarium oxysporum,* but excess expression of MKK6 causes lesion mimic phenotype. The cotton miR5727a was shown to target MKK6 by binding to its 3′ untranslated region (UTR). During *F. oxysporum* infection, miR5727a was induced to prevent excess expression of MKK6 by downregulation of MKK6. Overexpression of miR5727a via a virus vector impaired cotton resistance to *F. oxysporum* (Wang et al. 2017). These results established a case where plant immunity is fine-tuned through miRNA-mediated regulation of the MAPK cascade (Fig. 11.2a).

Reactive oxygen species (ROS) are important molecules in resistance signaling, and their levels are regulated via complex networks. In plants, Cu/Zn-superoxide dismutase (SOD) detoxifies O_2^- generated by photosystem I and generates H_2O_2. Cu/Zn-SOD is regulated by conserved miR398, and it was shown that downregulation of miR398 during oxidative stress led to the induction of Cu/Zn-SOD, which is important for oxidative stress tolerance (Sunkar et al. 2006). In Arabidopsis, infection by an avirulent pathogen reduced accumulation of miR398 and increased cytosolic Cu/Zn-SOD1 accumulation(Jagadeeswaran et al. 2009). In rice, *Magnaporthegrisea* and *Xanthomonasoryzaepv. Oryzae* infection also resulted in downregulation of miR398 and induction of SOD1 and SOD2 (Lu et al. 2011). These results indicated that miR398 negatively regulates plant immunity by repressing SOD and H_2O_2 production. A recent study in barley provided direct evidence of a role for miR398 in regulation of plant immunity through SOD. As observed in Arabidopsis, *Mla*-mediated ETI against powdery mildew fungus in barley downregulated miR398 and led to induction of SOD. Silencing SOD by VIGS inhibited *Mla*-mediated H_2O_2 production of HR (Xu et al. 2014). These results suggest that the miR398-SOD module may play a conserved role in the regulation of plant immunity (Fig. 11.2b).

The L-ascorbate oxidase (AO) positively regulates ROS by oxidizing ascorbic acid, which reduces ROS. In rice, AGO1-miR528 RISC represses *AO* expression by binding to its 3′ UTR. During RSV infection, miR528 is transcriptionally downregulated and *AGO18* is upregulated, which further sequestersmiR528 from AGO1 incorporation; thus, *AO* is upregulated and results in a higher level of ROS and resistance to RSV(Wu et al. 2017a) (Fig. 11.2b). Monodehydro-ascorbate reductase (MDHAR) regulates the ROS level via the ascorbate-glutathione (AsA-GSH) cycle. In wheat, knocking down *MDHAR* enhances resistance to wheat stripe rust in the seedling stage. Two wheat miRNAs (1136-P3 and PN-2013) were identified as negative regulators for *MDHAR*. During incompatible wheat-stripe rust interactions, miR-PN-2013 is initially induced and then starts to decline 24 h postinoculation, while *MDHAR* shows the opposite expression pattern, suggesting that miR-PN-2013 may contribute to wheat resistance to stripe rust via promoting ROS accumulation by downregulation of *MDHAR* (Feng et al. 2014) (Fig. 11.2b).

11.4 Future Aspects

In the past 30 years, our understanding of the molecular mechanisms of antiviral RNA silencing, PTI and ETI has advanced significantly. A plethora of literature has illustrated the core components in RNA silencing and their mode of action (Carbonell and Carrington 2015; Guo et al. 2019; Szittya and Burgyan 2013; Yang and Li 2018). A large volume of seminal works described the cloning and characterization of immune receptors in PTI and ETI pathways and signaling cascades, such as phosphorylation and ROS (Baker et al. 1997; Baxter et al. 2014; Macho and Zipfel 2014; Meng and Zhang 2013; Staskawicz et al. 1995). MiRNAs, as ubiquitous gene regulators in both plants and animals, have been shown to be deeply involved in all aspects of plant immunity. In this work, we focused on summarizing their roles in the regulation of the core components of RNA silencing, the immune receptors, the MAPK and ROS signaling in PTI and ETI. Within this scope, there are few regulatory modules showing different levels of conservation, such as the miR172-TOE1/2-FLS2, miR168-AGO1, miR398-SOD, miR403-AGO2, and miR482-NLR modules. For these conserved modules, it remains to be explored whether they play conserved roles in regulating plant immunity among different plant species. More miRNAs belong to species-specific immune regulators, such as the majority of the *NLR*-targeting miRNAs. More work is needed to comprehensively characterize a more complete set of genes in the core antiviral RNA silencing and the PTI and ETI pathway and miRNAs directly targeting them. Beyond the scope of this chapter, there is extensive cross-talk between core plant immunity pathways and other pathways, such as plant hormone signaling pathways (Alazem and Lin 2015). It has been shown that some conserved miRNAs, such as miR393, regulate plant immunity via auxin signaling pathways (Navarro et al. 2006). However, the direct connection between plant hormone signaling and the core plant immunity pathway remains largely unknown. As plant hormone signaling eventually functions through transcription factors, transcriptional regulation of core plant immunity genes and miRNA regulators may provide a link between them. Thus, elucidation of transcriptional control mechanisms for various antiviral RNA silencing genes, PTI and ETI components, and miRNA-mediated regulation of those transcription factors may provide a more comprehensive picture of the role of miRNAs in regulation of plant innate immunity.

Acknowledgements This work was supported by the Fundamental Research Funds for the Central Universities (2662014PY008) and the HZAU start fund (2013RC001).

References

Abedon BG, Tracy WF (1996) Corngrass1 of maize (*Zea mays* L.) delays development of adult plant resistance to common rust (*Puccinia sorghi Schw*) and European corn borer (*Ostrinia nubilalis hubner*). J Hered 87:219–223

Alazem M, Lin NS (2015) Roles of plant hormones in the regulation of host-virus interactions. Mol Plant Pathol 16:529–540

Allen E, Xie Z, Gustafson AM, Carrington JC (2005) microRNA-directed phasing during trans-acting siRNA biogenesis in plants. Cell 121:207–221

Axtell MJ, Jan C, Rajagopalan R, Bartel DP (2006) A two-hit trigger for siRNA biogenesis in plants. Cell 127:565–577

Baker B, Zambryski P, Staskawicz B, Dinesh-Kumar SP (1997) Signaling in plant-microbe interactions. Science 276:726–733

Baldrich P, Kakar K, Sire C, Beatriz Moreno A, Berger A, Garcia-Chapa M, Jose Lopez-Moya J, Luis Riechmann J, San Segundo B (2014) Small RNA profiling reveals regulation of Arabidopsis miR168 and heterochromatic siRNA415 in response to fungal elicitors. Bmc Genomics 15:1083

Bao D, Ganbaatar O, Cui X, Yu R, Bao W, Falk BW, Wuriyanghan H (2018) Down-regulation of genes coding for core RNAi components and disease resistance proteins via corresponding microRNAs might be correlated with successful Soybean mosaic virus infection in soybean. Mol Plant Pathol 19:948–960

Baxter A, Mittler R, Suzuki N (2014) ROS as key players in plant stress signalling. J Exp Bot 65:1229–1240

Boccara M, Sarazin A, Thiebeauld O, Jay F, Voinnet O, Navarro L, Colot V (2014) The Arabidopsis miR472-RDR6 silencing pathway modulates PAMP-and effector-triggered immunity through the post-transcriptional control of disease resistance genes. PLoS Pathogens 10:e1003883

Bouche N, Lauressergues D, Gasciolli V, Vaucheret H (2006) An antagonistic function for Arabidopsis DCL2 in development and a new function for DCL4 in generating viral siRNAs. EMBO J 25:3347–3356

Cai Q, Liang C, Wang S, Hou Y, Gao L, Liu L, He W, Ma W, Mo B, Chen X (2018) The disease resistance protein SNC1 represses the biogenesis of microRNAs and phased siRNAs. Nat Commun 9:5080

Canto-Pastor A, Santos B, Valli AA, Summers W, Schornack S, Baulcombe DC (2019) Enhanced resistance to bacterial and oomycete pathogens by short tandem target mimic RNAs in tomato. Proc Natl Acad Sci U S A 116:2755–2760

Cao YL, Ding XH, Cai M, Zhao J, Lin YJ, Li XH, Xu CG, Wang SP (2007) Expression pattern of a rice disease resistance gene Xa3/Xa26 is differentially regulated by the genetic backgrounds and developmental stages that influence its function. Genetics 177:523–533

Carbonell A, Carrington JC (2015) Antiviral roles of plant ARGONAUTES. Curr Opin Plant Biol 27:111–117

Century KS, Lagman RA, Adkisson M, Morlan J, Tobias R, Schwartz K, Smith A, Love J, Ronald PC, Whalen MC (1999) Short communication: developmental control of Xa21-mediated disease resistance in rice. Plant J 20:231–236

Chavez Montes RA, de Fatima Rosas-Cardenas F, De Paoli E, Accerbi M, Rymarquis LA, Mahalingam G, Marsch-Martinez N, Meyers BC, Green PJ, de Folter S (2014) Sample sequencing of vascular plants demonstrates widespread conservation and divergence of microRNAs. Nat Commun 5:3722

Chen HM, Chen LT, Patel K, Li YH, Baulcombe DC, Wu SH (2010) 22-Nucleotide RNAs trigger secondary siRNA biogenesis in plants. Proc Natl Acad Sci U S A 107:15269–15274

Chisholm ST, Coaker G, Day B, Staskawicz BJ (2006) Host-microbe interactions: shaping the evolution of the plant immune response. Cell 124:803–814

Chuck G, Cigan AM, Saeteurn K, Hake S (2007) The heterochronic maize mutant Corngrass1 results from overexpression of a tandem microRNA. Nat Genet 39:544–549

Couto D, Zipfel C (2016) Regulation of pattern recognition receptor signalling in plants. Nat Rev Immunol 16:537–552

Cuperus JT, Carbonell A, Fahlgren N, Garcia-Ruiz H, Burke RT, Takeda A, Sullivan CM, Gilbert SD, Montgomery TA, Carrington JC (2010) Unique functionality of 22-nt miRNAs in triggering RDR6-dependent siRNA biogenesis from target transcripts in Arabidopsis. Nat Struct Mol Biol 17:997–1003

Deleris A, Gallego-Bartolome J, Bao J, Kasschau KD, Carrington JC, Voinnet O (2006) Hierarchical action and inhibition of plant Dicer-like proteins in antiviral defense. Science 313:68–71

Deng Y, Liu M, Li X, Li F (2018a) microRNA-mediated R gene regulation: molecular scabbards for double-edged swords. Sci China-Life Sci 61:138–147

Deng Y, Wang J, Tung J, Liu D, Zhou Y, He S, Du Y, Baker B, Li F (2018b) A role for small RNA in regulating innate immunity during plant growth. PLoS Pathogens 14:e1006756

Diaz-Pendon JA, Li F, Li WX, Ding SW (2007) Suppression of antiviral silencing by cucumber mosaic virus 2b protein in Arabidopsis is associated with drastically reduced accumulation of three classes of viral small interfering RNAs. Plant Cell 19:2053–2063

Fang X, Qi Y (2016) RNAi in plants: an argonaute-centered view. Plant Cell 28:272–285

Fei QL, Xia R, Meyers BC (2013) Phased, secondary, small interfering RNAs in posttranscriptional regulatory networks. Plant Cell 25:2400–2415

Feng H, Wang X, Zhang Q, Fu Y, Feng C, Wang B, Huang L, Kang Z (2014) Monodehydroascorbate reductase gene, regulated by the wheat PN-2013 miRNA, contributes to adult wheat plant resistance to stripe rust through ROS metabolism. Biochim Biophys Acta 1839:1–12

Guo Z, Li Y, Ding SW (2019) Small RNA-based antimicrobial immunity. Nat Rev Immunol 19:31–44

Headrick JM, Pataky JK (1987) Expression of partial resistance to common rust in sweet corn hybrids at various host growth-stages. Phytopathology 77:454–458

Iki T, Clery A, Bologna NG, Sarazin A, Brosnan CA, Pumplin N, Allain FHT, Voinnet O (2018) Structural flexibility enables alternative maturation, ARGONAUTE sorting and activities of miR168, a global gene silencing regulator in plants. Mol Plant 11:1008–1023

Jagadeeswaran G, Saini A, Sunkar R (2009) Biotic and abiotic stress down-regulate miR398 expression in Arabidopsis. Planta 229:1009–1014

Jiang N, Meng J, Cui J, Sun G, Luan Y (2018) Function identification of miR482b, a negative regulator during tomato resistance to Phytophthora infestans. Hortic Res 5:9

Kasschau KD, Xie Z, Allen E, Llave C, Chapman EJ, Krizan KA, Carrington JC (2003) P1/HC-Pro, a viral suppressor of RNA silencing, interferes with Arabidopsis development and miRNA unction. Dev Cell 4:205–217

Li F, Ding SW (2006) Virus counter defense: diverse strategies for evading the RNA-silencing immunity. Annu Rev Microbiol 60:503–531

Li F, Orban R, Baker B (2012a) SoMART: a web server for plant miRNA, tasiRNA and target gene analysis. Plant J 70:891–901

Li F, Pignatta D, Bendix C, Brunkard JO, Cohn MM, Tung J, Sun H, Kumar P, Baker B (2012b) microRNA regulation of plant innate immune receptors. Proc Natl Acad Sci U S A 109:1790–1795

Li S, Castillo-Gonzalez C, Yu B, Zhang X (2017) The functions of plant small RNAs in development and in stress responses. Plant J 90:654–670

Lu YZ, Feng Z, Bian LY, Xie H, Liang JS (2011) miR398 regulation in rice of the responses to abiotic and biotic stresses depends on CSD1 and CSD2 expression. Funct Plant Biol 38:44–53

Macho AP, Zipfel C (2014) Plant PRRs and the activation of innate immune signaling. Mol Cell 54:263–272

Manavella PA, Koenig D, Weigel D (2012) Plant secondary siRNA production determined by microRNA-duplex structure. Proc Natl Acad Sci U S A 109:2461–2466

Meng X, Zhang S (2013) MAPK cascades in plant disease resistance signaling. In: Van Alfen NK (ed) Annual review of phytopathology, vol 51, pp 245–266

Mi S, Cai T, Hu Y, Chen Y, Hodges E, Ni F, Wu L, Li S, Zhou H, Long C et al (2008) Sorting of small RNAs into Arabidopsis argonaute complexes is directed by the 5′ terminal nucleotide. Cell 133:116–127

Mourrain P, Beclin C, Elmayan T, Feuerbach F, Godon C, Morel JB, Jouette D, Lacombe AM, Nikic S, Picault N et al (2000) Arabidopsis SGS2 and SGS3 genes are required for posttranscriptional gene silencing and natural virus resistance. Cell 101:533–542

Navarro L, Dunoyer P, Jay F, Arnold B, Dharmasiri N, Estelle M, Voinnet O, Jones JD (2006) A plant miRNA contributes to antibacterial resistance by repressing auxin signaling. Science 312:436–439

Ouyang S, Park G, Atamian HS, Han CS, Stajich JE, Kaloshian I, Borkovich KA (2014) microRNAs suppress NB domain genes in tomato that confer resistance to *Fusarium oxysporum*. PLoS Pathogens 10:e1004464

Pertermann R, Tamilarasan S, Gursinsky T, Gambino G, Schuck J, Weinholdt C, Lilie H, Grosse I, Golbik RP, Pantaleo V et al (2018) A viral suppressor modulates the plant immune response early in infection by regulating microRNA activity. Mbio 9:e00419

Pooggin MM (2013) How can plant DNA viruses evade siRNA-directed DNA methylation and silencing? Int J Mol Sci 14:15233–15259

Pruss GJ, Lawrence CB, Bass T, Li QQ, Bowman LH, Vance V (2004) The potyviral suppressor of RNA silencing confers enhanced resistance to multiple pathogens. Virology 320:107–120

Qu F, Ye X, Morris TJ (2008) Arabidopsis DRB4, AGO1, AGO7, and RDR6 participate in a DCL4-initiated antiviral RNA silencing pathway negatively regulated by DCL1. Proc Natl Acad Sci U S A 105:14732–14737

Raja P, Jackel JN, Li S, Heard IM, Bisaro DM (2014) Arabidopsis double-stranded RNA binding protein DRB3 participates in methylation-mediated defense against geminiviruses. J Virol 88:2611–2622

Rajagopalan R, Vaucheret H, Trejo J, Bartel DP (2006) A diverse and evolutionarily fluid set of microRNAs in *Arabidopsis thaliana*. Genes Dev 20:3407–3425

Saur IM, Kadota Y, Sklenar J, Holton NJ, Smakowska E, Belkhadir Y, Zipfel C, Rathjen JP (2016) NbCSPR underlies age-dependent immune responses to bacterial cold shock protein in *Nicotiana benthamiana*. Proc Natl Acad Sci U S A 113:3389–3394

Shivaprasad PV, Chen HM, Patel K, Bond DM, Santos BA, Baulcombe DC (2012) A microRNA superfamily regulates nucleotide binding site-leucine-rich repeats and other mRNAs. Plant Cell 24:859–874

Staskawicz BJ, Ausubel FM, Baker BJ, Ellis JG, Jones JD (1995) Molecular genetics of plant disease resistance. Science 268:661–667

Su Y, Li H-G, Wang Y, Li S, Wang H-L, Yu L, He F, Yang Y, Feng C-H, Shuai P et al (2018) Poplar miR472a targeting NBS-LRRs is involved in effective defence against the necrotrophic fungus *Cytospora chrysosperma*. J Exp Bot 69:5519–5530

Sunkar R, Kapoor A, Zhu JK (2006) Posttranscriptional induction of two Cu/Zn superoxide dismutase genes in Arabidopsis is mediated by downregulation of miR398 and important for oxidative stress tolerance. Plant Cell 18:2051–2065

Szittya G, Burgyan J (2013) RNA interference-mediated intrinsic antiviral immunity in plants. Curr Top Microbiol Immunol 371:153–181

Tang D, Wang G, Zhou JM (2017) Receptor kinases in plant-pathogen interactions: more than pattern recognition. Plant Cell 29:618–637

Taochy C, Gursanscky NR, Cao J, Fletcher SJ, Dressel U, Mitter N, Tucker MR, Koltunow AMG, Bowman JL, Vaucheret H et al (2017) A genetic screen for impaired systemic RNAi highlights the crucial role of DICER-LIKE 2. Plant Physiol 175:1424–1437

Varallyay E, Havelda Z (2013) Unrelated viral suppressors of RNA silencing mediate the control of ARGONAUTE1 level. Mol Plant Pathol 14:567–575

Varallyay E, Valoczi A, Agyi A, Burgyan J, Havelda Z (2010) Plant virus-mediated induction of miR168 is associated with repression of ARGONAUTE1 accumulation. EMBO J 29:3507–3519

Vaucheret H, Vazquez F, Crete P, Bartel DP (2004) The action of ARGONAUTE1 in the miRNA pathway and its regulation by the miRNA pathway are crucial for plant development. Genes Dev 18:1187–1197

Vaucheret H, Mallory AC, Bartel DP (2006) AGO1 homeostasis entails coexpression of MIR168 and AGO1 and preferential stabilization of miR168 by AGO1. Mol Cell 22:129–136

Wang XB, Wu Q, Ito T, Cillo F, Li WX, Chen X, Yu JL, Ding SW (2010) RNAi-mediated viral immunity requires amplification of virus-derived siRNAs in *Arabidopsis thaliana*. Proc Natl Acad Sci U S A 107:484–489

Wang XB, Jovel J, Udomporn P, Wang Y, Wu Q, Li WX, Gasciolli V, Vaucheret H, Ding SW (2011) The 21-nucleotide, but not 22-nucleotide, viral secondary small interfering RNAs direct potent antiviral defense by two cooperative argonautes in *Arabidopsis thaliana*. Plant Cell 23:1625–1638

Wang C, He X, Wang X, Zhang S, Guo X (2017) ghr-miR5272a-mediated regulation of GhMKK6 gene transcription contributes to the immune response in cotton. J Exp Bot 68:5895–5906

Wang T, Deng Z, Zhang X, Wang H, Wang Y, Liu X, Liu S, Xu F, Li T, Fu D et al (2018a) Tomato DCL2b is required for the biosynthesis of 22-nt small RNAs, the resulting secondary siRNAs, and the host defense against ToMV. Hortic Res 5:62

Wang Z, Hardcastle TJ, Pastor AC, Yip WH, Tang S, Baulcombe DC (2018b) A novel DCL2-dependent miRNA pathway in tomato affects susceptibility to RNA viruses. Genes Dev 32:1155–1160

Weaver LM, Swiderski MR, Li Y, Jones JDG (2006) The *Arabidopsis thaliana* TIR-NB-LRR R-protein, RPP1A; protein localization and constitutive activation of defence by truncated alleles in tobacco and Arabidopsis. Plant Journal 47:829–840

Wu L, Zhou H, Zhang Q, Zhang J, Ni F, Liu C, Qi Y (2010) DNA methylation mediated by a microRNA pathway. Mol Cell 38:465–475

Wu J, Yang Z, Wang Y, Zheng L, Ye R, Ji Y, Zhao S, Ji S, Liu R, Xu L et al (2015) Viral-inducible Argonaute18 confers broad-spectrum virus resistance in rice by sequestering a host microRNA. Elife 4:e05733

Wu J, Yang R, Yang Z, Yao S, Zhao S, Wang Y, Li P, Song X, Jin L, Zhou T et al (2017a) ROS accumulation and antiviral defence control by microRNA528 in rice. Nat Plants 3:16203

Wu YY, Hou BH, Lee WC, Lu SH, Yang CJ, Vaucheret H, Chen HM (2017b) DCL2- and RDR6-dependent transitive silencing of SMXL4 and SMXL5 in Arabidopsis dcl4 mutants causes defective phloem transport and carbohydrate over-accumulation. Plant J 90:1064–1078

Xia R, Xu J, Meyers BC (2017) The emergence, evolution, and diversification of the miR390-TAS3-ARF pathway in land plants. Plant Cell 29:1232–1247

Xie Q, Guo HS (2006) Systemic antiviral silencing in plants. Virus Res 118:1–6

Xie Z, Kasschau KD, Carrington JC (2003) Negative feedback regulation of Dicer-Like1 in Arabidopsis by microRNA-guided mRNA degradation. Curr Biol 13:784–789

Xu W, Meng Y, Wise RP (2014) Mla- and Rom1-mediated control of microRNA398 and chloroplast copper/zinc superoxide dismutase regulates cell death in response to the barley powdery mildew fungus. New Phytol 201:1396–1412

Yang Z, Li Y (2018) Dissection of RNAi-based antiviral immunity in plants. Curr Opin Virol 32:88–99

Yu Y, Jia T, Chen X (2017) The 'how' and 'where' of plant microRNAs. New Phytol 216:1002–1017

Zhai J, Jeong D-H, De Paoli E, Park S, Rosen BD, Li Y, Gonzalez AJ, Yan Z, Kitto SL, Grusak MA et al (2011) microRNAs as master regulators of the plant NB-LRR defense gene family via the production of phased, trans-acting siRNAs. Genes Dev 25:2540–2553

Zhang Y, Goritschnig S, Dong X, Li X (2003) A gain-of-function mutation in a plant disease resistance gene leads to constitutive activation of downstream signal transduction pathways in suppressor of npr1-1, constitutive 1. Plant Cell 15:2636–2646

Zhang R, Murat F, Pont C, Langin T, Salse J (2014) Paleo-evolutionary plasticity of plant disease resistance genes. BMC Genom 15:187

Zhang Y, Xia R, Kuang H, Meyers BC (2016) The diversification of plant NBS-LRR defense genes directs the evolution of microRNAs that target them. Mol Biol Evol 33:2692–2705

Zhao J, Fu J, Li X, Xu C, Wang S (2009) Dissection of the factors affecting development-controlled and race-specific disease resistance conferred by leucine-rich repeat receptor kinase-type R genes in rice. Theor Appl Genet 119:231–239

Zheng L, Zhang C, Shi C, Yang Z, Wang Y, Zhou T, Sun F, Wang H, Zhao S, Qin Q et al (2017) Rice stripe virus NS3 protein regulates primary miRNA processing through association with the miRNA biogenesis factor OsDRB1 and facilitates virus infection in rice. PLoS Pathogens 13:e1006662

Zou YM, Wang SF, Zhou YY, Bai JJ, Huang GZ, Liu XT, Zhang YY, Tang DZ, Lu DP (2018) Transcriptional regulation of the immune receptor FLS2 controls the ontogeny of plant innate immunity. Plant Cell 30:2779–2794

The manufacturer's authorised representative in the EU is Springer Nature Customer Service Centre GmbH, Europaplatz 3, 69115 Heidelberg, Germany. If you have any concerns regarding our products, please contact ProductSafety@springernature.com

Printed and bound by CPI Group (UK) Ltd, Croydon, CR0 4YY
24/07/2026
02175910-0010